Mind, Brain and Education

Vida Demarin • Leontino Battistin
Hrvoje Budinčević

Editors

Mind, Brain and Education

 Springer

Editors
Vida Demarin
International Institute for Brain Health
Zagreb, Croatia

Hrvoje Budinčević
Department of Neurology
Sveti Duh University Hospital
Zagreb, Croatia

Leontino Battistin
Department of Neuroscience
Università di Padova
Padova, Padova, Italy

ISBN 978-3-031-33015-5   ISBN 978-3-031-33013-1   (eBook)
https://doi.org/10.1007/978-3-031-33013-1

This Springer imprint is published by the registered company Springer Nature Switzerland AG
The registered company address is: Gewerbestrasse 11, 6330 Cham, Switzerland

Preface

The interrelation between the mind and the brain is fascinating and still rather challenging topic. Researchers are trying to contribute to its better understanding by their original and innovative approaches, and this debate is still open and ongoing.

After the successful contribution of our previous book *Mind and Brain, Bridging Neurology and Psychiatry* published by Springer 2 years ago, we decided to go on in the similar, but broader direction, pointing out the importance of this topic and our various ways of presenting it, to further intensify the education in this particular field. We invited our colleagues, mainly lecturers at our traditional "Mind & Brain Pula" Congress, to present their ideas and research. The first part of this book comprises of thoughts on neuroesthetics, arts, migraine, personality disorders, autoimmune psychosis and many other interesting topics.

The second part is mainly devoted to the brain, where one can find the ways of successful maintenance of the brain sharpness, functional neurological disorders, headaches, stroke, multiple sclerosis, neurodegenerative diseases, telemedicine, data on psychoneuroendocrinoimmunology and many more other interesting and challenging topics.

The third part of the book consists of several topics related to education, like "Psychopathology Summer School" in Pula Congresses, Education and Autism, Co-Creativity in the therapy with children and adolescents and many more.

With great privilege and estimation we express our sincere gratitude to our dear collaborators, experts in the field for their commitment and excellent contributions. The same words of gratitude go to devoted members of the team of our publisher Springer, for their precision and excellence in finalizing our project.

Zagreb, Croatia
Padova, Padova, Italy
Zagreb, Croatia

Vida Demarin
Leontino Battistin
Hrvoje Budinčević

Introduction

Our traditional Mind & Brain Congress in Pula has proved to be an excellent spot for sharing recent knowledge and innovative ideas, exchanging best practices and experiences, for interesting and challenging discussions and for helping young colleagues in their professional development. This interdisciplinary platform with experts acting as "happy teachers will change the world" provides the unique opportunity searching for details on the one hand and for broadening perspectives and comprehensive approach in decision-making on the other hand. Keeping up with the latest guidelines and best practices is crucial and live presentations and ongoing discussions with opinion leaders lead to a clearer understanding.

After the successful contribution of our previous book *Mind and Brain, Bridging the Neurology and Psychiatry* published by Springer 2 years ago, we decided to follow the same idea, but this time adding another chapter related to education.

We divided the contents of our book into three main parts. The first part is mainly dealing with research and ideas considering the mind, and the second is devoted to different topics and research of the brain and its importance, as the main subject. These topics are sometimes rather connected, and it was not always possible to put strict boundaries among them. All chapters are written by experts in the field, and each of them decided on the topic thus we got a wide platform of various topics and approaches.

The content of the third part is devoted to education. We present some ideas related to the possibilities of providing a kind of educational activities as well as emphasizing its importance in creating knowledgeable, educated and well-informed professionals.

One chapter is written by our long-standing lecturers, who participated in the "Pula" Congress for many years, witnessing their commitment and pleasure. On the other hand, the final chapter is written by young colleagues, expressing their attitude towards our congress as being the event which directs their future careers.

Vida Demarin
Leontino Battistin
Hrvoje Budinčević

Contents

Contributors

Ruth K. Abramson Hall Institute of Psychiatry, Department of Neuropsychiatry and Behavioral Science, University of South Carolina School of Medicine, Columbia, SC, USA

Vuk Aleksić Department of Neurosurgery, Clinical Hospital Center Zemun, Belgrade, Serbia

Rozi Andretć Waldowski University of Rijeka, Rijeka, Croatia

Silva Banović Medical Faculty and Faculty of Education and Rehabilitation, University of Tuzla, Tuzla, Bosnia and Herzegovina

Leontino Battistin Department of Neurosciences, University of Padova, Padova, Italy

Karl Bechter Clinic for Psychiatry and Psychotherapy II, Ulm University, Bezirkskrankenhaus Günzburg, Günzburg, Germany

University of Ulm, Department Psychiatry and Psychotherapy II, Günzburg, Germany

Raphael Béné Department of Psychiatry, University Hospital Centre Vaudois (CHUV), Lausanne, Switzerland

Vedran Bilić Department of Psychiatry and Psychological Medicine, University Hospital Center Zagreb, Zagreb, Croatia

Martin Brüne LWL University Hospital Bochum, Clinic for Psychiatry, Psychotherapy and Preventive Medicine, Division of Social Neuropsychiatry and Evolutionary Medicine, Ruhr University Bochum, Bochum, Germany

Hrvoje Budinčević Department of Neurology, Sveti Duh University Hospital, Zagreb, Croatia

Department of Neurology and Neurosurgery, Faculty of Medicine, J.J. Strossmayer University of Osijek, Osijek, Croatia

Department of Neurology, Stroke and Intensive Care Unit, Sveti Duh University Hospital, Zagreb, Croatia

Darko Chudy Department of Neurosurgery, Dubrava University Hospital, Zagreb, Croatia

Department of Surgery, University of Zagreb, School of Medicine, Zagreb, Croatia

Petra Črnac Žuna Department of Neurology, Department of Neurology, Zagreb, Croatia

Department of Neurology, Stroke and Intensive Care Unit, Sveti Duh University Hospital, Zagreb, Croatia

Miroslav Cuturic Department of Neurology, University of South Carolina School of Medicine, Columbia, SC, USA

Milica Dajević Department of Neurology, Clinical Center of Montenegro, Podgorica, Montenegro

Vida Demarin International Institute for Brain Health, Zagreb, Croatia

Filip Derke Department of Neurology, Referral Centre for Preoperative Assessment of Patients with Pharmacoresistant Epilepsy, University Hospital Dubrava, Zagreb, Croatia

Department of Psychiatry and Neurology, Faculty of Dental Medicine and Health, JJ Strossmayer University of Osijek, Osijek, Croatia

Department of Neurology, University Hospital Dubrava, Zagreb, Croatia

Department of Neurology, Dubrava University Hospital, Zagreb, Croatia

Hannes Deutschmann Department of Neuroradiology, Vascular and Interventional Radiology, Medical University of Graz, Graz, Austria

Vesna Đermanović Dobrota, MD, PhD Clinical Hospital Merkur – University Clinic Vuk Vrhovac, Zagreb, Croatia

Janice G. Edwards University of South Carolina School of Medicine, Columbia, SC, USA

Andreja Emeršič Laboratory for CSF Diagnostics, Department of Neurology, University Medical Centre Ljubljana, Ljubljana, Slovenia

Dominique Endres Department of Psychiatry and Psychotherapy, Medical Center-University of Freiburg, Faculty of Medicine, University of Freiburg, Freiburg, Germany

Luka Filipović-Grčić Department of Radiology, University Hospital Centre Zagreb, Zagreb, Croatia

University Hospital Center Zagreb, Department of Radiology, Zagreb, Croatia

Ana Filošević University of Rijeka, Rijeka, Croatia

Latica Friedrich Department of Neurology, Sveti Duh University Hospital, Zagreb, Croatia

Thomas Gattringer Department of Neurology, Vascular and Interventional Radiology, Medical University of Graz, Graz, Austria

Department of Neuroradiology, Vascular and Interventional Radiology, Medical University of Graz, Graz, Austria

Anton Glasnović School of Medicine, University of Zagreb, Zagreb, Croatia

Nenad Jakšić Department of Psychiatry and Psychological Medicine, University Hospital Center Zagreb, Zagreb, Croatia

Marta Jeremić Faculty of Medicine, University of Belgrade, Belgrade, Serbia

Neurology Clinic, University Clinical Center of Serbia, Belgrade, Serbia

Stefan Jerotić Faculty of Medicine, University of Belgrade, Belgrade, Serbia

Clinic for Psychiatry, University Clinical Center of Serbia, Belgrade, Serbia

Tatjana Kapuralin University Graduate Social Pedagogue, Integrative Psychotherapist ECP, Family Counseling Center and Private Practice, Pula, Croatia

Guenther E. Klein Department of Radiology, Medical University Graz, Graz, Austria

Markus Kneihsl Department of Neurology, Vascular and Interventional Radiology, Medical University of Graz, Graz, Austria

Department of Neuroradiology, Vascular and Interventional Radiology, Medical University of Graz, Graz, Austria

Maja Lačković Faculty of Medicine, University of Belgrade, Belgrade, Serbia

Clinic for Psychiatry, University Clinical Center of Serbia, Belgrade, Serbia

Günther Lanner Clinic of Neurosurgery Klagenfurt, Klagenfurt, Austria

Slaven Lasić Department of Neurology, University Hospital Dubrava, Zagreb, Croatia

Milan Latas Faculty of Medicine, University of Belgrade, Belgrade, Serbia

Clinic for Psychiatry, University Clinical Center of Serbia, Belgrade, Serbia

Maja Šeparović Lisak Department of Psychiatry and Psychological Medicine, University Hospital Center Zagreb, Zagreb, Croatia

Vesna Lukinović-Škudar, MD, PhD School of Medicine, University of Zagreb, Zagreb, Croatia

Ivanka Maduna Health Center of Osijek-Baranja County, Osijek, Croatia

Francesco Maffessanti Maria Cecilia Hospital, GVM Care and Research, Cotignola, RA, Italy

Anna Maria Malagoni Maria Cecilia Hospital, GVM Care and Research, Cotignola, RA, Italy

Darko Marčinko Department of Psychiatry and Psychological Medicine, University Hospital Center Zagreb, Zagreb, Croatia

School of Medicine, University of Zagreb, Zagreb, Croatia

Milija Mijajlović Faculty of Medicine, University of Belgrade, Belgrade, Serbia

Neurology Clinic, University Clinical Center of Serbia, Belgrade, Serbia

Nenad Milošević Faculty of Medicine, University of Pristina, Kosovska Mitrovica, Serbia

Clinical-Hospital Center Pristina, Gracanica, Serbia

Ninoslav Mimica School of Medicine, University of Zagreb, Zagreb, Croatia

Ivana Mitić Department of Neurology, General Hospital Zaječar, Zaječar, Serbia

Sandra Morovic Juraj Dobrila University of Pula, Zagreb, Croatia

Arsano Medical Group, Zagreb, Croatia

Igor Mosic NAM Emotional Hygiene Technology, Rijeka, Croatia

Mirsad Muftić Sarajevo Medical School of Medicine, University of Sarajevo School of Science and Technology, Sarajevo, Bosnia and Herzegovina

Faculty for Health Sciences, University of Sarajevo, Sarajevo, Bosnia and Herzegovina

Norbert Müller Department of Psychiatry and Psychotherapy, Ludwig Maximilian University, Munich, Germany

Filip Mustač Department of Psychiatry and Psychological Medicine, University Hospital Center Zagreb, Zagreb, Croatia

Department of Psychiatry, University Hospital Centre Zagreb, Zagreb, Croatia

Kurt Niederkorn Department of Neurology, Vascular and Interventional Radiology, Medical University of Graz, Graz, Austria

Department of Neurology, Medical University Graz, Graz, Austria

Darko Orešković Department of Neurosurgery, University Hospital Dubrava, Zagreb, Croatia

Draginja Petković Institute for Cardiovascular Diseases "Dedinje", Belgrade, Serbia

Milan Petrović University of Rijeka, Rijeka, Croatia

Valentino Rački Department of Neurology, UHC Rijeka, Rijeka, Croatia

Department of Neurology, Faculty of Medicine, University of Rijeka, Rijeka, Croatia

Marina Raguž Department of Neurosurgery, University Hospital Dubrava, Zagreb, Croatia

Catholic University of Croatia, School of Medicine, Zagreb, Croatia

Vladimir Rendevski University Clinic for Neurosurgery, Medical Faculty, Ss. Cyril and Methodius University in Skopje, Skopje, North Macedonia

Franka Rigo University of Rijeka, Rijeka, Croatia

Uroš Rot Laboratory for CSF Diagnostics, Department of Neurology, University Medical Centre Ljubljana, Ljubljana, Slovenia

Faculty of Medicine, University of Ljubljana, Ljubljana, Slovenia

Duško Rudan Department of Psychiatry and Psychological Medicine, University Hospital Center Zagreb, Zabok, Croatia

Maida Seferović Department of Neurology, Zabok General Hospital and Croatian Veterans Hospital, Zagreb, Croatia

Souvik Sen Department of Neurology, University of South Carolina School of Medicine, Columbia, SC, USA

Osman Sinanović Medical Faculty and Faculty of Education and Rehabilitation, University of Tuzla, Tuzla, Bosnia and Herzegovina

Sarajevo Medical School of Medicine, University of Sarajevo School of Science and Technology, Sarajevo, Bosnia and Herzegovina

University of Tuzla, Tuzla, Bosnia and Herzegovina

Emina Sinanović Psychotherapy Training Institute for Integrative Psychotherapy, Sarajevo, Bosnia and Herzegovina

Ana Sruk Department of Neurology, Sveti Duh University Hospital, Zagreb, Croatia

Natasa Stojanovski Faculty of Medicine, University of Belgrade, Belgrade, Serbia

Danijela Tiosavljević Faculty of Medicine, University of Belgrade, Belgrade, Serbia

Clinic for Psychiatry, University Clinical Center of Serbia, Belgrade, Serbia

Sanja Toljan Clinics for Otorhinolaryngology, Surgery and Anesthesiology "Orlando", Zagreb, Croatia

Dorotea Vidaković Health Center of Osijek-Baranja County, Osijek, Croatia

Vladimira Vuletić Department of Neurology, UHC Rijeka, Rijeka, Croatia

Department of Neurology, Faculty of Medicine, University of Rijeka, Rijeka, Croatia

Claudia Wallner Department of Neurology, Medical University Graz, Graz, Austria

Marjan Zaletel Department of Vascular Neurology and Intensive Therapy, Neurology Clinic, University Medical Centre, Ljubljana, Slovenia

University Clinical Center Ljubljana, Department of Vascular Neurology, Ljubljana, Slovenia

University Clinical Center Ljubljana, Pain Clinic, Ljubljana, Slovenia

Department of Vascular Neurology and Intensive Therapy, Neurology Clinic, University, Medical Centre, Zaloska 2, Ljubljana, Slovenia

Sanela Zukić Medical Faculty and Faculty of Education and Rehabilitation, University of Tuzla, Tuzla, Bosnia and Herzegovina

Petra Črnac Žuna Department of Neurology, Sveti Duh University Hospital, Zagreb, Croatia

Matija Zupan Department of Vascular Neurology and Intensive Therapy, Neurology Clinic, University Medical Centre, Zaloska 2, Ljubljana, Slovenia

Bojana Žvan Department of Vascular Neurology and Intensive Therapy, Neurology Clinic, University Medical Centre, Zaloska 2, Ljubljana, Slovenia

University Clinical Center Ljubljana, Department of Vascular Neurology, Ljubljana, Slovenia

Chapter 1
Neuroaesthetics: How We Like What We Like

Filip Derke, Luka Filipović-Grčić, Marina Raguž, Slaven Lasić, Darko Orešković, and Vida Demarin

Introduction

Neuroaesthetics is a young field of neuroscience that studies the neurobiological and evolutionary basis of experiencing and creating art and rests on the combination of cognitive and affective neuroscience. Before the development of technology, especially functional magnetic resonance imaging (fMRI), research was mostly based on the study of the impact of various brain damage (from epilepsy, and cerebral infarction to neurodegenerative diseases) on artistic creation, i.e. the experience of works of art. As fMRI technology has developed that it can accurately register brain activity signals during exposure to a stimulus, research aimed at discovering the brain regions involved in the so-called aesthetic experience, mainly of fine art,

F. Derke (✉)
Department of Neurology, University Hospital Dubrava, Zagreb, Croatia

Department of Psychiatry and Neurology, Faculty of Dental Medicine and Health, JJ Strossmayer University of Osijek, Osijek, Croatia
e-mail: filip@mozak.hr

L. Filipović-Grčić
Department of Radiology, University Hospital Centre Zagreb, Zagreb, Croatia

M. Raguž
Department of Neurosurgery, University Hospital Dubrava, Zagreb, Croatia

Catholic University of Croatia, School of Medicine, Zagreb, Croatia

S. Lasić
Department of Neurology, University Hospital Dubrava, Zagreb, Croatia

D. Orešković
Department of Neurosurgery, University Hospital Dubrava, Zagreb, Croatia

V. Demarin
International Institute for Brain Health, Zagreb, Croatia

and the interpretation of the findings in the light of previous knowledge about the processes that take place in certain regions of the brain. However, over time the field of research has expanded and the reductionist approach of researching exclusively aesthetic experience as beautiful and not-beautiful has often been criticized, and it is proposed that it includes the study not only of the neurological but also of the evolutionary bases of cognitive and affective processes involved in the aesthetic or artistic approach to other to forms of artistic expression—music, dance, film, theatre, poetry, literature, architecture, etc. and not only to them but also to non-artistic objects and natural phenomena [1–5].

The History of the Perception of Beauty

Archaeological findings indicate that the early beginnings of 'art', along with the realized prerequisite of the existence of cognitive abilities for symbolic, abstract, and referential thinking, began with the stratification of society and the need for mutual distinction through the use of jewellery and ornaments. Practically serving as a system of communication between artists and viewers, showing experiences and ideas, it has become an integral part of society. Starting from this assumption, art had a social role in the very beginning—it reflected social behaviour and rituals and created better social cohesion. The latter determined the adaptive role of art, and thus its transmission to other generations. In addition to the above, the artist himself in this sense showed biological signals of good genetic material in his works—talent, skill and cognitive abilities that were worth transferring to future generations [6].

However, the question arises as to how the specific regions in the brain responsible for our experience of art have developed. Neuroimaging studies on humans have shown that activity in parts of the brain involved in creating an aesthetic response to works of art overlaps with an activity that occurs when evaluating objects of evolutionary importance—the desirability of certain foods or the attractiveness of potential partners. Therefore, it is considered that the processes involved in the creation of the aesthetic experience actually 'upgraded' to the adaptive evaluation processes related to food, i.e. mate selection [7]. In addition, some also mention the connection between the preference for visual aesthetic objects and the choice of habitat, i.e., the assessment of lines suitable for settlement [8].

The Impact of Illness on Creativity and the Experience of Art

There are studies on how neurodegenerative diseases such as Alzheimer's and frontotemporal dementia and brain infarction in a certain region of the brain, especially the right hemisphere, affect the creation and experience of art. However, due to the insufficient number of studies, we can rely more on their anecdotal than comprehensive meaning and the creation of only a temporary and conditional picture of the role of certain areas of the brain in certain aesthetic categories [5]. There are cases of people with a special type of frontotemporal dementia, semantic dementia in which degenerative changes occur in the front part of the temporal lobe, in which a tendency to create art appeared. Their creation is obsessive-compulsive to the extent that they forget about basic life needs such as food. Typical is a realistic representation without symbolism or abstract elements, focusing on details with frequent use of recurring motifs [3]. In three patients with frontotemporal dementia, at the onset of the disease, there was a compulsive and constant listening to music that they did not particularly like until then [6]. In the case of artists suffering from Alzheimer's dementia, the works became more and more abstract and symbolic, moving away from a realistically accurate representation [5]. One of the most famous artists who continued to create after the onset of Alzheimer's dementia is Willem de Kooning. His 'late' phase is characterized by simplicity and purity of form down to the very 'essence' and the use of vivid colours. Another is the case of the painter William Utermohlen, whose— self-portraits also became more and more simplified, and distorted [3].

Research by the US researcher Halpern [9] showed that in patients with frontotemporal and Alzheimer's dementia, the taste in visual art remains consistent, only that patients with Alzheimer's dementia do not remember the works they have seen before [4]. Another research studied the consequences of cerebral infarction of the right hemisphere. Damage in different parts of the frontal-parietal and lateral temporal cortex led to a changed judgment of the conceptual characteristics of the image—abstraction, symbolism, realism, and vividness. On the other hand, damage to the lower prefrontal cortex changed only the formal characteristics and the experience of depth. However, neither area of brain damage was associated with changes in judgments of interest or preference for a particular work [2].

An example of a change in style after a stroke in the left hemisphere is the American painter, Katherine Sherwood. Before the stroke, she described her style as 'too cerebral', and her paintings contained esoteric depictions of people dressed in clothing of the opposite sex and medieval stamps, after which she began to paint with her left hand in a new 'raw' and 'intuitive' style with ease and tenderness that her right hand never had [1].

On the other hand, another group of researchers led by Griffiths [10] the consequences of a stroke that left damage to the left insula, frontal cortex, and amygdala in one patient left him unable to experience music emotionally, although he enjoyed other activities without hindrance and even successfully recognized other musical characteristics. This leads to the assumption that the pathways of perceptual and emotional processing of music are separate, and that the amygdala is crucial in creating an emotional response when listening to music [3].

Furthermore, the case of two patients with bilateral damage to the amygdala is described, who showed a greater preference for three-dimensional objects, landscapes, and colour arrangements than a healthy control group, especially for those stimuli that the control group liked the least. Also, the case of a patient who, after the same damage, was unable to recognize sad and frightening, but not happy music, was presented, although she was able to process other characteristics of music such as tempo and tonality [4].

There is also one interesting case of a patient with epilepsy in whom the left temporal lobe was resected, leaving the parahippocampal gyrus and amygdala [11]. In the first year after the operation, the patient's taste in music changed, he no longer enjoyed listening to rock music but much more enjoyed Celtic or Corsican polyphonic singing. In addition, his literary taste shifted from science fiction to Kafkaesque novels. He also began to enjoy more realistic pictures, especially small details that he had not noticed before. Despite all these changes, his taste in food and clothes and facial preferences remained unchanged [4].

Brain Mechanisms in the Processing of the Beautiful

It is assumed that the aesthetic experience arises when evaluating an object. The authors of the early works relied on Kant's concept of aesthetics [12], according to which "beautiful" is considered to be what is observed without interest, i.e. an object that e we want to acquire, control or manage it. To explain the neurobiological mechanism of such a mental state of 'lack of interest, a hypothesis was put forward that it could be a consequence of the activity of the 'liking' system (tendency towards something) without the activity of the 'desire' system (for something) with the corresponding experience of pleasant 'aesthetic' emotions. Aesthetic emotions themselves, such as pleasure or disgust, are triggered by the experience of the object, unlike emotions triggered by the outcome (happiness, disappointment). Furthermore, research on rodents has shown that there is a difference in the neurotransmitter activity of these two systems – the 'liking' system is mediated by opiates and cannabinoids, and the 'wanting' system is mediated by dopamine [1]. Our experience of art is based on the dynamic interaction of processes that activate different parts of the brain in different time frames, and are linked to perceptive, cognitive (memory, attention, decision-making) and affective abilities [4, 5]. Considering the previous findings from neuroimaging studies, an overview of brain regions and their functions in the processing of aesthetic experience follows.

Brain Plasticity in Artists

Fine art is a complex and unique human phenomenon. The creation of works of art has historically been a poorly understood and very mysterious process, often even to the artists themselves [13]. Neuroscience is still investigating exactly how the brain supports the skills necessary to create a work of art. Art appears in many forms and styles throughout history, and the quality that art has achieved is often difficult to define. Although there are different styles of fine artworks, one thing is common to all artists who create them, and that is that they depend on motivation and experience through which they develop many of the ideas and skills necessary to create their works. To study this phenomenon, it is difficult to isolate all external influences that affect the very motivation and creation of the work, such as the influence of society, culture and the like [14–17].

Can painting or drawing be learned? We will focus on three factors in the process of realizing a work of art, namely: creative inspiration, visual perception and motor skills. To find out if there are differences in brain structures between artists and those who are not involved in art, a study was done in which 35 volunteers were collected, of which 17 volunteers were sent to a three-month painting lesson and drawing [18, 19]. Creative cognition is a mental action by which the artist creates and develops new forms, thoughts and phenomena and links between them. It is very difficult to define it in a scientific context, partly because creativity manifests itself in many different areas (e.g. in speech, dance, and music...). It remains an open question what makes an artist creative while considering that the brain plays a major role in this. Many neuroscientific researchers have had different approaches and knowledge about the neural basis of creativity, but very few consensuses have been reached. For example, in one study [20] a difference was found between artists and laymen in short and long-range neural synchronized EEG waves when painting a given object. It was also found by fMRI [21] that creativity is correlated with the dominance of right over left prefrontal cortical activity. It has been observed that the activity of the parietal cortex decreases in experienced painters when drawing portraits [20]. The very few consensuses reached in scientific research on creativity shows how complex a process it is, and scientists should avoid narrowing down its conceptual construct because of its complexity. Today's knowledge is not enough to say which regions and processes in the brain are crucial for artists. It is most likely that it is an integration of individual experience of the world and improvement of motor skills. It has been found that practice and experience improve and increase creativity, while visual perception is unaffected. Research has shown that the brain of an artist is different from an "ordinary" brain in terms of creative inspiration, different perception, a more precise visual system as well as more precise fine motor skills, but what is crucial in the brain of an artist that makes him an artist in the first place has not yet been proven, but it is certainly unique to artists.

The Activity of Cortical Areas Involved in the Experience of Art

Dorsolateral Prefrontal Cortex

In one study, led by Spanish researchers, the participants were shown different artistic and non-artistic, abstract, and representational pictures [22]. When evaluating a visual stimulus as 'beautiful', there was significant activation in the dorsolateral prefrontal cortex. It is believed that in this area decisions are made based on the wider external context of the visual stimulus—style, explicit content, and level of skill [5].

Anterior Medial (Frontomedian) Prefrontal Cortex

The frontomedian prefrontal cortex shows greater activation in assessing the beauty of a geometric shape than its symmetry [5, 6, 13]. In addition, according to research by Kirk [13], increased activation is also present when evaluating images of unusual or unexpected figure-background combinations that were more aesthetically appealing [5, 6, 15]. Considering that this region is related to the evaluation of what is related to oneself, it is correlated with the nature of the aesthetic evaluation, which is subjective in contrast to the evaluation of symmetry, which is an objective judgment [5].

The activation of the latter brain regions most likely reflects the formation of the first impression of visual support that affects other processes related to attention, perception and response selection and may be responsible for the further outcome of interaction with the stimulus [5, 6].

Ventrolateral Prefrontal Cortex

The increased activity of this area is also present in the assessment of the beauty of geometric shapes compared to their symmetry, and more aesthetically appealing images with an unusual and unexpected figure-background combination. It is believed that this is a consequence of increased demands on attention when converting a non-dichotomous assessment into a binary response (nice or not), that is, deciphering and processing complex visual stimuli [6, 15].

Temporal Lobe

In the research of Jacobsen et al. [8] when assessing the beauty of a geometric shape, a stronger activity of the left temporal lobe was found when assessing symmetry. The authors believe that this indicates the recall of information from memory to create a semantically and emotionally coloured context for evaluating the visual stimulus [6, 13]. Also, Kirk's study [13] found increased activity in the left temporal lobe when evaluating aesthetically appealing photographs of abnormal figure-background pairings. It is assumed to indicate the use of prior knowledge to organize affectively salient figure-background combinations into comprehensible and relatable representations [6, 15].

In addition, in the left hemisphere in the area where the temporal and parietal lobes touch, in the posterior part of the Sylvian fissure, activation was observed during a viewing of 'beautiful' and not 'neutral' images. Activity in this area is thought to be responsible for recognizing meaningful, familiar content in representational, but not abstract, images [5].

Posterior Cingulate Cortex and Praecuneus

The research of Kirk [15] showed that in the aesthetic evaluation of scenes showing matching figure-background pairings, there is an activation in the posterior cingulate cortex. This is explained using semantic memory and familiarity with the content in generating a response to a stimulus [6]. Furthermore, in the research of Jacobsen et al. [13] increased activity of the posterior cingulate cortex and praecuneus was present when evaluating the beauty of geometric shapes. It is thought to reflect processes of recall from memory either because they considered shapes seen before during research or they relied on those shapes from everyday life that they were familiar with. In addition, architects who evaluate the aesthetic appeal of buildings show the greater activity of the precuneus than non-professionally educated people in this sense. It is assumed that this activity reflects the recall of information stored in memory to create a context appropriate for evaluation [6].

Insula

According to the meta-analysis conducted by S. Brown et al. [16], the anterior part of the insula of the right hemisphere coincided the most with a positive aesthetic assessment in four sensations. That paralimbic area relates to interoceptive awareness/insight, or sense of self. He believes that interoceptive processing is the key to assigning valence to the observed object. The assumption is that the determination of valence takes place by referring to the visceral or motivational state caused by the observation of that object [16].

Another very important finding is increased activation during the subjective experience of a work of art and the mood and feelings evoked. It is the activation of the insular cortex that is considered responsible for the emotional experience when viewing images of aesthetic value [6].

Orbitofrontal Cortex (OFC)

The orbitofrontal cortex is considered the most important brain region for multisensory integration because it receives input from the five major sensory pathways and the visceral afferent system. In this area, there are secondary gustatory and olfactory cortices, and it is the place of integration of these two senses, where the perception of 'aroma' is created. In addition, like the insula, it is connected to visceromotor regions (anterior cingulate cortex) and the hypothalamus, provides a response to the reward valence of stimuli and is an important region of the brain for emotions. However, the OFC also has important connections with learning and memory, possibly due to its connection with the dorsolateral prefrontal cortex. It is also considered that in addition to the representation of the object's valence, it maintains it in working memory and thus influences decision-making and behaviour. It is even thought to serve to store memories of the stimuli that led to pleasure, which is also important in decision-making [16].

As the OFC is part of the reward system in which the polysensory convergence of stimuli takes place, among which is the perception of the quality of the food source, including taste, smell, and visual and textural components, it is considered that other modalities have evolved from this basis of food evaluation, from partner evaluation to sublime works of art [16].

Anterior Cingulate Cortex

It is a common finding that participants experience increased activation in the anterior cingulate cortex when they experience works of art that they like. Its role is connected with monitoring one's affective state, which serves in further assessment and decision-making related to the observed object [6].

Nucleus Accumbens and Ventral Striatum

The nucleus accumbens and ventral striatum are subcortical structures that play an important role in various complex processes related to learning, anticipation and expectation of reward, emotions, and pleasure. In a study by Kirk and colleagues [15] when participants looked at photographs of faces and buildings they liked, regardless of their level of expertise in architectural content, increased activity in the nucleus accumbens was observed. In another study, there was increased activation of the ventral striatum when looking at supports that they considered to be works of art. Furthermore, it is the ventral striatum that is considered the "hedonic hotspot" of the brain [6] and is responsible for the enjoyment of art [6].

Occipital Cortex

The occipital cortex consists of the primary visual cortex and extrastriate visual areas (areas of the association visual cortex). In the research, regardless of whether it was an abstract or a representative image, the activation of the visual cortex of the participants was greater the more they liked a certain image. This increase in activation could be related to the positive valence of the preferred image or the elicitation of greater attention. In addition, increased activation of the occipital cortex was also observed when exposed to displays of congruent figure-background pairs, and to visual stimuli that were said to be works of art [6].

Parietal Cortex

According to research by Cela-Conde [22], greater activity in the bilateral angular gyri during the observation of images, photographs and designs considered beautiful is associated with an increase in their spatial processing strategies. In another study, during the aesthetic experience of images with soft transitions (an example of Monet's water lilies), increased activity of the superior parietal cortex was noticed, which was interpreted as the participant's effort to create a coherent representation from undefined shapes [6].

Auditory Cortex

Research by Koelsch et al. [23] showed that pleasant music leads to greater bilateral activity in the area of transverse temporal gyrus, the area of the primary auditory cortex and the processing of constant pitch. The authors suggest that the latter higher activity of the auditory cortex is stimulated by the activation of attention mechanisms in response to the positive affective valence of pleasant music and consequently caused an increased perceptual analysis of pleasant musical fragments [6].

Movement Representation Bodies

Body perception takes place in the occipital and premotor cortex, according to research by Calvo-Merino [24, 25]. Extrastriate visual areas (dorsal visual pathway) are involved in processing specific details about body posture, and the ventral premotor cortex in processing body configuration [5, 6]. TMS (transcranial magnetic stimulation) was used to examine the influence of these regions on the aesthetic assessment of body movements. It was shown that the aesthetic sensitivity to the dancer-like posture was significantly reduced when TMS was applied in the area of extrastriatal visual areas, which leads to the conclusion that early perceptual processes in that area have a significant contribution to the aesthetic assessment of body movements [6].

However, although explanations have been proposed as to what the increased activity of sensory cortices is for, the question arises as to how and why it occurs and what role it plays in the experience of art. It is known that attention can affect various brain activities, from feelings to decision-making. It is assumed that the modulation of neuronal activity in the sensory cortices during the selection of spatial locations, specific characteristics or even entire visual objects results from an increase in the sensitivity of the neurons involved in these processes. Furthermore, it is thought that the increased activity of the sensory cortices serves to stimulate activity in the ventral striatum leading to the sensation of pleasure associated with the preferred stimulus. Namely, the results of the research by Lacey et al. [26] showed that the activity of the ventral striatum is stimulated by the activity of the caloric sulcus and presupplementary motor cortex in the left, and by the activity of the hypothalamus, posterior frontal gyrus and lateral occipital complex in the right hemisphere. In addition, the activity of the occipital cortex appears to be predominantly driven by the activity of the cingulate cortex and the posterior frontal gyrus suggesting a top-down regulation of attention [6].

Criticism of this Approach to Aesthetics

One of the first unresolved questions is whether the field of neuroaesthetics should be limited only to aesthetic experience, or whether it can elucidate artistic creation as well. (3.5) Among the reasons given why neuroaesthetics cannot deal with art is the reductionist focus only on the 'aesthetic' response, that is, the experience of the 'beautiful', which does not correspond to everything that art represents. Among other things, beauty is not necessary if the purpose of the work of art was to frighten, to sadden, to display the opulence of a community, to contemplate one's existence, etc. Moreover, a work of art has a different purpose depending on the context, time, and people [1, 5, 7].

Another objection is the search for general relationships between cognitive processes and neurological mechanisms when observing art, rather than looking at an individual work of art [3]. Therefore, it is said that neuroaesthetics bypasses everything that art criticism deals with—it does not provide any basis for evaluating great, good or bad art, nor does it provide insight into the concrete nature of objects and experiences of art and the individual contribution of certain artists.

Furthermore, the methodology of most research roughly limits artworks, specifically visual ones, to 'stimuli', ignoring their cultural, 'natural', context and possible personal significance to the individual who observes them [7]. Therefore, a legitimate question arises as to whether the activity in certain brain regions shows the processes involved in the aesthetic assessment or is involved in the performance of tasks related to attention or affective discrimination. Although, the scientists tried to get around the problem of semantic context by telling the participants that the visual stimulus was from an art gallery. Of course, this is far from creating a historical and cultural context when measuring aesthetic assessment, but it is still an attempt in that direction [5, 8].

Lastly, it is considered that the aesthetic experience cannot be explained by neurological mechanisms that are involved in other activities that are not at all related to art, some of which can be found in animals as well. Moreover, such an approach does not consider the uniqueness of the human experience of great works of art concerning everyday perception [1, 27–34].

Conclusion

Taking all the above into account, it is suggested that further research focus more on the study of the dynamics of neurological processes than on their localization [5] and construct different contextual conditions that are more like 'real' ones to gain a better insight into how they influence to the aesthetic assessment of different works of art [3].

Moreover, Steven Brown and Ellen Dissanayake propose that neuroaesthetics be renamed 'neuroartology' which will place greater emphasis on the behavioural function of art creation and perception. In addition, it would include the study of cognitive and behavioural mechanisms that do not have direct aesthetic functions or consequences. Also, they believe that the experience of art cannot be reduced only to aesthetic emotions based on the evaluation of objects, because they do not describe the entire spectrum of emotions that arise when creating or experiencing art, especially the feeling of satisfaction provided by social belonging and togetherness—and they are one of the most important drivers of art.

References

1. Demarin V, Bedeković MR, Puretić MB, Pašić MB. Arts, Brain and Cognition. Psychiatr Danub. 2016;28(4):343–8.
2. Zeki S, Bao Y, Pöppel E. Neuroaesthetics: the art, science, and brain triptych. Psych J. 2020;9(4):427–8. https://doi.org/10.1002/pchj.383.
3. Chatterjee A. Neuroaesthetics: a coming of age story. J Cogn Neurosci. 2011;23(1):53–62. https://doi.org/10.1162/jocn.2010.21457.
4. Bromberger B, Sternschein R, Widick P, Smith W 2nd, Chatterjee A. The right hemisphere in esthetic perception. Front Hum Neurosci. 2011;5:109. Epub 20111014. https://doi.org/10.3389/fnhum.2011.00109.
5. Cela-Conde CJ, Agnati L, Huston JP, Mora F, Nadal M. The neural foundations of aesthetic appreciation. Prog Neurobiol. 2011;94(1):39–48. https://doi.org/10.1016/j.pneurobio.2011.03.003.
6. Nadal M, Chatterjee A. Neuroaesthetics and art's diversity and universality. Wiley Interdiscip Rev Cogn Sci. 2019;10(3):e1487. https://doi.org/10.1002/wcs.1487.
7. Nadal M, Pearce MT. The Copenhagen Neuroaesthetics conference: prospects and pitfalls for an emerging field. Brain Cogn. 2011;76(1):172–83. https://doi.org/10.1016/j.bandc.2011.01.009.
8. Brown S, Gao X, Tisdelle L, Eickhoff SB, Liotti M. Naturalizing aesthetics: brain areas for aesthetic appraisal across sensory modalities. Neuroimage. 2011;58(1):250–8. Epub 20110615. https://doi.org/10.1016/j.neuroimage.2011.06.012.
9. Halpern AR, Ly J, Elkin-Frankston S, O'Connor MG. "I know what I like": stability of aesthetic preference in Alzheimer's patients. Brain Cogn. 2008;66(1):65–72. https://doi.org/10.1016/j.bandc.2007.05.008.
10. Tischler V, Howson-Griffiths T, Jones CH, Windle G. Using art for public engagement: reflections on the dementia and imagination project. Arts Health. 2020;12(3):270–7. https://doi.org/10.1080/17533015.2019.1608565.
11. Sellal F, Andriantseheno M, Vercueil L, Hirsch E, Kahane P, Pellat J. Dramatic changes in artistic preference after left temporal lobectomy. Epilepsy Behav. 2003;4(4):449–50.; author reply 451. https://doi.org/10.1016/s1525-5050(03)00146-x.

12. Zaidel DW. Art and brain: the relationship of biology and evolution to art. Prog Brain Res. 2013;204:217–33. https://doi.org/10.1016/b978-0-444-63287-6.00011-7.
13. Jacobsen T, Schubotz RI, Höfel L, Cramon DY. Brain correlates of aesthetic judgment of beauty. Neuroimage. 2006;29(1):276–85. Epub 20050808. https://doi.org/10.1016/j.neuroimage.2005.07.010.
14. Ishizu T, Zeki S. The brain's specialized systems for aesthetic and perceptual judgment. Eur J Neurosci. 2013;37(9):1413–20. Epub 20130203. https://doi.org/10.1111/ejn.12135.
15. Kirk U. The neural basis of object-context relationships on aesthetic judgment. PLoS One. 2008;3(11):e3754. https://doi.org/10.1371/journal.pone.0003754.
16. Skov M, Vartanian O, Martindale C, Berleant A. Neuroaesthetics. 1st ed. Routledge; 2009. https://doi.org/10.4324/9781315224091.
17. Chatterjee A, Vartanian O. Neuroaesthetics. Trends Cogn Sci. 2014;18(7):370–5. https://doi.org/10.1016/j.tics.2014.03.003.
18. Cinzia DD, Vittorio G. Neuroaesthetics: a review. Curr Opin Neurobiol. 2009;19(6):682–7. https://doi.org/10.1016/j.conb.2009.09.001.
19. Schlegel AA, Rudelson JJ, Tse PU. White matter structure changes as adults learn a second language. J Cogn Neurosci. 2012;24(8):1664–70. https://doi.org/10.1162/jocn_a_00240.
20. Bhattacharya J, Petsche H. Drawing on Mind's canvas: differences in cortical integration patterns between artists and non-artists. Hum Brain Mapp. 2005;26(1):1–14. https://doi.org/10.1002/hbm.20104.
21. Kowatari Y, Lee SH, Yamamura H, Nagamori Y, Levy P, Yamane S, Yamamoto M. Neural networks involved in artistic creativity. Hum Brain Mapp. 2009;30(5):1678–90. https://doi.org/10.1002/hbm.20633.
22. Cela-Conde CJ, Ayala FJ. Art and brain coevolution. Prog Brain Res. 2018;237:41–60. https://doi.org/10.1016/bs.pbr.2018.03.013.
23. Koelsch S, Fritz T, V. Cramon DY, Müller K, Friederici AD. Investigating emotion with music: an fMRI study. Hum Brain Mapp. 2006;27(3):239–50. https://doi.org/10.1002/hbm.20180.
24. Calvo-Merino B, Jola C, Glaser DE, Haggard P. Towards a sensorimotor aesthetics of performing art. Conscious Cogn. 2008;17(3):911–22. https://doi.org/10.1016/j.concog.2007.11.003.
25. Calvo-Merino B, Urgesi C, Orgs G, Aglioti SM, Haggard P. Extrastriate body area underlies aesthetic evaluation of body stimuli. Exp Brain Res. 2010;204(3):447–56. https://doi.org/10.1007/s00221-010-2283-6.
26. Lacey S, Hagtvedt H, Patrick VM, Anderson A, Stilla R, Deshpande G, Hu X, Sato JR, Reddy S, Sathian K. Art for reward's sake: visual art recruits the ventral striatum. NeuroImage. 2011;55(1):420–33. https://doi.org/10.1016/j.neuroimage.2010.11.027.
27. Conway BR, Rehding A. Neuroaesthetics and the trouble with beauty. PLoS Biol. 2013;11(3):e1001504. https://doi.org/10.1371/journal.pbio.1001504.
28. Nadal M. The experience of art: insights from neuroimaging. Prog Brain Res. 2013;204:135–58. https://doi.org/10.1016/b978-0-444-63287-6.00007-5.
29. Li R, Zhang J. Review of computational neuroaesthetics: bridging the gap between neuroaesthetics and computer science. Brain Inform. 2020;7(1):16. https://doi.org/10.1186/s40708-020-00118-w.
30. Chatterjee A, Coburn A, Weinberger A. The neuroaesthetics of architectural spaces. Cogn Process. 2021;22(Suppl 1):115–20. https://doi.org/10.1007/s10339-021-01043-4.
31. Iigaya K, O'Doherty JP, Starr GG. Progress and promise in neuroaesthetics. Neuron. 2020;108(4):594–6. https://doi.org/10.1016/j.neuron.2020.10.022.
32. Osaka N. Emotional neuroaesthetics of color experience: views from single, paired, and complex color combinations. Psych J. 2022;11:628–35. https://doi.org/10.1002/pchj.577.
33. Baroncini M, Jissendi P, Balland E, Besson P, Pruvo JP, Francke JP, et al. MRI atlas of the human hypothalamus. NeuroImage. 2012;59(1):168–80. https://doi.org/10.1016/j.neuroimage.2011.07.013.
34. Herbet G, Maheu M, Costi E, Lafargue G, Duffau H. Mapping neuroplastic potential in brain-damaged patients. Brain. 2016;139(Pt 3):829–44. https://doi.org/10.1093/brain/awv394.

Chapter 2
Neuroimaging and Art

A Short Introduction to Neuroimaging Techniques Employed in Neuroaesthetics Research

Luka Filipović-Grčić and Filip Derke

Introduction

Like most writings concerning neuroaesthetics, this too will begin with an attempt to outline what neuroaesthetics and aesthetic is. The neuro part is easy, it defines the organ this scientific field explores, and that is the brain. Aesthetic experience like any other unfolds in the brain and may or may not engulf the rest of the body. The aesthetic part is the tricky one as there is no general consensus of what comprises an aesthetic experience. The definition I clung to, which may seem as an oversimplification is that aesthetic process is about what we like and dislike. Indeed, aesthetic experience can occur while hiking, visiting a museum, watching a football game, listening to music, dancing or enjoying a bath, a key aspect being immersing oneself in an activity and deriving pleasure from it [1]. However, as people strive to explore and explain the world around us, so too do we question and analyze ourselves and our interaction with the world. This is where the science of everything, including aesthetics begins. In order to analyze something and to draw conclusions we need a scientific method, involving hypothesis tested by a structured experiment that is reproducible and confirms or discards the hypothesis. But how did we come to associate the experience of beauty, or any experience for that matter, with the brain? It seems obvious to us today, but it has been more or less obvious for only the last 200

L. Filipović-Grčić (✉)
Department of Radiology, University Hospital Centre Zagreb, Zagreb, Croatia

F. Derke
Department of Neurology, Dubrava University Hospital, Zagreb, Croatia

Department of Psychiatry and Neurology, Faculty of Dental Medicine and Health, JJ Strossmayer University of Osijek, Osijek, Croatia
e-mail: filip@mozak.hr

© The Author(s), under exclusive license to Springer Nature Switzerland AG 2023
V. Demarin et al. (eds.), *Mind, Brain and Education*,
https://doi.org/10.1007/978-3-031-33013-1_2

to 300 years [2]. Therefore it would be worthwhile to take a short recap of the history of neurophysiology.

During the greater part of human history, brain was considered a rather uninteresting organ in the skull, with functions such as cooling the blood from the heart [2]. Although there were some notable physicians which correctly ascribed certain functions to the brain and located the mind within it such as Galen, Vesalius and Willis [3] to name a few, the concept of specialized functions within clearly delineated brain regions did not catch up until early nineteenth century. At that time a strange misconception called phrenology helped to usher in the modern brain research and to divide the brain into functional areas. Phrenology was a brainchild of Franz Joseph Gall, a German-Austrian neuroanatomist who devised and, more importantly, popularized the concept of localization of certain mental functions in certain areas of the brain. His teaching of brain as a cluster of specialized mental organs, although fundamentally wrong, provoked criticism from other experts like Jean Pierre Flourens, and, more notably, Paul Broca and Karl Wernicke. The latter two will correlate the lesions of certain brain areas with various speech disabilities and demonstrate it as a proof of function localization within the brain [4, 5]. In the second half of the nineteenth century the work of Camillo Golgi and Santiago Ramon y Cajal brought about deeper and more comprehensive understanding of neurohistology, culminating in the cerebral cytoarchitectonic maps by Korbinian Brodmann and Constantin von Economo. These maps delineate 43 (Brodmann) areas distinct from one another by neuronal organization (architecture), and, in some cases, by function. Despite many shortcomings of Brodmann's maps, they and their many versions are still in use today for localizing activation in neuroimaging studies [6]. An investigation into neurophysiology developed parallel to neuroanatomy, discovering the electrical nature of neural signal and a pivotal role of neurotransmitters. All of this gave a tremendous contribution to the understanding of brain functions and their localization. However, virtually none of these discoveries was made on living patients. No matter how valid and important, they still lacked the ability to directly assess a living, functioning brain. At that point (in the first half of the twentieth century) the neurophysiology stepped in with the help of Electroencephalography (EEG). However, although we could detect activity within the brain with the help of EEG, the exact location of the activity proved hard to determine since there wasn't any method that could provide images of brain morphology. That changed in 1971 with the invention of computed tomography (CT) by Godfrey Hounsfield. It took scientists almost 80 years from Roentgen's discovery of x-rays to penetrate inside the skull and obtain useful pictures of the brain. Around the same time, a method for detecting brain activity (amongst other applications) was finally developed; positron emission tomography (PET). It detects the gamma radiation emitted by radiotracers (radioactive substances) injected into the body assessing their accumulation in certain organs and tissues and giving information about metabolic activity, blood flow, etc. This is where the neuroimaging stood before the development of magnetic resonance imaging (MRI).

Neuroimaging

According to Wikipedia, neuroimaging encompasses computed tomography (CT), magnetic resonance imaging (MRI), positron emission tomography (PET), single photon emission computed tomography (SPECT), cranial ultrasound, functional magnetic resonance imaging (fMRI), diffuse optical imaging (DOI), event-related optical signal (EROS), magnetoencephalography (MEG), functional ultrasound imaging (fUS), functional near-infrared spectroscopy (fNIRS) and quantum optically-pumped magnetometer [7]. Due to its availability, safety, versatility and precision, fMRI is by far the most widely used in modern neuroaesthetics research albeit often combined with other techniques. Methods as CT and PET have been used to a certain extent, but given their radiation and ionizing effect, to subject healthy volunteers to them seems unnecessary. On the other hand, EEG measures the voltage created by the postsynaptic potentials, mostly from cortical pyramidal neurons. It detects the voltage employing a number of electrodes positioned on the subject's scalp and produces a recording of brain waves. EEG is the oldest of above mentioned techniques, with concept dating to the end of the nineteenth century, and clinical application from the first half of the twentieth century. Its simplicity, robustness, noninvasive nature, mobility and low cost made it a popular clinical tool for assessing a multitude of brain disorders, most notably epilepsy. However it is not without downsides, the most prominent being a poor spatial resolution and significant errors in analyzing subcortical activities as it mainly detects voltage from brain surface. Another method, MEG, relies on the same effect as EEG, namely the electrical current running through neurons, but, instead of measuring the voltage, it measures the magnetic field created by the electrical current using magnetometers. With a temporal resolution of 1 ms it is somewhat better than EEG, but it surpasses it greatly in the spatial resolution since magnetic fields are less distorted by the skull. However, it doesn't come close to fMRI due to the same reasons as EEG. Other downsides include the fact that it has to be in a magnetically shielded room to avoid distortion by external magnetic field. This obstacle is being tackled through the integration of quantum sensors, enabling the movement of the subjects, making more complex and less restrained research possible [8]. Another similar method is fNIRS; similar in a sense it is a positioned on the scalp and measures mostly cortical activity, but different as it doesn't utilize the electrical activity but blood flow to generate its results. It uses near infrared light in such way that it penetrates the skull and brain tissue, but is absorbed by the hemoglobin, allowing for the estimate of blood flow and inference of brain activity. As EEG and MEG, it has a great temporal resolution and mobility, but low spatial resolution, detecting changes predominantly in the cortex. Likewise its spatial resolution is significantly lower than that of fMRI. All of the above described techniques can be combined with fMRI, adding to the quality of the assessment. Never the less, fMRI remains the most popular technique used in neuroimaging, particularly in neuroaesthetics and hence this section will focus on the basic principles of fMRI.

As stated by Bandettini in his book, fMRI was more discovered than invented since all the necessary technology was already at hand in conventional MRI machines [9]. Therefore, let us take a brief look in to the working principles of MRI. The physics behind the MRI is quite extensive and complicated so it will be grossly simplified here. To those interested in specific physical principles of MRI, I suggest excellent lectures available on YouTube [10]. In essence, once put in the MR machine, our body is exposed to a strong magnetic field (in most clinical settings it is 1.5 T to 4 T, but in research facilities devices as strong as 11 T are used). This field acts upon hydrogen atoms in our body, namely on their nuclei consisting of a single proton which has a distinct property of spin. Think of it as a direction, which is random at any given time. The field forces most of these protons to align their spins in the direction of the field. As they are aligned with the field, protons are bombarded by radiofrequency energy from radiofrequency coil, another part of the MRI machine. Having absorbed this energy, some of the protons aligned with the magnetic field turn in opposite direction becoming high energy protons. As the radiofrequency coil is turned off, the protons realign with magnetic field, emitting energy in the process. This energy is then detected by receiver coil that transforms it into electrical signal, sends it to the computer, where the picture is reconstructed using different algorithms. Based on their density and molecules around them which are tissue specific, different protons will take longer time to realign with the magnetic field than others, producing a contrast between different tissues. By combining different times of signal acquisition, radiofrequency pulses and magnetic gradients (different sequences) MRI can produce images with differently accentuated tissues, providing a multitude of clinical applications. One such application is functional magnetic resonance imaging. We should go back in history for a moment to appreciate that this technology, the MRI, was created almost at the same time as CT. Indeed, Paul Lauterbur and Peter Mansfield are credited with the invention of MRI already in the early 70-s. However, the first such machine suitable for clinical application appeared in the 80-s. Their development was dictated by the computational power of the contemporary computers since the reconstruction of the picture from obtained signals required an immense number of mathematical and statistical operations [9]. Then, in the early 90-s Seji Ogawa coined the term blood-oxygenation-level-dependent contrast (BOLD) contrast. It is an observation that oxygenated blood has a higher susceptibility than deoxygenated blood which causes the decrease of MRI signal. He postulated that in the case of brain activation blood consumption in a given area would increase, following an increase in deoxygenated blood and decrease of MRI signal [11]. However, for still unclear reasons, the opposite happens; in active areas the flow does increase, but the blood is instead oxygenated and therefore the signal actually increases. At the time, a new sequence was developed capable of faster image acquisition, the echo-planar imaging (EPI). Kenneth Kwong used this together with BOLD contrast to obtain a firs dynamic depiction of brain activity responding to visual stimuli [12]. That was the beginning of fMRI. Fortunately, it did not take long for the technique to spread and become immensely popular since the existing MRI machines were already capable of performing this function. The main problem of fMRI is the relatively poor temporal resolution compared to EEG and MEG (today tenth of a second). Further issues concern the question whether the activated area really takes part and in which sense in the activity we investigate. These could be improved by better study design and combining fMRI with other techniques as EEG, MEG, etc. Another concern is the fact that any experiment requiring MRI doesn't allow free movement, thus prohibiting experiments in natural environments for art experience like museums, concert halls, etc. In modern clinical settings fMRI is chiefly used to identify and delineate the regions involved in motor, somatosensory, visual, auditory and language functions prior to neurosurgical procedure [9].

Notable Contributions to the Field

Amongst many applications of fMRI ranging from clinic to academia it also provided significant contribution to neuroaesthetics research. Here we look at some of the most important studies performed with the aid of fMRI. Early experiments into beauty and its processing in the brain using fMRI were not focused on the works of art but on other stimuli, most notably human faces. One such study undertaken by Aharon et al. demonstrated that viewing beautiful faces activates the reward circuit, in particular the nucleus accumbens [13].

The first experiments focusing on art were conducted by Zeki and Kawabata in 2003 demonstrated the role of specialized visual areas in perception of different categories of paintings. They also detected that categorizing a picture as beautiful or ugly also correlates with specific brain structures, namely the orbitofrontal cortex, known to be engaged during the perception of rewarding stimuli, but also in motor cortex. They also postulated that due to the intricate connections these areas have with other regions their activation and thus judgment can be influenced by the activity in those regions, most prominently, the anterior cingulate cortex and left parietal cortex. Anterior cingulate cortex has already been known to play a role of pleasurable response to music from the earlier PET study by Blood and Zatorre [14, 15].

Another almost simultaneously published study was that of Vartanian and Goel, who observed that the differential patterns of activation observed in right caudate nucleus, bilateral occipital gyri, left cingulate sulcus, and bilateral fusiform gyri in response to preference ratings are specific examples of their roles in evaluating reward-based stimuli [16].

Jacobsen demonstrated in 2006 that aesthetic judgment shares the same neural substrate like other modes of judgment (social and moral) [17]. A series of studies performed by Calvo-Merino, some of them using fMRI, focused their attention on our processing of dancing (while not dancing ourselves) revealing which aspects of dancing play a role in the aesthetic appreciation [18]. Another important aspect of the perception of art is the difference between experts and non-experts in assessing certain works of art. This matter was the subject of a study conducted by Kirk and Skov that showed increased activation of hippocampus of experts apprising a work of arts, regardless whether they liked it or not, a finding in favor of the hypothesis that there is a strong influence of memory in aesthetic appraisal [19]. In 2009 Lacey et al. tackled an interesting problem in neuroaestheticss, namely, how do we know if a piece of art really is art. They found that images representing art activated specific brain regions associated with the reward circuit regardless whether one liked them or not or considered them beautiful or aesthetically pleasing. This region in turn, did not show activation when viewing non-artistic images [20].

An insight into why do we like certain works of art was given by Iigaya et al. as they found activity patterns in ventral visual pathway associated with the presence of higher level features in the picture making artistic preference predictable [21].

Of course, like in any other field, it helps to sometimes take a step back and try to appreciate the bigger picture, or, as they say in academia, to do a meta-analysis. One of the first such studies done by Brown in 2011 searched for the common brain areas involved in valuation of different stimuli across a large number of studies, irrespective of their design and methods used. What became clear is that these common areas belong to the reward circuit: striatal structures, orbitofrontal cortex, anterior cingulate cortex, insula and amygdala [22]. His and most other meta-analyses involved only the studies which used fMRI or PET. Brown went on to hypothesize, like some have before him, that there are no specialized brain areas for the assessment of art, but rather that it works "piggybacked" on the system for evaluating objects and actions of biological importance. Skov would expand on that by stating that the role of aesthetic appreciation with all of its connections to other brain areas is to help steer our behavior thus constituting a neurobiological phenomenon not fundamentally different from our appreciation of food, social interaction, gaining wealth or sex [23]. Naturally, this kind of conceptualization is not solely a product of neuroimaging findings, but their integration with knowledge gained from evolutionary science, psychology, psychiatry, endocrinology, etc. The story of neuroaestheticss has thus seemingly come a full circle; from purely biological face attractiveness appreciation, through assessment of art works, food, odors and others, to just another biological response helping us move through life.

Why Research in Neuroaesthetics Matters?

One may perceive the neuroaesthetics research and experiments here mentioned as purely academic and even lartpourlartistic pursuit lacking practical application; however, that is not the case. These investigations are not solely focused on museum like environment and experience of art, but explore a more broad, all-human experience of creativity and enjoyment, an experience not limited to art historians and aficionados. It is a part of a great and eternal pursuit to understand ourselves as species and as individuals. Analyzing these phenomena has already yielded considerable insight into how art may be used to help treat various medical conditions. From stroke rehabilitation to Parkinson's disease, early introduction of art into therapeutic process proved to bring considerable benefit [24]. It is through research of these basic principles that we gain insight into the working mechanisms and physiology underlying this very important aspect of our everyday life, how we may preserve it, advance it and mobilize it to alleviate pain and suffering. Hopefully the readers of this chapter will not require the assistance of these insights to help them recover from stroke, but they may find some satisfaction in the fact that it doesn't matter which kind of art or which genre of given art they like. No matter whether you like Dolly Parton, Tokio Hotel, Brahms or your local band, your brain likes them in the same way someone else's likes Iron maiden, Britney Spears, Beatles or Beethoven. As we learn more and more about perceiving beauty one thing seems to become ever more apparent: beauty is in the mind of the beholder [25].As when writing on any other topic I am compelled to believe the neuroimaging techniques in neuroaesthetics deserve a book of their own.

References

1. Chatterjee A, Cardilo E (2022) Brain, beauty, and art
2. Finger S. Origins of neuroscience. A history of explorations into brain function; 2001. https://doi.org/10.1604/9780195146943.
3. Mazzarello P. A Vesalian guide to neuroscience. Stud History Philos Sci Part C Stud History Philos of Biol Biomed Sci. 2016;55:121–3.
4. Broca P. Perte de la parole, ramollissement chronique et destruction partielle du lobe antérieur gauche du cerveau. Bulletin de la Société d'anthropologie de Paris. 1861;2:235–8.
5. Eling P (2014) Reader in the History of Aphasia: from Franz Gall to Norman Geschwind.
6. Zilles K. Brodmann: a pioneer of human brain mapping—his impact on concepts of cortical organization. Brain. 2018;141:3262–78.
7. (2013) Neuroimaging - Wikipedia. In: Neuroimaging - Wikipedia. https://en.wikipedia.org/wiki/Neuroimaging. Accessed 16 Sep 2022
8. Boto E, Holmes N, Leggett J, et al. Moving magnetoencephalography towards real-world applications with a wearable system. Nature. 2018;555:657–61.
9. Bandettini PA (2020) FMRI.
10. (2018) How MRI Works - Part 1 - NMR Basics. In: YouTube. https://www.youtube.com/watch?v=TQegSF4ZiIQ. Accessed 12 Nov 2022.
11. Ogawa S, Lee TM, Kay AR, Tank DW. Brain magnetic resonance imaging with contrast dependent on blood oxygenation. Proc Natl Acad Sci. 1990;87:9868–72.
12. Kwong KK, Belliveau JW, Chesler DA, Goldberg IE, Weisskoff RM, Poncelet BP, Kennedy DN, Hoppel BE, Cohen MS, Turner R. Dynamic magnetic resonance imaging of human brain activity during primary sensory stimulation. Proc Natl Acad Sci. 1992;89:5675–9.
13. Aharon I, Etcoff N, Ariely D, Chabris CF, O'Connor E, Breiter HC. Beautiful faces have variable reward value. Neuron. 2001;32:537–51.
14. Kawabata H, Zeki S. Neural correlates of beauty. J Neurophysiol. 2004;91:1699–705.
15. Blood AJ, Zatorre RJ. Intensely pleasurable responses to music correlate with activity in brain regions implicated in reward and emotion. Proc Natl Acad Sci. 2001;98:11818–23.
16. Vartanian O, Goel V. Neuroanatomical correlates of aesthetic preference for paintings. Neuroreport. 2004;15:893–7.
17. Jacobsen T, Schubotz RI, Höfel L, Cramon DY. Brain correlates of aesthetic judgment of beauty. NeuroImage. 2006;29:276–85.
18. Calvo-Merino B, Glaser DE, Grèzes J, Passingham RE, Haggard P. Action observation and acquired motor skills: an fMRI study with expert dancers. Cereb Cortex. 2004;15:1243–9.

19. Kirk U, Skov M, Christensen MS, Nygaard N. Brain correlates of aesthetic expertise: a parametric fMRI study. Brain Cogn. 2009;69:306–15.
20. Lacey S, Hagtvedt H, Patrick VM, Anderson A, Stilla R, Deshpande G, Hu X, Sato JR, Reddy S, Sathian K. Art for reward's sake: visual art recruits the ventral striatum. NeuroImage. 2011;55:420–33.
21. Iigaya K, Yi S, Wahle IA, Tanwisuth K, O'Doherty JP. Aesthetic preference for art can be predicted from a mixture of low- and high-level visual features. Nat Hum Behav. 2021;5:743–55.
22. Brown S, Gao X, Tisdelle L, Eickhoff SB, Liotti M. Naturalizing aesthetics: brain areas for aesthetic appraisal across sensory modalities. NeuroImage. 2011;58:250–8.
23. Skov M. Aesthetic appreciation: the view from neuroimaging. Empir Stud Arts. 2019;37:220–48.
24. Lo TLT, Lee JLC, Ho RTH. Creative arts-based therapies for stroke survivors: a qualitative systematic review. Front Psychol. 2018;9 https://doi.org/10.3389/fpsyg.2018.01646.
25. Hume D (2010) Four Dissertations.

Chapter 3
How Things Exist: Interdependence Whose Essence Is Compassion

Igor Mosic

Text

"At the spacious, open meadow, bountiful, colorful spring flowers - primroses, lilies and daisies are tenderly bathing the air with gentle perfumes. The flowers communicate their kindness by generously offering their essence while the gentle breeze casts a web of their seductive scents through the fields and the forest nearby. The perfect peace and stillness of this place is interrupted by the sound of nightingales' wings soaring above the meadow. While singing cantabile songs, he starts the journey across the sunny sky with only a few clouds. He rises further and further, and nobody can really tell how far will he reach on his journey of celebrating life and everything that is wholesome [1]".

For but a brief moment, this description captures the attention of an interested reader because it takes one to the meadow through the rich, textured sensory experience.

Each readers' subjective experience of reading about the nightingale as he flies off, can be different. We can get more specific by asking the questions: While reading about it: Did you see the nightingale? Did you hear the melody of his song? Did you see the clouds in the sky? Do you see the sun? Did you feel the scent of the flowers?

Or did you maybe experience vast space which encompasses all sensory representations presented in this short description?

It appears as if the nightingale is being completely free and unrelated with the observer or the sky around him. But does the nightingale really exist in such an independent way?

I. Mosic (✉)
NAM Emotional Hygiene Technology, Rijeka, Croatia
e-mail: igor@n2sed.hr

V. Demarin et al. (eds.), *Mind, Brain and Education*,
https://doi.org/10.1007/978-3-031-33013-1_3

If it would exist in an independent way, it would mean that his flying would not depend on gravity and dynamic buoyancy. If the nightingale would exist as an independent entity, he would not need any causes to be born. No mother, no father, no food to grow and develop. If the nightingale would be an independent entity, he would not have any parts like wings, beak, feathers, or vocal cords to support his singing. Also, if there would be no observer, there would be no-one to give him the name "the nightingale".

The nightingale, just like any other phenomena can exist only in one of two ways: independently or dependently.

The nightingale exists depending on many causes and conditions like his mother and father, egg, nest, his parents' care and food, weather conditions which enabled his development.

The nightingale also depends on parts of his body like beak, vocal cords, wings, heart, inner organs, brain.

His ability to sing depends on his vocal cords, lungs, and air. His ability to fly depends on his wings, sky, gravity, and dynamic buoyancy.

None of those causes, conditions, or parts are "a nightingale". For him, to become a nightingale, a name "nightingale" has to be designated from the side of the observer upon the basis of those causes, conditions, and parts.

Therefore, it is logical to conclude that the nightingale exists in an interdependent way, as opposed to an independent way of existence.

At the first glance, nightingale flying across the sky looks so independent and unrelated to the environment. But it is because of all those causes, conditions, and parts that the nightingale came to existence. There is no nightingale apart from that. Just as there is no wave apart from the sea.

Interdependent existence does not undermine the existence of the nightingale, but the way he exists is not independent or solid as it appears without analysis.

The nightingale is empty of independent existence because it exists in an interdependent way.

In addition to that, nightingale's flight also implies the existence of the sky - vast space within which the flying becomes possible. Just as the wave becomes possible because of existence of the sea.

If with the narrow view we focus on the nightingale as an independent entity, we disregard the continuous process of arising, abiding, and dissolving of countless causes, conditions, parts, and labels which are the womb of nightingales' existence - a vast ocean of possibilities.

Continuously ignoring interdependence as the way things exist, directs attention to the appearance of an independently existent objects and their significance for or against us. That kind of approach promotes limited view which obscures the complete view and understanding of causes, conditions, and our own personal responsibility in the world.

Just as with the nightingale, the existence of a human being implies the existence of countless causes, conditions, parts, and name given to a specific person. The person does not exist independently, which means it is empty of independent existence, because it exists in an interdependent way.

Each person exists depending on many causes and conditions like: mother and father, parents' care and food, environment within which is brought up, education, interaction with others. Every person depends on parts of his body and mind. All good in every persons life, all the benefits received during the course of ones life are the result of the affection and care in direct and indirect way, received from countless people like teachers, doctors, nurses, family members, coworkers, neighbours, and friends.

Persons' ability to walk depends on legs, but also on the earth and the gravity. Ability to talk, apart from depending on vocal cords, depends on air. Each persons' ability to see, apart from depending on eyes, depends on sunlight. Each persons' ability to think depends on the mind.

And also, for each person a specific name is designated on the basis of those causes, conditions, and parts, from the side of the observer, usually parents.

Even though at the first glance, with the limited view each person appears as an independent, solid entity, not related to the environment, parts, causes, or conditions. It also appears as an independent entity completely separated from our subjective experience of it.

But looking deeply, persons' parts, causes, conditions, and name are the person, and it is because of all those causes, conditions, parts, and name that the specific person comes to existence, but apart from that there is no separate, independent, solid person existing unrelated to all those causes, conditions, parts, and name.

The person does not exist independently, which means it is empty of independent existence because it exists in an interdependent way.

There is no nightingale apart from all elements which make the nightingale possible. Just as there is no wave apart from the sea.

"There is no phenomenon that is not dependently arisen. Therefore, there is no phenomenon which is not empty." [2] (of independent existence)

Believing in an independent, solid person, nightingale, or the wave is believing in an impossible way of existence.

Impossible way of existence believes things do not exist in accordance with reality.

When our way of thinking believes in an impossible way of existence, stream of countless different problems will follow in our life, just as if we jump from the plane without the parachute. If we ignore the natural law of gravity, our landing on the ground will not turn to our advantage.

Limited view focused on self as the main object of observation and judging everything through the lens of its significance for or against us, is following an impossible way of existence as true, because that kind of view ignores interdependent way of existing.

Limited view of impossible way of existence is focused on self, which is unrelated to causes, conditions, parts, and the environment which made it possible. That is limited because it is focused on our own benefit, being indifferent, or disregarding the benefit of all elements that made and are making "the person" possible.

It is as if the person would be interested in his or her own well-being, but not being interested when his or her arm was wounded. That kind of self-absorbed view is completely against the logic and intelligence. It is the view which harms the one who holds it, because that person does not comprehend the complete view of interdependent existence.

If we care for the well-being of our own body, it is logical that the left hand will take genuine care of the right hand when the latter is wounded. If we care for the well-being of self, it is logical to take genuine care for the elements that make us possible - other people and the environment.

Because a simple, yet profound truth states: "Just like me, all human beings want to be happy. Just like me, all human beings want to be free from suffering."

Therefore, focusing just on the benefit for self and the significance of others' behaviour for or against self, would be logical if human beings would exist in an independent, unrelated, solid way, but it is completely against logic and intelligence if we understand the interdependent way of existence.

Even if the baby throws up on the mother, mother will not harm the baby, but lovingly take care for it and make sure the baby is safe and calm, so the baby would not repeat the unwanted behaviour. Because the well-being of the baby completely depends on the kindness of the mother.

The well-being of every human being and the humanity completely depends on our kindness and genuine care for each other.

Not because someone said so, or it is imposed on you by some kind of authority, but because you deeply understand it to be the way things exist.

The more we work for just for our own happiness, the more frustration, disturbance, irritation, and problems we have in life. Because the unrestrained mind which believes in an impossible way of existence and is not in accordance with reality, can never be satisfied. It is like quenching thirst with salty water, which has no end. All those problems become a fertile ground for various physical and mental sicknesses.

Once the complete view of existence which is in accordance with reality is established, limited view of impossible way of solid, independent existence cannot operate because those are contradictory ways of thinking. Developing complete view breaks the axle of perpetual human mental suffering and physical sickness.

"Consciousness pervades the experience of its objects and the experience of the organs, if the consciousness is transformed, or ones' mode of experience of consciousness is transformed into the pure realisation of final mode of abiding or the ultimate truth, then the appearances of the objects, and also the organs themselves, will become pure and sacred." [3]

In adjunction to developing of the complete view of existence, the more our efforts are directed on using available resources to work for the benefit of others, for all those elements that make us possible, the more happiness and joy will instantly enter our lives.

> "Whatever joy there is in this world, all comes from desiring others to be happy,
> Whatever suffering there is in the world, all comes from desiring myself to be happy [4]".

The moment we extend our self-absorbed view from our own benefit to the benefit of others, the view becomes more complete, because it is in accordance with the wishes of others, and it is completely in accordance with reality. As a consequence, even if we do not want it, more peace, joy, happiness, and success will immediately enter our life.

> "Whatever degenerations there are in the world,
>   The root of all this is ignorance,
>   You thought that it is dependent origination,
>   The seeing of which will undo this ignorance [5]".

Practical model of deeper understanding and familiarising with the way of thinking of dependent origination which is in accordance with reality is presented through Neuroplasticity Activation Modeling (NAM) - technology of emotional hygiene. NAM is an elegant and practical model for transforming destructive states of mind and cultivating inner-peace and well-being for self and others. NAM is developed based on Tibetan neuroscience and yogic meditation techniques. NAM is tailored for busy people living "western" type of lifestyle. From 2017. it is regularly thought online and in-person to various groups of interested participants, as well as specific support groups.

NAM has proven to be very beneficial in numerous ways because learning about human mind and how to use human intelligence for breaking the axle of perpetual mental and physical suffering becomes the source of inner-strength and this confidence overflows to all other areas of life.

"Once the conventional nature of mind - the mere clear knower - has been identified, then, through analyzing its nature, finally we will gradually be able to identify the ultimate nature of the mind, if that is done, there is great progress unlike anything else [6]".

The intention of this article is to present the mode of abiding of all things, using human intelligence and deductive logic as the basis of analysis and understanding. The intention of this article is not to impose any conclusion, but to inspire well-intended, interested reader to check whether the presented ideas are valid. If they are valid, what could be their specific benefit in the process of treating mental, as well as physical sickness?

It is my wish that this article serves as the spur for the rest of the scientific community to use available resources and explore more the potential of human consciousness in breaking the axle of human mental suffering and physical sickness, as opposed to merely treating the symptoms.

In conclusion, you as a person, all other beings, and our environment came into being in an interdependent way and all will cease to exist in that same manner, and during that process, all things, events, humans, and other living beings communicate to us the wisdom of interdependence whose essence is compassion.

And that is why skilfully cultivating genuine, kind heart filled with compassion is a sign of superior intelligence and most meaningful way of life which is in accordance with reality.

Because of that, a life without wise, kind, compassionate heart is like a forest without nightingales' songs - all objects are there, but something is missing.

References

1. Mošič I. Magnificent the Tale of Love, Compassion, and Reality. Rijeka, Croatia: N2SED Ltd.; 2022. p. 25.
2. Guru Arya Nagarjuna Mulamadhyamakakarika – Chapters 26, 18, 23, 24, 22, and 1" [Internet]. In: Translator and editor Geshema Kelsang Wangmo. Chapter 24, Verse 19. Available from: http://tushita.info/wp-content/uploads/2018/07/Nagarjunas-Mulamadhyamakakarika-Translation-by-Geshe-Kelsang-Wangmo-2018.pdf
3. Khenchen Thrangu Rinpoche. Medicine Buddha Teachings. Ithaca, NY: Snow Lion Publications; 2004. p. 6.
4. Guru Arya Shantideva. A Guide to the Bodhisattva's Way of Life. Dharamsala: Library of Tibetan Works and Archives; 1979. p. 110.
5. Guru Je Tsongkhapa, In Praise of Dependent Origination n.p., verse 2
6. His Holiness the Dalai Lama, The Buddhism of Tibet and The Key to the Middle Way, Boston: Snow Lion; 2002. n.sp.

Chapter 4
Disrupted Sleep and Brain Functioning

Darko Orešković, Marina Raguž, and Filip Derke

Sleep is a recurrent, physiological phenomenon, which consists of many measurable factors [1] and is ubiquitous throughout the natural world [2–5]. It is a highly active, easily reversible process, which is crucial not only for the physical and mental well-being of virtually all living organisms, but also for the very concepts we as humans have of ourselves and the world around us [6]. There are many theories regarding the possible function of sleep, ranging from the physiological explanations such as rest of individual cells [7] to behavioral explanations of why a biological system needs periodic inactivity [8].

Sleep can be impaired in many ways. Indeed, there is a growing understanding of how the modern lifestyle disrupts the natural circadian rhythm in humans, consequences of which are still not sufficiently explored [9]. The current classification of sleep disorders consists of several clinical entities such as insomnia, parasomnia, hyper-somnolence, sleep-related movement disorders, etc. [10]. However, we will refer to all of this broad pathology as "sleep disruption," primarily for clarity and simplicity sake. In this chapter we will explore how a single specific pathology (a malignant brain tumor) can disrupt sleep, after which we will list some of the many known consequences of such sleep disruption.

D. Orešković (✉) · M. Raguž
Department of Neurosurgery, University Hospital Dubrava, Zagreb, Croatia

Catholic University of Croatia, School of Medicine, Zagreb, Croatia

F. Derke
University Hospital Dubrava, Zagreb, Croatia

Department of Psychiatry and Neurology, Faculty of Dental Medicine and Health, JJ Strossmayer University of Osijek, Zagreb, Croatia
e-mail: filip@mozak.hr

© The Author(s), under exclusive license to Springer Nature Switzerland AG 2023
V. Demarin et al. (eds.), *Mind, Brain and Education*,
https://doi.org/10.1007/978-3-031-33013-1_4

Sleep and a Malignant Disease of the Brain

The annual prevalence of tumors of the central nervous system (CNS) is little over 22 per 100,000 [11]. Around a third of these lesions are malignant. Among the malignant tumors, gliomas are by far the most common type, constituting over 80% of the number. Among gliomas, the most aggressive type (glioblastoma) is the most common one, making up over a half of all newly diagnosed gliomas [12, 13]. The five-year survival of patients with malignant CNS tumors is around 30%, with patients being diagnosed a glioblastoma having a five-year survival rate of less than 5%. All this goes to show how malignant CNS tumors are some of the most aggressive human malignancies today. It also shows how the vast accumulated knowledge on the disease origin and progression still has not translated into significant improvement of the overall survival of these patients. New treatment modalities are therefore desperately needed. Besides the devastating diagnosis of a malignant brain tumor, these patients often experience a wide variety of severe symptoms, which significantly diminish their quality of life [15–17]. There has been an increasing awareness of the importance of supportive and palliative care in patients suffering from malignant brain tumors, especially those in whom other treatment modalities have been exhausted [15–17]. And one of the most commonly reported symptoms is sleep disturbance [1, 14, 18–21].

Indeed, an emerging field of interest in sleep disruption in patients suffering from malignant diseases has recently gained much attention [21, 22]. It has been shown that disrupted sleep is one of the most common complaints in patients undergoing oncological treatment, with patients suffering from malignant brain tumors being especially susceptible [14, 20]. Furthermore, the risk of developing several different neoplasms can be directly correlated with different sleep disturbances [23–26]. On the other hand, patients suffering from various types of neurological disorders very often also suffer from some sort of disrupted sleeping pattern [27, 28], indicating that patients with CNS pathology are very susceptible to sleep impairment. Research into the complex relationship which malignant brain tumors could have with disrupted sleep is quite scarce. It usually addresses the severity and frequency with which disrupted sleep occurs as a symptom in these patients [18, 19, 29] as well as the effects of the oncological treatment on sleeping patterns [18], but usually does not explore in detail the possible pathophysiological mechanisms by which disrupted sleep could actually detrimentally affect patients'ability to fight off the disease, or the ways in which malignant brain lesions themselves could actually impair sleep. We therefore explore this issue in greater detail, first by listing the many ways in which a malignant brain lesion can disrupt sleep.

1. The first possible mechanism is pretty straightforward. A malignant lesion can directly disrupt and destroy the neural projections or the structures involved in the sleep-wake circuitry (the hypothalamus for example) which would then, in turn disrupt the sleeping patterns of a patient. It has indeed been reported that patients with a malignant tumor in these regions could have severely impaired sleeping cycles [30].
2. The second possibility is that a malignant lesion could disrupt sleeping patterns even if not directly located in the sleep-wake circuitry. This could happen since the brain function inevitably relies on its structure. And, as mentioned earlier,

one of the most important brain functions is sleep. Therefore, it is likely that the complex changes occurring during sleep are not limited to certain cerebral areas, but are instead function of the entire brain, at least to some extent. It would follow then that any disruption in the cerebral structure can impair sleep to a greater or lesser degree. More specifically, even malignant lesions in the brain regions that are not usually considered to be crucial for sleep could mechanically disrupt the complex cerebral structure and its function. Indeed, sleeping is disrupted in many other neurologic pathological conditions that compromise the general brain structure [28, 31]. Also of note is that the growing intracranial mass can cause many other non-specific symptoms (such as headaches) during sleep and thus further impairing the sleeping schedule.

3. Another possibility is an indirect physical effect. Indeed, a growing intracranial mass can cause a variety of nonspecific symptoms. This is especially true for highly proliferative malignant intracranial tumors, where the compensatory mechanisms of an organism are rapidly rendered insufficient due to the fast and infiltrative tumor growth. The most commonly reported indirect physical effect of a malignant lesion on brain function is disruption in a patient's breathing patterns. Breathing is a highly complex physiological phenomenon. It is tightly controlled in several distinct control points, namely the central control at the level of the brainstem, effector control (respiratory muscles for example), and sensory control. The central breathing control is primarily performed by three large neuron groups in the pons and medulla [32, 33]. Disruption or damage in these neurons and neuron groups can lead to severe breathing disorders and even death (Ondine's curse). Beside malignant lesions, such damage can occur in other types of neuropathology as well, such as in multiple sclerosis [34]. Breathing is also controlled at a higher level, namely through corticobulbar and corticospinal pathways [32]. This suprapontine control allows for voluntary respiration modification. Destruction of these pathways can also lead to severe breathing disorders [35]. Another possible mechanism by which malignant brain tumors can compromise breathing is through elevated intracranial pressure, which arises due to the growing lesion. This in turn can cause an indirect compression of the breathing centers, similarly to what occurs in patients with a Chiari malformation [36]. All of the mechanisms described above can compromise breathing and breathing patterns of a patient, causing various types of dyspneas and apneas, especially during sleep. Indeed, it has already been recognized that disordered breathing has a much higher prevalence in patients with neurologic disorders such as stroke and epilepsy [37]. While these sleeping disorders have been labeled highly prevalent and grossly under-recognized [38], they have also not been sufficiently investigated in patients with malignant brain tumors. This seems remarkable since it has already been shown that apneas and insomnias lead to a functional reorganization of the brain [39, 40]. Moreover, this type of sleep disruption has been connected to a worse outcome and prognosis in several diseases, including malignant ones [41]. Furthermore, when different effects of acute, chronic, and cyclic hypoxia were investigated with regards to tumor aggressiveness, it has been shown that cyclic hypoxia (such as the one occurring in patients with sleep apneas) actually significantly enhances tumor

cell aggressiveness by altering various cancer hallmarks, such as angiogenesis, metastasis, cell proliferation, and/or inflammation. Although these findings have not yet been tested on brain tumor cells, it seems possible that similar effects could take place in them as well. Knowing how common and often undetected various breathing disorders are in the general population [37, 42], and taking into account the pathophysiological mechanisms described above, it is likely that patients with brain tumors are also susceptible to this type of pathology. Unfortunately, research is still fairly limited and warrants further effort into deciphering the complex relationship which brain tumors, sleeping, and breathing have with each other.

4. Finally, sleep could be even disrupted chemically. Alongside all of the aforementioned mechanisms by which malignant brain tumors can disrupt sleeping patters of a patient through their physical interaction with normal brain tissue, there is also another important way in which this disruption can occur. The disruption in question is a chemical one, through glutamate. Glutamate is not only the predominant excitatory neurotransmitter in the central nervous system, but it also has a crucial role in regulating sleep-wake cycles [43]. More specifically, it has a significant excitatory role in promoting and maintaining wakefulness. This is in fact true whether the molecule is located in the intra-synaptic [44] or extrasynaptic space [45]. Besides this physiological role, cellular glutamate metabolism has gained much interest recently due to its apparent crucial role in the survival of malignant cells, especially in their cellular growth and proliferation [46]. Glutamate is mostly secreted into the synaptic cleft by the cysteine-glutamate transporter (system xc), which exchanges it with extracellular cysteine. Glutamate cannot passively diffuse back to the intracellular space, nor can it be metabolized by extracellular enzymes. It is therefore transported into the intracellular space primarily by molecules known as the excitatory amino acid transporters (EAATs) [47]. Glutamate release and uptake to and from the synaptic cleft are both tightly regulated through the aforementioned molecules. If this tight control is disrupted, it leads to glutamate accumulation and causes detrimental excitotoxicity [48, 49]. This type of damage can also occur in the presence of necrosis, which causes the intracellular glutamate to leak into the extracellular space, damaging cells, and causing a cascade of neurotoxocity [49] and further cellular decay [51]. This over-abundance of extracellular glutamate is detrimental not only to the surrounding cells, but also to the cellular ability to cope with reactive oxygen radicals. This happens since the abundance of extracellular glutamate impairs the system xc, which normally exchanges it with extracellular cysteine. The deficiency of this transporter leads to an intracellular lack of cysteine, which in turn causes a complex cascade in which cellular cysteine and subsequent glutathione production are impaired [49], therefore disabling the anti-oxidative properties of glutathione (see earlier). The abundance of glutamate in malignant brain tumors is well-known [52–54]. And even though in the past it has been proposed that this abundance is caused primarily by tumor cells necrosis (thus being merely a side-effect of necrosis), it has since been repeatedly shown that malignant cells express a significant upregulation of sys-

tem xc as well as a significant downregulation of EAAT molecules. Both of these changes of genetic expression allow for a higher extracellular glutamate concentration [49]. This would therefore imply an important role that extracellular glutamate has in the survival of malignant cells. Knowing the positive effect that disrupted sleep has on tumor cells, it seems likely that at least one of the functions of this active glutamate secretion is actually chemically disrupting sleep.

Consequences of Sleep Disruption

Consequences of sleep disruption in general and by a malignant lesion in particular are numerous. First of all, for the normal functioning of the brain itself [50]. These effects can be classified into moment-to-moment disruption and broader, more general changes of the brain structure and function. The former are fluctuations in brain activity that occur across minutes and are especially important for example in attention and working memory. The latter are on the other hand a complex set of changes which highly depend on the specific function and anatomical region. The mesolimbic reward system for example (which includes the midbrain ventral tegmental area, the striatum and regions of the prefrontal cortex), has repeatedly been shown to be very sensitive to sleep disruption. Such changes have been observed, to a varying degree, in virtually all of the studied neural and behavioral brain systems. And all of these different changes in the brain structure and function then in turn lead to various disruptions in virtually all aspects of human behavior, cognition and affect through various mood disorders, memory impairment, increased dementia risk, etc. [9]. Beside this devastating effects on the normal brain functioning, sleep disruption has a well-known detrimental role for the entire organism. We will list several of these systemic changes here.

Phase Shifts

Disrupted sleeping patterns have been found to significantly influence the transcription of the so-called clock genes [55]. These genes, alongside the circadian "master clock" in the suprachiasmatic nuclei of the brain, govern the rhythmic circadian synchronization of almost all of the physiological processes within the body [12]. The physiological circadian clock functions as a tumor suppressor at the systemic, molecular, and cellular levels. Indeed, these circadian rhythms have been found to be so important that their various disruptions lead to the so-called phase shifts, which have been linked to both tumorigenesis and tumor progression [12, 56]. In fact, there is a growing awareness of how chronotherapy could improve the efficacy of cancer treatment and the quality of patients' lives [56, 57]. Malignant brain tumors have also been found to rely heavily on the expression of clock genes, namely in their growth [58], cellular proliferation [59], and migration [60].

Reduced Antioxidant Levels

Excess production of free radicals (or oxidative stress) plays an important role in the metabolism of all living aerobic organisms, including humans. These free radicals, also called reactive oxygen species, induce oxidative damage to certain cellular macromolecules, and this damage has been linked to many common human diseases including cancer [61]. Several protective cellular mechanisms have evolved to counter this damage, namely in the form of various antioxidant molecules and antioxidant enzymes [62]. Of these, glutathione has been found to be the most important mammalian cellular antioxidant molecule, with a crucial role in cell protection against oxidative stress [62]. Glutathione is a well-known antioxidant molecule with a significant protective role against oxidative free-radicals and carcinogens [63]. However, glutathione levels are not constant. They actually strongly depend on the circadian rhythm. More specifically, they are significantly elevated during sleep [64]. Thus, due to the disruption of sleep cycles, physiologically elevated levels of glutathione are diminished, making cells more susceptible to oxidative damage. On the other hand, the therapeutic potential of glutathione is highly complex and controversial. Indeed, beside its protective role, glutathione also significantly influences the response to therapy of the tumor cells themselves, namely it allows these cells to suffer less damage from oncological therapy [65, 66]. Bansal and Simon [63] offer a more in-depth analysis of the immensely complex dual role which glutathione plays in cancer patients. The knowledge of these complex systems and mechanisms is still insufficient, and further research is needed.

Melatonin Depletion

Melatonin is a pineal hormone that is involved in the circadian regulation and facilitation of sleep [67]. Besides in the pineal gland, melatonin is also synthesized in various other organs, tissues, and cells, also in a circadian fashion, with high rhythm amplitude and a prominent nocturnal maximum. Melatonin is also often called a hormone of darkness since all of the body's melatonin is secreted at night-time [68]. The modern, industrialized lifestyle with its dependency on light disrupts significantly the synthesis and the secretion of melatonin [68, 69]. Even different diets have been found to affect the melatonin levels in the organism [70]. It has also been found that disrupted sleeping schedules significantly further diminish melatonin secretion [68, 71]. And while melatonin is currently successfully administered in the treatment of restoring the diurnal rhythm [72], its complex metabolism has repeatedly been suggested as a possible therapeutic target in oncological treatment [73, 74]. This type of research is still unfortunately in its infancy and further investigation is needed.

Metabolic Changes

Even short-term sleep loss has been shown not only to disrupt the physiological functioning of various metabolic processes such as glucose regulation or cortisol and insulin secretion, but also to lead to an increased appetite and caloric intake [9]. Chronic sleep disruption has also been linked to severe complications such as cortisol and insulin dysregulation, obesity, and diabetes mellitus [9, 75].

Beside these, there are many more systemic effects which have been linked to sleep disruption such as cardiovascular disorders [76], systemic and local inflammation [77, 78], immunosuppression [79, 80] and epigenetic changes [81]. All this goes to show how complex these changes are, and how future research is needed.

Conclusions

Sleep disruption has many detrimental effects on an organism. It impairs many, if not all of the normal functioning of the brain itself, while also having many devastating effects on the organism as a whole. In this chapter we tried to look at just one of the many common pathologies affecting the brain, that of a malignant disease and how it can severely disrupt sleeping patterns and cause many systemic effects. The topics raised here have been explored in greater detail in our previous work [82]. The vast majority of research into sleep disruption usually looks at acute sleep disruption and leaving out the more chronic changes. These changes however are especially important in the modern industrialized world where the chronic sleep deficiency is already well-documented. Also, understanding the exact ways in which the brain recovers from chronic and acute sleep disruption is still unfortunately lacking. All this goes to show that research into the complex nature of sleep is still unexplored with many interesting findings waiting to be discovered.

References

1. Redeker NS, Pigeon WR, Boudreau EA. Incorporating measures of sleep quality into cancer studies. Support Care Cancer. 2015;23:1145–55. https://doi.org/10.1007/s00520-014-2537-0.
2. Cirelli C, Tononi G. Is sleep essential? PLoS Biol. 2008;6:e216. https://doi.org/10.1371/journal.pbio.0060216.
3. Kim JA, Kim HS, Choi SH, Jang JY, Jeong MJ, Lee SI. The importance of the circadian clock in regulating plant metabolism. Int J Mol Sci. 2017;18:18. https://doi.org/10.3390/ijms18122680.
4. Miyazaki S, Liu CY, Hayashi Y. Sleep in vertebrate and invertebrate animals, and insights into the function and evolution of sleep. Neurosci Res. 2017;118:3–12. https://doi.org/10.1016/j.neures.2017.04.017.

5. Villafuerte G, Miguel-Puga A, Rodríguez EM, Machado S, Manjarrez E, Arias-Carrión O. Sleep deprivation and oxidative stress in animal models: a systematic review. Oxidative Med Cell Longev. 2015;2015:234952. https://doi.org/10.1155/2015/234952.

6. Hong CC, Fallon JH, Friston KJ, Harris JC. Rapid eye movements in sleep furnish a unique probe into consciousness. Front Psychol. 2018;9:2087. https://doi.org/10.3389/fpsyg.2018.02087.

7. Vyazovskiy VV, Harris KD. Sleep and the single neuron: the role of global slow oscillations in individual cell rest. Nat Rev Neurosci. 2013;14:443–51. https://doi.org/10.1038/nrn3494.

8. Field JM, Bonsall MB. The evolution of sleep is inevitable in a periodic world. PLoS One. 2018;13:e0201615. https://doi.org/10.1371/journal.pone.0201615.

9. McEwen BS, Karatsoreos IN. Sleep deprivation and circadian disruption: stress, allostasis, and allostatic load. Sleep Med Clin. 2015;10:1–10. https://doi.org/10.1016/j.jsmc.2014.11.007.

10. Sateia MJ. International classification of sleep disorders-third edition. Chest. 2014;146:1387–94. https://doi.org/10.1378/chest.14-0970.

11. Barnholtz-Sloan JS, Ostrom QT, Cote D. Epidemiology of brain tumors. Neurol Clin. 2018;36:395–419. https://doi.org/10.1016/j.ncl.2018.04.001.

12. Fritschi L, Glass DC, Heyworth JS, Aronson K, Girschik J, Boyle T, et al. Hypotheses for mechanisms linking shiftwork and cancer. Med Hypotheses. 2011;77:430–6. https://doi.org/10.1016/j.mehy.2011.06.002.

13. Wirsching HG, Galanis E, Weller M. Glioblastoma. Handb Clin Neurol. 2016;134:381–97. https://doi.org/10.1016/B978-0-12-802997-8.00023-2.

14. Armstrong TS, Vera-Bolanos E, Acquaye AA, Gilbert MR, Ladha H, Mendoza T. The symptom burden of primary brain tumors: evidence for a core set of tumor- and treatment-related symptoms. Neuro-oncol. 2016;18:252–60. https://doi.org/10.1093/neuonc/nov166.

15. Batash R, Asna N, Schaffer P, Francis N, Schaffer M. Glioblastoma multiforme, diagnosis and treatment; recent literature review. Curr Med Chem. 2017;24:3002–9. https://doi.org/10.2174/0929867324666170516123206.

16. Braun K, Ahluwalia MS. Treatment of glioblastoma in older adults. Curr Oncol Rep. 2017;19:81. https://doi.org/10.1007/s11912-017-0644-z.

17. Perrin SL, Samuel MS, Koszyca B, Brown MP, Ebert LM, Oksdath M, et al. Glioblastoma heterogenity and the tumour microenvironment: implications for preclinical research and development of new treatments. Biochem Soc Trans. 2019;47:625–38. https://doi.org/10.1042/BST20180444.

18. Armstrong TS, Shade MY, Breton G, Gilbert MR, Mahajan A, Scheurer ME, et al. Sleep-wake disturbance in patients with brain tumors. Neuro-oncol. 2017;19:323–35.

19. Jeon MS, Dhillon HM, Agar MR. Sleep disturbance of adults with a brain tumor and their family caregivers: a systematic review. Neuro-oncol. 2017;19:1035–46. https://doi.org/10.1093/neuonc/nox019.

20. Kim BR, Chun MH, Han EY, Kim DK. Fatigue assessment and rehabilitation outcomes in patients with brain tumors. Support Care Cancer. 2012;20:805–12. https://doi.org/10.1007/s00520-011-1153-5.

21. Rha SY, Lee J. Symptom clusters during palliative chemotherapy and their influence on functioning and quality of life. Support Care Cancer. 2017;25:1519–27. https://doi.org/10.1007/s00520-016-3545-z.

22. Howell D, Oliver TK, Keller-Olaman S, Davidson JR, Garland S, Samuels C, et al. Sleep disturbance in adults with cancer: a systematic review of evidence for best practices in assessment and management for clinical practice. Ann Oncol. 2014;25:791–800. https://doi.org/10.1093/annonc/mdt506.

23. Chen Y, Tan F, Wei L, Li X, Lyu Z, Feng X, et al. Sleep duration and the risk of cancer: a systematic review and meta-analysis including dose–response relationship. BMC Cancer. 2018;18:1149. https://doi.org/10.1186/s12885-018-5025-y.

24. Kakizaki M, Inoue K, Kuriyama S, Sone T, Matsuda-Ohmori K, Nakaya N, et al. Sleep duration and the risk of prostate cancer: the Ohsaki cohort study. Br J Cancer. 2008;99:176–8. https://doi.org/10.1038/sj.bjc.6604425.

25. Kakizaki M, Kuriyama S, Sone T, Matsuda-Ohmori K, Hozawa A, Nakaya N, et al. Sleep duration and the risk of breast cancer: the Ohsaki cohort study. Br J Cancer. 2008;99:1502–5. https://doi.org/10.1038/sj.bjc.6604684.
26. Thompson CL, Larkin EK, Patel S, Berger NA, Redline S, Li L. Short duration of sleep increases risk of colorectal adenoma. Cancer. 2011;117:841–7. https://doi.org/10.1002/cncr.25507.
27. Castriotta RJ, Murthy JN. Sleep disorders in patients with traumatic brain injury: a review. CNS Drugs. 2011;25:175–85. https://doi.org/10.2165/11584870-000000000-00000.
28. Provini F, Lombardi C, Lugaresi E. Insomnia in neurological diseases. Semin Neurol. 2005;25:81–9. https://doi.org/10.1055/s-2005-867074.
29. Yavas C, Zorlu F, Ozyigit G, Gurkayanak M, Yavas G, Yuce D, et al. Health-related quality of life in high-grade glioma patients: a prospective single-center study. Support Care Cancer. 2012;20:2315–25. https://doi.org/10.1007/s00520-011-1340-4.
30. Stahl SM, Layzer RB, Aminoff MJ, Townsend JJ, Feldon S. Continuous cataplexy in a patient with a midbrain tumor: the limp man syndrome. Neurology. 1980;30:1115–8. https://doi.org/10.1212/WNL.30.10.1115.
31. Tesoriero C, Del Gallo F, Bentivoglio M. Sleep and brain infections. Brain Res Bull. 2019;145:59–74. https://doi.org/10.1016/j.brainresbull.2018.07.002.
32. Corne S, Bshouty Z. Basic principles of control of breathing. Respir Care Clin N Am. 2005;11:147–72. https://doi.org/10.1016/j.rcc.2005.02.011.
33. Newton K, Malik V, Lee-Chiong T. Sleep and breathing. Clin Chest Med. 2014;35:451–6. https://doi.org/10.1016/j.ccm.2014.06.001.
34. Braley TJ, Boudreau EA. Sleep disorders in multiple sclerosis. Curr Neurol Neurosci Rep. 2016;16:50. https://doi.org/10.1007/s11910-016-0649-2.
35. Discolo CM, Akst LM, Schlossberg L, Greene D. Anterior cranial fossa gliolastoma with sleep apnea as initial manifestation. Am J Otolaryngol. 2005;26:327–9. https://doi.org/10.1016/j.amjoto.2005.01.014.
36. Leu RM. Sleep-related breathing disorders and the Chiari 1 malformation. Chest. 2015;148:1346–52. https://doi.org/10.1378/chest.14-3090.
37. Foldvary-Schaefer NR, Waters TE. Sleep-disordered breathing. Continuum (Minneap Minn). Sleep Neurol. 2017;23(4):1093–116.
38. Ramar K, Olson EJ. Management of common sleep disorders. Am Fam Physician. 2013;88:231–8.
39. Khazaie H, Veronese M, Noori K, Emamian F, Zarei M, Ashkan K, et al. Functional reorganization in obstructive sleep apnoea and insomnia: a systematic review of the resting-state fMRI. Neurosci Biobehav Rev. 2017;77:219–31. https://doi.org/10.1016/j.neubiorev.2017.03.013.
40. Urrila AS, Artiges E, Massicotte J, Miranda R, Vulser H, Bézivin-Frere P. Sleep habits, academic performance, and the adolescent brain structure. Sci Rep. 2017;7:41678. https://doi.org/10.1038/srep41678.
41. Bonsignore MR, Baiamonte P, Mazzuca E, Castrogiovanni A, Marrone O. Obstructive sleep apnea and comorbidities: a dangerous liaison. Multidiscip Respir Med. 2019;14:8. https://doi.org/10.1186/s40248-019-0172-9.
42. Young T, Peppard PE, Gottlieb DJ. Epidemiology of obstructive sleep apnea: a population health perspective. Am J Respir Crit Care Med. 2002;165:1217–39. https://doi.org/10.1164/rccm.2109080.
43. Holst SC, Landolt HP. Sleep-wake neurochemistry. Sleep Med Clin. 2018;13:132–46. https://doi.org/10.1016/j.jsmc.2018.03.002.
44. Pedersen NP, Ferrari L, Venner A, Wang JL, Abbott SB, Vujovic N, et al. Supramamillary glutamate neurons are a key node of the arousal system. Nat Commun. 2017;8:1405. https://doi.org/10.1038/s41467-017-01004-6.

45. Pal B. Involvment of extrasynaptic glutamate in physiological and pathophysiological changes of neuronal excitability. Cell Mol Life Sci. 2018;75:2917–49. https://doi.org/10.1007/s00018-018-2837-5.

46. Maus A, Peters GJ. Glutamate and α-ketoglutarate: key players in glioma metabolism. Amino Acids. 2017;49:21–32. https://doi.org/10.1007/s00726-016-2342-9.

47. Danbolt NC. Glutamate uptake. Prog Neurobiol. 2001;65:1–105. https://doi.org/10.1016/S0301-0082(00)00067-8.

48. Nicholls D, Attwell D. The release and uptake of excitatory amino acids. Trends Pharmacol Sci. 1990;11:462–8. https://doi.org/10.1016/0165-6147(90)90129-V.

49. Noch E, Khalili K. Molecular mechanisms of necrosis in glioblastoma: the role of glutamate excitotoxicity. Cancer Biol Ther. 2009;8:1791–7. https://doi.org/10.4161/cbt.8.19.9762.

50. Krause AJ, Simon EB, Mander BA, et al. The sleep-deprived human brain. Nat Rev Neurosci. 2017;18(7):404–18. https://doi.org/10.1038/nrn.2017.55.

51. Louis DN. Molecular pathology of malignant gliomas. Annu Rev Pathol. 2006;1:97–117. https://doi.org/10.1146/annurev.pathol.1.110304.100043.

52. Corsi L, Mescola A, Alessandrini A. Glutamate receptors and glioblastoma multiforme: an old "route" for new perspectives. Int J Mol Sci. 2019;20:20. https://doi.org/10.3390/ijms20071796.

53. Majos C, Alonso J, Aguilera C, Serrallonga M, Coll S, Acebes JJ, et al. Utility of proton MR spectroscopy in the diagnosis of radiologically atypical intracranial meningiomas. Neuroradiology. 2003;45:129–36. https://doi.org/10.1007/s00234-002-0933-5.

54. Monleon D, Morales JM, Gonzalez-Darder J, Talamantes F, Cortes O, et al. Benign and atypical meningioma metabolic signatures by high-resolution magic-angle spinning molecular profiling. J Proteome Res. 2008;7:2882–8. https://doi.org/10.1021/pr800110a.

55. Heyde I, Kiehn JT, Oster H. Mutual influence of sleep and circadian clocks on physiology and cognition. Free Radic Biol Med. 2018;119:8–16. https://doi.org/10.1016/j.freeradbiomed.2017.11.003.

56. Fu L, Lee CC. The circadian clock: pacemaker and tumour suppressor. Nat Rev Cancer. 2003;3:350–61. https://doi.org/10.1038/nrc1072.

57. Sulli G, Lam MTY, Panda S. Interplay between circadian clock and cancer: new frontiers for cancer treatment. Trends Cancer. 2019;5:475–94. https://doi.org/10.1016/j.trecan.2019.07.002.

58. Dong Z, Zhang G, Qu M, Gimple RC, Wu Q, Qiu Z, et al. Targeting glioblastoma stem cells through disruption of the circadian clock. Cancer Discov. 2019;9:1556–73. https://doi.org/10.1158/2159-8290.CD-19-0215.

59. Yu M, Li W, Wang Q, Wang Y, Lu F. Circadian regulator NR1D2 regulates glioblastoma cell proliferation and motility. Oncogene. 2018;37:4838–53. https://doi.org/10.1038/s41388-018-0319-8.

60. Li A, Lin X, Tan X, Yin B, Han W, Zhao J, et al. Circadian gene clock contributes to cell proliferation and migration of glioma and is directly regulated by tumor-suppressive miR-124. FEBS Lett. 2013;587:2455–60. https://doi.org/10.1016/j.febslet.2013.06.018.

61. Cacciapuoti F. Oxidative stress as "mother" of many human diseases at strong clinical impact. J Cardiovasc Med Cardiol. 2016;3(1):1–6.

62. Matschke V, Theiss C, Matschke J. Oxidative stress: the lowest common denominator of multiple diseases. Neural Regen Res. 2019;14:238–41. https://doi.org/10.4103/1673-5374.244780.

63. Bansal A, Simon MC. Glutathione metabolism in cancer progression and treatment resistance. J Cell Biol. 2018;217:2291–8. https://doi.org/10.1083/jcb.201804161.

64. Aaling NN, Nedergaard M, DiNuzzo M. Cerebral metabolic changes during sleep. Curr Neurol Neurosci Rep. 2018;18:57. https://doi.org/10.1007/s11910-018-0868-9.

65. Kumar A, Dhull DK, Gupta V, Channana P, Singh A, Bhardwaj M, et al. Role of glutathione-S-transferases in neurological problems. Expert Opin Ther Pat. 2017;27(3):299–309.

66. Zhu Z, Du S, Du Y, Ren J, Ying G, Yan Z. Glutathione reductase mediates drug resistance in glioblastoma cells by regulating redox homeostasis. J Neurochem. 2018;144:93–104. https://doi.org/10.1111/jnc.14250.

67. Costello RB, Lentino CV, Boyd CC, O'Connell ML, Crawford CC, Sprengel ML, et al. The effectiveness of melatonin for promoting healthy sleep: a rapid evidence assessment of the literature. Nutr J. 2014;13:106. https://doi.org/10.1186/1475-2891-13-106.
68. Blask DE. Melatonin, sleep disturbance and cancer risk. Sleep Med Rev. 2009;13:257–64. https://doi.org/10.1016/j.smrv.2008.07.007.
69. Touitou Y, Reinberg A, Touitou D. Association between light at night, melatonin secretion, sleep deprivation, and the internal clock: health impacts and mechanisms of circadian disruption. Life Sci. 2017;173:94–106. https://doi.org/10.1016/j.lfs.2017.02.008.
70. Chaput JP. Sleep patterns, diet quality and energy balance. Physiol Behav. 2014;134:86–91. https://doi.org/10.1016/j.physbeh.2013.09.006.
71. Huang CT, Chiang RP, Chen CL, Tsai YJ. Sleep deprivation aggravates median nerve injury-induced neuropathic pain and enhances microglial activation by suppressing melatonin secretion. Sleep (Basel). 2014;37:1513–23. https://doi.org/10.5665/sleep.4002.
72. Pfeffer M, Korf HW, Wicht H. Synchronizing effects of melatonin on diurnal and circadian rhythms. Gen Comp Endocrinol. 2018;258:215–21. https://doi.org/10.1016/j.ygcen.2017.05.013.
73. Reiter RJ, Mayo JC, Tan DX, Sainz RM, Alatorre-Jimenez M, Qin L. Melatonin as an antioxidant: under promises but over delivers. J Pineal Res. 2016;61:253–78. https://doi.org/10.1111/jpi.12360.
74. Reiter RJ, Tan DX, Galano A. Melatonin: exceeding expectations. Physiology (Bethesda). 2014;29:325–33.
75. Deng T, Lyon CJ, Bergin S, Caligiuri MA, Hsueh WA. Obesity, inflammation, and cancer. Annu Rev Pathol. 2016;11:421–49. https://doi.org/10.1146/annurev-pathol-012615-044359.
76. Tobaldini E, Constantino G, Solbiati M, Cogliati C, Kara T, et al. Sleep, sleep deprivation, autonomic nervous system and cardiovascular diseases. Neurosci Biobehav Rev. 2017;74(Pt B):321–9.
77. Irwin MR. Why sleep is important for health: a psychoneuroimmunology perspective. Annu Rev Psychol. 2015;66:143–72. https://doi.org/10.1146/annurev-psych010213-115205.
78. Irwin MR, Olmstead R, Carroll JE. Sleep disturbance, sleep duration, and inflammation: a systematic review and metaanalysis of cohort studies and experimental sleep deprivation. Biol Psychiatry. 2016;80:40–52. https://doi.org/10.1016/j.biopsych.2015.05.014
79. Aguirre CC. Sleep deprivation: a mind-body approach. Curr Opin Pulm Med. 2016;22:583–8. https://doi.org/10.1097/MCP.0000000000000323.
80. Hurtado-Alvarado G, Domínguez-Salazar E, Pavon L, VelázquezMoctezuma J, Gómez-González B. Blood-brain barrier disruption induced by chronic sleep loss: low-grade inflammation may be the link. J Immunol Res. 2016;2016:4576012. https://doi.org/10.1155/2016/4576012.
81. Gaine ME, Chatterjee S, Abel T. Sleep deprivation and the epigenome. Front Neural Circuits. 2018;12:14. https://doi.org/10.3389/fncir.2018.00014.
82. Orešković D, Kaštelančić A, Raguž M, et al. The vicious interplay between disrupted sleep and malignant brain tumors: a narrative review. Croat Med J. 2021;62(4):376–86. https://doi.org/10.3325/cmj.2021.62.376.

Chapter 5
Migraine and Interoception

Marjan Zaletel

Introduction

Migraine is a chronic neurological disorder characterized by recurrent moderate to severe headaches associated with several autonomic nervous system symptoms. In addition, disorders of exteroception such as phono and photophobia are combined with interception phenomena. Shortly, we can recognize migraine as a neurologic sensory disturbance. Why does such a complex reaction occur, and why is the reason for migraine episodes?

Migraine is a poorly explainable and understandable neurology condition that prevents effective communication between patient and physician. In addition, the patient's understanding of migraine is crucial for desirable therapeutic outcomes. The modern science of computational models and theoretical neurobiology of interoception makes it possible to understand changes in the nervous system and whole organism during a migraine episode. Moreover, it enables us to better understand the influence of the social environment on migraineurs. Also, the comorbid conditions, especially mood disorders, could be consequences of deeper interoception and its inability to adapt successfully to environmental changes. From this point, a migraine can be considered a maladaptive or hyper-protective response for survival. Thus, migraine represents an allostatic load for an organism, making disability an economic burden for society. The modern approach using calcitonin gene-related (CGRP) antagonism is highly effective in preventing and terminating migraine episodes. Nevertheless, understanding the deeper mechanism of information processes during migraine episodes in the central nervous system offers the

M. Zaletel (✉)
Department of Vascular Neurology, University Clinical Center Ljubljana, Ljubljana, Slovenia

University Clinical Center Ljubljana, Pain Clinic, Ljubljana, Slovenia
e-mail: marjan.zaletel@kclj.si

V. Demarin et al. (eds.), *Mind, Brain and Education*,
https://doi.org/10.1007/978-3-031-33013-1_5

possibility to cope with migraine by improving the internal state model. This paper trying to explain how interoception is linked to migraine and how migraineurs can cope with it.

Interoception

Interoception is an emerging physiologic concept that links human mental activity with body homeostatic and allostatic systems [1]. Interoception describes the perception of symptoms and sensations that originate within the body [2]. Internal states and their regulation are largely unconsciously. Changes in body state cause automatic physiological reactions and mental feelings such as hunger, thirst, pain or fear, and different kinds of pleasure and pain. Central to such interoceptive inference perspectives is the notion that interoceptive signals encode representations of the internal, physiological state of the body. Evidence suggests that body state changes are mapped topographically in the central nervous system [3]. Changes recorded in these neural maps serve as triggers for corrective responses. Interoception refers to the afferent signaling and central representation of the body's internal physiological state [4]. It is the material basis of self, constructed from early development through a continuous representation of biological data from the body affected by cognitive and emotional processes [5].

Clinical data and functional neuroimaging provide evidence of interoception involvement in migraine. In the prodromal state, typical homeostatic feelings include cravings for specific foods, the need to pee more often, and thirst. Emotional changes such as irritability belong to interoceptive feelings. In the headache phase, headache pain is a predominant interoceptive phenomenon because pain is referred to as interoceptive feeling [6]. The functional changes to the insula, which is supposed to be the interoceptive hub, were observed in migraine subjects compared to healthy controls. There is an increase in insular activity that corresponds to the default mode network [7]. In addition, increased functional alterations to the default mode network were found [8]. The data indicate that migraine may alter inzular activity and interoception, resulting in chronic migraine.

Interoceptive Model of Internal State and Migraine

The central part of interoception is the intrinsic brain model of an internal state, which is still an enigma in its functioning. According to the good regulator theorem, every good regulator of a system must be a model of that system [4]. The model is complex and supposed to be hierarchical in structure. This can be achieved through updating the brain's generative model. The model assumes forecasting future states and monitoring prediction error at the metacognition level. The goal of the Bayesian brain is to minimize prediction error. The intrinsic model incorporates prior knowledge about the structure of the environment and the body, a model of the body in the world. It is the source of predictions about the causes of interoceptive signals in the external environment [4].

The internal state model contains the representation of the external world, implemented in forecasting and metacognition [4, 9]. The generative model is hierarchically built. Each level of hierarchy holds beliefs or predictions about the state of the level below. This prediction is transmitted to a lower level where it is compared against to actual condition. The result of the comparison of interoceptive signal and prediction is an interoceptive error (IE). This is transmitted back to the upper level. The result is an updated prediction in the upper level of the generative model. In turn, this process minimizes the IE in the lower level. The dynamic model works instantaneously in updating the prediction model using the predictive coding principle.

The inference which updated the intrinsic model is perceptual inference: more complex regulation involving inferential, anticipatory, and forecasting [10]. The brain predicts how internal states change as a consequence of actions and forecasts future conditions as a function of their internal dynamics. The forecasting process indicates future internal and external states in the context of the intrinsic dynamics of the body and the environment. Switching from perceptual inference to forecasting requires ignoring sensory precision and abolishing model updates. The new predictions due to forecasting could be signaled from higher levels in the hierarchical model and implemented forecasting in the intrinsic generative model of the internal state. Finally, metacognition [4] is connected to the generative model through an additional layer that holds expectations about the level of IE. Metacognition is the awareness and understanding of one's cognitive processes. It enables us to process, monitor, and evaluate our knowledge and makes us aware of our strengths and weaknesses. This layer infers the performance of the inference-control loop, enabling updating the intrinsic model and consequently mastery or self-efficacy beliefs. Metacognitive beliefs about the efficacy of one's body regulation may influence both cognitive processes through learning. With self-reflection, metacognition is likely to be linked to perceptual awareness and consciousness.

In migraine, the generative model might include cortical and subcortical structures. It is supposed that the first-order integrated maps of an interoceptive generative model from the whole body are located in the brainstem [2]. Functional studies showed that increased activity in the pons is linked explicitly to the pain of the migraine attack [11]. In addition, the results of the fMRI study have shown within-individual alterations in brain activity immediately preceding migraine onset [12]. Therefore, it supports the hypothesis that altered regional brainstem function before a migraine attack is involved in underlying migraine neurobiology and might be important in producing IE, which is not resolved subcortically but conveys to higher order generative model units such as the inzula, which makes conscious experiences of migraine symptoms during an attack.

Interoceptive Perceptions and Migraine

From the perspective of selfhood, the interoceptions provide information about the global physiological condition of the body in the context of the social and physical environment through forecasting. The response involves error awareness, the process in which the subject becomes consciously aware of IE [13]. From this perspective, a migraine episode could be considered conscious awareness of IE. More migraineurs are adaptive fewer IEs appear at higher levels and become aware. Anatomically, the region for IE awareness is pointed to the inzula and anterior cingulum, which are supposed to be strategically located as hubs between the crossroads of interoceptive sensations, limbic system, exteroceptive input, and metacognitive location prefrontal. Although, some studies argued subcortically regions as leading sites of conscious awareness of interoceptive states [3]. Impairment of the error awareness network might produce further maladaptation resulting in chronic migraine, an allostatic load incorporated into behavior response, and disabled the subjects from functioning normally. Also, inzula-the anterior cingulum complex, is an important source of autonomic response in migraine episodes.

Some symptoms during migraine episodes can be considered error awareness. The others might be connected with compensation as an active inference response during predictive coding. Presumably, forecasting might change the intrinsic model, which directed priors or interoceptive predictions to match interoceptive sensations in the central nervous system inducing IE in the generative system. Evidence of a high rate of psychologic stress, where individuals perceive their environment requires more resources than the person is capable of adjusting [14], serve as a trigger of migraine episode preceding the onset of migraine by several days was suggested to play an essential role in the occurrence of migraine [15] supporting the role of forecasting in the interoceptive model. Indeed, psychological stress is prevalent among patients who suffer from migraines, with about 70% reporting it initiates migraine episodes [16].

In the prodromal period, a typical sign of interoceptive symptoms included cravings for specific foods, changes in thirst, and urination. The symptoms might be the consequence of IE due to mismatch interoceptive predictions and sensations at the highest level of the generative interoceptive system and conscious experiences. During the prodromal state emotional feelings may appear. According to bodily marker theory, emotion is the representation and regulation of the complex array of homeostatic changes that occur in different levels of the brain and body in given situations [3]. Therefore, emotional instability in the prodroma phase appears as part of the activation program. The headache phase could be regarded as a continuum of the prodromal phase. Headache or pain as the most prominent symptoms can be considered interoceptive feeling, that includes nociception [6]. It might result from high precision altered predictions, which can produce IE error.

Nevertheless, it could arise because of a compensatory increase of nociceptive input due to sensitization during migraine episodes that is a more probable cause of headache. The sensitization of the nervous system is a generalized phenomenon that includes vision and hearing. That is a reason for phono and photophobia during a migraine episode. In the headache phase, the parasympathetic symptoms appear, which is part of the action program due to IE error. Also, the behavior change during migraine episodes follows the signs as part of the action program. The aura symptoms, such as flashes of light and blurred vision, might be considered excessive sensitization in the occipital lobe. Thus, sensitization of the central nervous system might be the consequence of IE and represents compensatory activities in the nervous system due to unresolved IE.

Treatment Decisions Based on the Interoceptive Concept of Migraine

Interoception presumes an internal model or representation of the body and internal environment, which should be neuroplastic and influenced by the environment. The model represents systematically organized knowledge of the internal environment. Therefore, it could be changed consciously or unconsciously throught the mental processes of cognitive and meditative methods. Indeed, current studies and experiences confirm the such approach to migraine management. Maladaptive thinking and behaviors may negatively affect the course of several medical disorders, including migraine [17]. Overall, these findings were mixed, with some studies supporting the suggestion that people experiencing migraines can benefit from cognitive behavior therapy (CBT) and that CBT can reduce the physical symptoms of migraine [18]. For people with migraine who are currently experiencing a large number of headache days per month, behavioral treatment strategies that have demonstrated efficacy for migraine day reduction may be more appropriate than Mindfulness-Based Cognitive Therapy for Migraine (MBCT-M). However, for people reporting high levels of migraine-related disability despite relatively less frequent headache days, a mindfulness-based treatment such as MBCT-M may be an acceptable and appropriate treatment choice [19]. MBCT-M is a promising treatment for reducing headache-related disability, with more significant benefits for episodic than chronic migraine. Mindfulness-based stress reduction did not improve migraine frequency more than headache education, as both groups had similar decreases; however, MBCT-M improved disability, quality of life, self-efficacy, pain catastrophizing, and depression with decreased experimentally induced pain suggesting a potential shift in appraisal process [20]. Spiritual Meditation was found not to affect pain sensitivity, but it does improve pain tolerance with reduced headache-related analgesic medication usage [21].

Shortly, mental models are related to the extensively researched personality traits, locus of control, and attribution or explanatory style. Those with an internal locus of control tend to believe they play a significant role in affecting the events that influence their lives. Internal locus of control is linked to self-efficacy, the belief in one's ability to succeed in a particular situation [22]. Conversely, those with an external locus of control view themselves as victims of life. They believe rewards and punishments are controlled by outside forces, other people, or some higher power. Thus, the goal of psychotherapy is to move from an external to an internal locus of control. From this point, metacognition helps us decrease the burden of migraine, which should be paramount in the treatment strategy for some patients with migraine.

In an acute migraine attack or episode characterized by the peripheral response to a significant cause, peripheral sensitization with an inflammatory response exists in the first phase. Still, in advanced stages, the central sensitization and its clinical correlates predominate. Therefore, analgesics that affect peripheral mechanisms are effective at the beginning of migraine headaches, but later on, their effects diminish [23]. Among them, the most important are non-steroid analgesics and specific ones such as triptans and gepants. The triptans could stop peripheral sensitization by their action on CGRP, which supposes to be the most crucial molecule in producing sensitization during a migraine episode. Neurogenic inflammation is a secondary phenomenon due to trigeminal sensitization [24]. This explains why they are less effective compared to triptans. The role of acetaminophen is exciting because it has peripheral and central effects. It is well known that it can be effective in the first 2 h but could help later in the attack, especially in the parenteral form, which implicates its central effects through TRPV1 and cannabinoid receptors [25]. Therefore, in an acute migraine attack, during the headache phase, peripheral interoceptive – nociceptive signaling is crucial for sensory input, which updates the existing internal homeostatic model and representation through the predictive coding process and triggers protective behavior and autonomic response that could be lessened by antiemetics. Shortly, treatment and interceptive responses tend to diminish nociceptive input. After normalization, the generative body model was updated again to a non-painful body representation.

The state of high frequency or chronic migraine represents permanent sensitization of the nervous system. The development and maintenance of central sensitization depend mainly on activating the glutamate N-methyl-D-aspartate (NMDA) receptors [26]. CGRP may contribute to central sensitization in migraine since CGRP receptors were identified, presynaptically, on nerve terminals of glutaminergic neurons [27]. Nevertheless, sensitization should be a generalized process involving the central or peripheral part of the nervous system, wherever it starts. For instance, complex regional pain syndrome (CRPS) is a model for the top-down developing sensitization process [28]. Although the mechanism is poorly understood sympathetic autonomic system may be involved. Also, migraine is appeared to be top-down spreading sensitization with involvement of the peripheral nervous system in the head and trigeminovascular system. Therefore, prevention should be focused on the excitability of central networks, including glia. Thus, antiepileptics, beta-blockers, and tricyclics remain the therapy of choice. Recently, monoclonal anti-CGRP antibodies have been acceptable and well tolerated prophylactic agents for migraine.

Conclusion

Interoception enables us to understand better mechanisms of migraine, which is essential for communication with patients and treatment planning. Accordingly, migraine is a consequence of interaction internal and external human environment. It might represent the organism's response to an anticipatory appraisal of potential threats and reactions to them, including interoceptive feelings, including headache pain. In addition, the corrective responses are accurate in the form of emotional feelings, autonomic signs, and behavior leading to disability. Therefore, migraine could be considered a maladaptive response to environmental changes. Thus, migrainours should develop appropriate coping strategies to functionally adapt to external changes for long-term beneficial effect of anti-CGRP therapy.

References

1. Quadt L, Critchley HD, Garfinkel SN. The neurobiology of interoception in health and disease. Ann N Y Acad Sci. 2018;1428(1):112–28.
2. Carvalho GB, Damasio A. Interoception and the origin of feelings: a new synthesis. BioEssays. 2021;43(6):e2000261.
3. Damasio A, Carvalho GB. The nature of feelings: evolutionary and neurobiological origins. Nat Rev Neurosci. 2013;14(2):143–52.
4. Petzschner FH, Weber LAE, Gard T, Stephan KE. Computational psychosomatics and computational psychiatry: toward a joint framework for differential diagnosis. Biol Psychiatry. 2017;82(6):421–30.
5. Chen WG, Schloesser D, Arensdorf AM, Simmons JM, Cui C, Valentino R, Gnadt JW, Nielsen L, Hillaire-Clarke CS, Spruance V, Horowitz TS, Vallejo YF, Langevin HM. The emerging science of Interoception: sensing, integrating, interpreting, and regulating signals within the self. Trends Neurosci. 2021;44(1):3–16.
6. Craig AD. How do you feel? Interoception: the sense of the physiological condition of the body. Nat Rev Neurosci. 2002;3(8):655–66.
7. Xue T, Yuan K, Zhao L, Yu D, Zhao L, Dong T, Cheng P, von Deneen KM, Qin W, Tian J. Intrinsic brain network abnormalities in migraines without aura revealed in resting-state fMRI. PLoS One. 2012;7(12):e52927.
8. Schwedt TJ, Schlaggar BL, Mar S, Nolan T, Coalson RS, Nardos B, Benzinger T, Larson-Prior LJ. Atypical resting-state functional connectivity of affective pain regions in chronic migraine. Headache. 2013;53(5):737–51.
9. Seth AK, Suzuki K, Critchley HD. An interoceptive predictive coding model of conscious presence. Front Psychol. 2012;2:395.
10. Aguilera M, Millidge B, Tschantz A, Buckley CL. How particular is the physics of the free energy principle? Phys Life Rev. 2022;40:24–50.
11. Stankewitz A, May A. Increased limbic and brainstem activity during migraine attacks following olfactory stimulation. Neurology. 2011;77(5):476–82.
12. Meylakh N, Marciszewski KK, Di Pietro F, Macefield VG, Macey PM, Henderson LA. Brainstem functional oscillations across the migraine cycle: a longitudinal investigation. Neuroimage Clin. 2021;30:102630.
13. Borsook D, Aasted CM, Burstein R, Becerra L. Migraine mistakes: error awareness. Neuroscientist. 2014;20(3):291–304.

14. Al-Quliti K. Stress and its correlates in migraine-headache patients with a family history of migraine. Behav Sci (Basel). 2022;12(3):65.
15. Hashizume M, Yamada U, Sato A, Hayashi K, Amano Y, Makino M, Yoshiuchi K, Tsuboi K. Stress and psychological factors before a migraine attack: a time-based analysis. Biopsychosoc Med. 2008;2:14.
16. Theeler BJ, Kenney K, Prokhorenko OA, Fideli US, Campbell W, Erickson JC. Headache triggers in the US military. Headache. 2010;50(5):790–4.
17. Borsook D, Maleki N, Becerra L, McEwen B. Understanding migraine through the lens of maladaptive stress responses: a model disease of allostatic load. Neuron. 2012;73(2):219–34.
18. Harris P, Loveman E, Clegg A, Easton S, Berry N. Systematic review of cognitive behavioural therapy for the management of headaches and migraines in adults. Br J Pain. 2015;9(4):213–24.
19. Seng EK, Conway AB, Grinberg AS, Patel ZS, Marzouk M, Rosenberg L, Metts C, Day MA, Minen MT, Buse DC, Lipton RB. Response to mindfulness-based cognitive therapy differs between chronic and episodic migraine. Neurol Clin Pract. 2021;11(3):194–205.
20. Wells RE, O'Connell N, Pierce CR, Estave P, Penzien DB, Loder E, Zeidan F, Houle TT. Effectiveness of mindfulness meditation vs headache education for adults with migraine: a randomized clinical trial. JAMA Intern Med. 2021;181(3):317–28.
21. Wachholtz AB, Malone CD, Pargament KI. Effect of different meditation types on migraine headache medication use. Behav Med. 2017;43(1):1–8.
22. Grinberg AS, Seng EK. Cross-sectional evaluation of the psychometric properties of the headache-specific locus of control scale in people with migraine. Headache. 2019;59(5):701–14.
23. Mayans L, Walling A. Acute migraine headache: treatment strategies. Am Fam Physician. 2018;97(4):243–51.
24. Ramachandran R. Neurogenic inflammation and its role in migraine. Semin Immunopathol. 2018;40(3):301–14.
25. Ohashi N, Kohno T. Analgesic effect of acetaminophen: a review of known and novel mechanisms of action. Front Pharmacol. 2020;11:580289.
26. Woolf CJ. Central sensitization: implications for the diagnosis and treatment of pain. Pain. 2011;152(3 Suppl):S2–S15.
27. Marvizón JC, Perez OA, Song B, Chen W, Bunnett NW, Grady EF, Todd AJ. Calcitonin receptor-like receptor and receptor activity modifying protein 1 in the rat dorsal horn: localization in glutamatergic presynaptic terminals containing opioids and adrenergic alpha2C receptors. Neuroscience. 2007;148(1):250–65.
28. Littlejohn G. Neurogenic neuroinflammation in fibromyalgia and complex regional pain syndrome. Nat Rev Rheumatol. 2015;11(11):639–48.

Chapter 6
Shame in the Context of Personality and Eating Disorders Patients

Darko Marčinko, Vedran Bilić, Maja Šeparović Lisak, Duško Rudan, Filip Mustač, and Nenad Jakšić

Introduction

Shame is usually conceptualized as a self-debilitating emotion that that encompasses a wide range of negative beliefs about onself. It is also an emotion with often visible physical manifestations, making it particularly disruptive in social circumstances. In recent years shame is described and empirically investigated as a transdiagnostic entity in the fields of clinical psychology and psychiatry, but much of the literature focuses on individuals suffering from personality and eating disorders. In order to better understand the role of shame in these and other clinical populations, one needs to go back to the roots of historical understandings of shame and explore different theoretical perspectives of this disturbing psychological entity. In order to make the concept of shame more scientific, discussion about its measurement is also warranted.

Shame and Narcissism

Narcissism

It is essential to mention that 'narcissism' is often used as a shortening for narcissistic personality disorder. A narcissist or narcissistic personality imposes a pejorative tone on the whole concept of narcissism. We must remember the existence of a continuum from healthy narcissism to its various psychopathological

D. Marčinko (✉) · V. Bilić · M. Š. Lisak · D. Rudan · F. Mustač · N. Jakšić
Department of Psychiatry and Psychological Medicine, University Hospital Center Zagreb, Zagreb, Croatia

© The Author(s), under exclusive license to Springer Nature Switzerland AG 2023
V. Demarin et al. (eds.), *Mind, Brain and Education*,
https://doi.org/10.1007/978-3-031-33013-1_6

manifestations. The same has happened with other mental states, such as hysteria. It is also good to emphasize how contemporary psychodynamic theory no more favors one standpoint in understanding the psyche to the exclusion of all others. Therefore, looking at the psyche only from one standpoint is no longer acceptable. On the contrary, a modern psychodynamic theory is an integration of various standpoints that have proved their validity. Some of them are: causal and final understanding of psychic events, drives, their representative affects and emotions, and their vicissitudes, guilt and shame as dominant regulatory emotions, the dominance of unconscious conflicts or psychic defects, forms and development of object relations and narcissism, the intersubjective processes and more.

It is possible to recognize manifestations of these components in different people. In individual cases, perhaps some of them dominate, but others are also present. Contemporary psychodynamic theory no longer considers that object relations replace narcissism. Narcissism and object relations have separate developments, with lower and higher forms. The development of narcissism and object relations starts at the infantile archaic level, which gradually, but never wholly, transforms into mature forms of narcissism and object relations.

Shame

Shame indicates problems of impaired self-esteem for several reasons. First, shame is associated with dissatisfaction with oneself. That is, it indicates a negative self-evaluation. It can be an unfavorable comparison with others or a comparison of the real with the idealized representation of oneself, the present with the former, and now significantly reduced and disappeared representation. Shame can also arise due to loss of self-control, awareness of lack of independence, and dependence on others. On a social level, shame signals a departure from a person's socially desirable representation, indicating a conflict with accepted social norms. Therefore, adapting to the norms of the community or other people we interact with is an unavoidable prerequisite for socialization. However, this prerequisite is also tiring because it implies giving up some personal aspirations that conflict with and do not fit the interests of the environment. Moreover, in adapting to the other, one loses a part of oneself.

Shame and Narcissism

Shame and narcissism are inextricably linked. Shame is an emotion that, like narcissism, occurs when dealing with the Self. Narcissism and shame focus on various representations and manifestations of the Self. However, focusing on oneself and dealing with oneself motivated by shame is a very unpleasant experience in most cases. Nathanson believes that narcissism's essence is avoiding shame in various ways [1, 2]. In its broadest sense, narcissism means focusing on the Self, dealing with oneself rather than anyone or anything else. Focusing attention on various representations, aspects, and manifestations of the Self, which deviate from the expected or even ideal, can cause narcissistic injury, with painful emotions of confusion, shame, humiliation, and loss of value [3].

At the root of oversensitivity to shame is infantile narcissism. The stunted development of narcissism, which has remained at an archaic level, means that only perfection ensures the absence of negative self-evaluation. Comparing the ideal self-image with the actual state is the basis for a particular sensitivity to shame and sensitivity to critical evaluation of the Self by others. One can never achieve the perfection that would ensure the complete absence of a negative evaluation of Self by others or, more importantly, negative self-evaluation. Therefore, with narcissism and shame, there is always covert or explicit discomfort, if not suffering. The discomfort from shame can be of various intensities, from mild confusion and self-consciousness to extreme and blocking shame. Although it may begin as a confrontation with some unacceptable aspect of the Self, shame spreads rapidly and irrationally and engulf the entire Self. A person whose dominant narcissistic needs remained at the infantile level can, for a moment, experience delight when he considers his adult achievements and shows them to the environment, which mirrors them positively. Nevertheless, behind this appearance lurks discomfort and insecurity. Unresolved infantile narcissistic needs generate unrealistic demands that nothing can satisfy. Such a person, in vain, tries to deal with unsatisfying unconscious self-experience by pointing out his successes and virtues to himself and the environment. The perfections of the Ego that Ego-deal demands are impossible to achieve [4].

Self-assessment stems from the assessment of self-images, that is, mental images of oneself that can be conscious, pre-conscious, and unconscious. If the self-images are negative, the impact on self-esteem is strong, and the person can hardly bear them. Parting with former representations of the Self is painful. It can cause shame and low self-esteem when current capabilities and features are significantly reduced compared to what they were at their peak. For example, loss of good looks, former sexual potency, social position, and change of circumstances in the family due to the departure of adult children or the death of a partner are all blows to self-esteem. Furthermore, if a person does not integrate them and process them through mourning, they can cause an unfavorable comparison of the present and past Self and lead to shame. Comparing oneself to something or someone, whether it is other people, their achievements, or something they have, or comparing an ideal image of oneself with an actual state, is a characteristic of narcissism and mostly a sure path to discomfort and suffering. Comparing oneself and one's achievements with unrealistic unconscious demands or with the achievements or possessions of others to determine one's value and positively regulate self-esteem can easily lead to the violation of narcissism, causing an emotion of shame, discomfort, and suffering.

Everything experienced remains in the unconscious, remembered as it was. Earlier experiences of oneself and once essential persons that are no longer relevant persist. However, petrified insistence on the validity of former self-representations indicates an incorrect understanding of current reality, which can go as far as losing contact with reality. One has to change Self-representations, 'images,' and attitudes about oneself throughout life. When we look at old photos, we can be surprised at what changes have occurred on a purely physical level. We can also become aware of how many psychological changes are now present concerning the past. Parting with former representations of the Self is painful. It can cause shame and diminish self-esteem. Negative self-evaluation destroys the good feeling about one's existence and endangers self-esteem and satisfaction with one's existence. However, regardless of all the transformations, there is also a sense of identity throughout time in normalcy. This sense of Self, which we could express in words: "I am I," is unchanging.

Shame is an unpleasant emotion that cannot become pleasant and desirable. It is difficult to imagine a person who consciously and deliberately wants to experience shame. Shame can function as a signal emotion that warns of danger. The danger that shame warns of is a negative self-evaluation, and the social danger is rejection from the community. The psychological dangers that shame warns of are an attack on the Self, a negative evaluation of the Self, a drop in self-esteem, and the social danger is isolation and rejection from the community. However, shame has essential psychological and social functions. An unfavorable evaluation of the environment in extreme cases leads to excluding a person from all interactions with the environment. Both narcissism and shame can be mature or immature. In mature manifestations, they play a very positive role because they encourage self-awareness and the elimination of negativity, thereby improving genuine self-esteem. Mature narcissism enables a person's interest, motivation, and active participation in essential aspects of life connected with the integrity and coherence of the Self and realistic self-esteem. Mature shame makes it possible to maintain social connection, protect the integrity of the Self, and preserve self-esteem. It can be about protection from external or internal factors that pose a danger. The external danger is loss of connection with others, and the internal is condemnation and devaluing of oneself, i.e., turning aggression on oneself. The intensity of shame, which functions as a danger signal, must still be tolerable for the person. Otherwise, shame becomes a traumatic, overwhelming emotion that completely numbs and blocks the ego's functioning. Although shame may begin with a confrontation in a social situation with some hidden unacceptable behavior or aspect of the Self, shame-prone persons with weak resilience can spread shame rapidly and irrationally and engulf the entire Self and all its representations. In such cases, the intense shame, which a person cannot bear, is no longer a signal emotion but causes psychological trauma that completely numbs and blocks functioning.

Avoidance of Shame

The immediate or first defense against shame is withdrawal from public exposure and interactions with the environment to avoid further narcissistic injury. The public exposure of one's socially unacceptable traits or behaviors usually induces shame, and a series of defenses activate to alleviate the pain of shame. As shame results from exposure, a person first tries to hide from public exposure. However, external criticism activates internal one. There is a loss of harmonious psychological integration of oneself. A critical part of the psyche attacks the Self and directs aggression and devaluation of the whole Self. However, better still is to avoid shaming behaviors, and best is to develop oneself by transforming one's infantile needs and improving one's moral standards. Such changes in the person that the evaluation of the Self becomes positive are radical changes because shame indicates an evaluation of the whole Self, not just correcting some inappropriate behavior.

To become conscious of one's previously unconscious but shaming features is also painful. Therefore, a person tries to avoid such an experience by using everything available. There are many ways to avoid the awareness of shame, which is possible through various defense mechanisms, defensive behaviors, emotions, and ways of thinking. However, all these ways only prevent awareness of shame and do not influence the cause of shame. Moreover, the intensity of defenses against shame seldom ultimately succeeds in abolishing shame. The complete absence of shame is rare because experiencing shame in socially embarrassing situations is a typical human trait. A person's inability to experience any shame points to the lack of necessary psychological structures that enable the experience of shame. Such psychological defects mean moving away from humanity.

The possibility of experiencing shame is also a developmental achievement. The complete absence of shame indicates the underdevelopment of the psychological structures that make it possible to experience shame. Living without any possibility of shame is a departure from humanity. One such constellation is present in persons suffering from psychopathy and sociopathy [5], in whom there is a lack of the limiting psychic structure of the Superego [6]. A personality disorder known as psychopathy is not a unique disorder, but a defective superego is one of its most noticeable and common characteristics. "Pathological identifications and the failure of internalization, including the failure of internalized values, also influence the development of Superego" [7]. Due to the lack of meaningful and solid identifications with others, these individuals use pseudo-identification processes, including conscious and unconscious simulations and imitations of attitudes and behaviors. However, they do not lead to corresponding changes in the Self. Kernberg [8] points out that the Superego is generally poorly developed in narcissistic pathology. "Aggressive introjects and identifications have replaced those based on parental ideals, resulting in psychopathic attribution of negative values to concepts such as attachment, morality, love, and empathy" [7]. People who have fallen behind on an infantile level in the development of narcissism, which is especially true of people with psychopathic personality disorder, have not developed a mature Superego. In milder cases of psychopathy and sociopathy, some rudiments of the Superego exist. However, the Superego is insufficiently narcissistically invested and idealized to function correctly. If the psychic structure, the Superego, which has a critical and rewarding part, is not sufficiently narcissistically invested and idealized, its 'attitude' and assessment of the Self have no significant impact on the person. Therefore, the Superego does not significantly impact self-evaluation and can not cause anxiety, shame, or guilt that could influence behavior and stop aggression and destruction of the environment.

Defenses from Shame

Narcissism can be a defense against shame. Freud described the 'stone wall of narcissism' by which a person tries to block knowledge about himself that would cause shame if he were aware of it [9]. One tries to avoid negative self-images in various ways. Nevertheless, narcissistic persons form and continuously maintain a usually unconscious, perfect image of themselves - the Grandiose Self [10]. The Grandiose Self is a compensatory opposite self-image and a method of avoiding or defending against the shame of narcissistic devaluation. The False Self is another defensive psychic constellation against devaluation and shame. The False Self is exposed and can receive an attack from the environment and devaluation without danger to the genuine but concealed True Self [11]. In an adverse interaction with a shame-inducing environment, withdrawal and isolation are usually the first defenses. However, external criticism can activate internal criticism that directs aggression and devaluation toward the Self. One can avoid the negative evaluation of the social environment by changing the subjective experience of the environment. If the environment ceases to be valued, its assessment is also unimportant. Devaluation results from the unconscious psychological processes by which a person dehumanizes the environment.

It can be a family community or other less close but essential communities such as, for example, professional communities. Such an experience of others is a defense against shame, fueled by the negative evaluation of others and oneself. A mature relationship with others includes understanding others as equal (with us) subjects with genuine aspirations that they strive to realize in their interactions with us, just as we strive to do the same. However, such relationships can be impossible due to the effort they require. First, one can take away from others their subjectivity and thus their 'right' to authenticity and autonomy and perceive them as tools, even as things in the environment that are available for use. One uses others as long as they are helpful and can discard them, completely ignore them, or even destroy them after use. In the field of sexuality, a dehumanized attitude toward others is a characteristic of pornography [12]. If others do not psychologically exist, there is no danger of their negative evaluation.

With dehumanization, the experience of others fundamentally changes. Others 'lose' their status as complete and autonomous persons whose opinion is important. By 'taking away' subjectivity, people become dehumanized objects. Such defense removes an essential element of shame-inducing interaction. For shame, a person must perceive the environment that is critical of him as valuable. If one devalues the environment, its opinion and attitudes are irrelevant. They do not require consideration, i.e., in dealing with them, reckless manipulation is appropriate. People become things in the environment available for use. One can reject, completely ignore or even destroy them. The dominant mode of psychic functioning shifts to a paranoid-schizoid position, as described by M. Klein, in which the experience of integration of Self and object disappears, and the partial experience of Self and object dominates [13].

Envy, a very unpleasant emotion, is also associated with a paranoid-schizoid position and shame. A person notices something precious in the environment or others that he lacks. As a result, he compares himself unfavorably with others, which causes envy, narcissistic injury, and shame. The "solution" is destroying the other or what the person perceives as extremely valuable [13].

Shame plays a vital role in conflicts and violence. Conflicts and violence often arise as a defense against humiliation. Narcissistic hurt causes shame, and one defense against shame is anger. Nevertheless, anger makes it possible to carry out destructive actions, and while the anger lasts, the usual moral inhibitions and consideration for the environment disappear. Directing aggression and devaluing the environment, facilitated by dehumanizing the environment, is a method of suppressing shame. Directing aggression based on extensive projection is dangerous for the environment. In extreme cases, we find it in sadistic bullies [1, 2].

One defense against shame is turning the passive into the active, which stems from narcissistic omnipotence. The person now actively places others in his former passive position in which the environment treats him inconsiderately [14]. Shame plays a vital role in conflicts and violence. Conflicts and violence often arise as a defense against humiliation. Katz found the prevalence of humiliation as a proximate motive for violent crime and homicide [15]. Retzinger states that the anger arising from a narcissistic injury is also a defense against unbearable shame [16]. Anger makes it possible to carry out destructive actions, and while the anger lasts, the usual moral inhibitions and consideration for the environment disappear. People in violent relationships tend to be shame-prone [17].Much shame is present in marriages dominated by physical violence [17]. Marital arguments and conflicts are often a defense against shame. Mutual shaming involves mutual shaming and shame defense. Unrecognized shame is a common and crucial hidden factor that disrupts relationships in psychotherapy. A common motive for prison riots and war events is a defense against conscious and unconscious shame [16, 18].

Shame in the Context of Personality and Eating Disorders: Transactional Analysis Perspective

"Shame begins as a largely wordless experience" [19]. As Kaufman states [20] "The experience of shame is a fundamental sense of being defective as a person, accompanied by fear of exposure and self-protecting rage."

From TA perspective, a sense of shame is profoundly related to the core script belief that "something is wrong with me" (I am not Ok, You are OK), while we experience guilt when we have done something wrong. When we are feeling guilty, we know exactly what we have done wrong and what we need to do to fix it and make amends. The original shame comprises pre-verbal experience within interpersonal dynamics and is involving person/s of great importance (e.g., parents) rupturing emotional bond. The shame develops as a result of the misattunements and internalization of unfulfilled wishes, as well as a result of trauma and abusive behaviors.

According to Fanita English [21] shame reactions, compared to guilt, are at their origin more primitive and non-verbal. The first sensation we experience when feeling ashamed is physiological (the blushes, the perspiration, the faster heartbeat, etc.), and requires translation into words, whereas guilt reactions are formed later, when the child possesses linguistic ability. Shame invades and inhabits a person's sense of being, and guilt is associated with what a person has done.

From the Ego state model (PAC) perspective, "Archaic shame is an internal expression of an intrapsychic conflict between a reactive Child ego state and an influencing Parent ego state" (Erskine, 1993). Considering the Parent subdivisions "Critical Parent" and the "Nurturing Parent", as well as the Child subdivisons, the "Natural Child" and the "Adaptive Child", the most people with an eating disorder typically have a large Critical Parent (criticizing, shame-inducing, and controlling) and consequently a large Adaptive Child. Typically, people who have eating disorders frequently have Please (you) and Be Perfect or Be Strong as their Drivers, based on underlying belief: "As long as I can keep other people happy, I am a worthwhile person"; or: "As long as I can do things perfectly I am an OK person" [22]. The Be Strong driver indicates a need to hide vulnerability, by not expressing feelings and needs, which is in the long run impossible demand to achieve and leads to a failure that results in low self-esteem and guilt. Some of people with an eating disorder seem to develop a parallel belief "I can stay OK if I am slim." or more dangerous "anorexic driver" "I can stay OK if I continue to lose weight" [22].

A person (mostly women) with an eating disorder switches positions from I'm not OK, You're OK (I-U+) to I'm OK, You're not OK (I + U-), wherein "frequently experiences a feeling of superiority in her self-control while pitying those whose will is weak enough to keep them fat" [22]. "It is control but not autonomy, a desperate control aimed at creating a conditional OKness" [22].

Persons with BPD also have the predominant ego state not OK Adaptive Child, and besides, they could manifest rapid switches to Vengeful Child or negative Critical Parent with highly risky and aggressive behaviour. Script beliefs and feelings represent contamination of Adult, and sometimes there may be periods of Adult exclusion when they experience brief psychotic episodes. According to Transactional Analysis theory, a personality disorder is characterised by a script with self-destructive messages that impact the functioning with negative consequences especially within interpersonal relationships.

The Child makes "adaptations" or script decisions based on Injunctions (Don't be you, Don't grow up, Don't feel, etc.) or Drivers (Be perfect, Be strong, Please other, Try hard and Hurry up).

The complex origin of the core belief "something is wrong with me" within the Child ego state can be explained from three perspectives [23]: "messages with compliant decisions" (parental/caregiver criticism); "conclusions in response to an impossibility" (impossible demands that result in failure); and "defensive reactions of hope and control" (taking the responsibility/blame for other to preserve the relationship).

"The Child ego state fixations occurred when critical childhood needs for contact were not met, and the child's use of defenses against the discomfort of the unmet needs became habitual" [24].

An introjection as a defense mechanism is frequently used within circumstances when there is a lack of adequate psychological contact between a child and the parents/caregivers. "The function of introjection (introjected Parent ego state) is to reduce the external conflict between the child and the person on whom the child depends for need fulfillment" [25], while introjected shame in the Parent ego state may be misidentified as self [26].

Eric Berne [27] suggested that dysfunctional behavior is the result of self-limiting (unconscious) decisions about how life should be lived, made in childhood in order to survive (Script). Erskine [24] defines Script as a "life plan based on introjections and/or defensive reactions made under pressure, at any developmental age, which inhibits spontaneity and limits flexibility in problem-solving and in relating to people" [24], while Susan A. Adams [28] concludes "Shamed individuals hide their pain behind masks and create scripts to get others to react in prescribed ways". Accordingly "Life Scripts are designed to protect individuals from three types of shame: social shame (What will people think?); competence shame (based on the gap between what people can do and what they think they should be able to do); and existential shame (induced by the internal message "I am worthless")" ([29], Potter-Efron & Potter-Efron 1999).

To maintain the stability of script beliefs and confirm life positions people are playing Games (Berne, 1964). Everybody plays games sometimes. We play Games to get strokes without risking intimacy and confirm our basic beliefs about ourselves, others, and the world (Kick me, Now I've got you son of a bitch, Stop me if you can, If it weren't for you, Poor me, See what you made me do…). "People who have decided they are helpless may play some version of "Poor Me" or "Kick Me", while "Kick me" game is often played by people who have decided to be rejected; they set themselves up to be mistreated by others so that they can play the role of the victim whom nobody likes" [30].

"Adult ego state approaches scenarios by gathering the facts from either external sources or other ego states and maintains balance among the multiple Parent and Child ego states" [28]. The Adult ego state's task is achieving autonomy, that is, script free model of functioning, wherein we have the ability to spontaneously choose the best possible state and reaction to the circumstances.

Shame, Mentalization, and Eating Disorders

Shame is also an unavoidable topic in the context of eating disorders. It is believed that in eating disorders, shame plays an important role in several aspects. First of all, shame is in a way the origin of problems in the sense of shame towards the very experience of pleasure and love, as well as one's own image and perception of one's own body. A deficit in mentalizing abilities, which can be reflected and further embarrass the person and the person can constantly feel inadequate, ashamed is why we should look mentalization based treatment more closely.

When the environment perceives that someone often refuses food and consequently has a low body mass index, external shame also develops which can have a degree of self-disgust [31, 32]. In general, the appearance of a regressive symptom as the final expression of emotion can be explained by a deficit in the personality structure [33]. The aforementioned is often the same through the emotion of shame, since it is a strong emotion that overwhelms a person. All of this is then connected to the experience of one's own body image, but also the relationship with other people, since eating also has an important sociocultural role [34].

Also, recent research [35, 36] also indicates major cognitive changes in eating disorders, especially in the restrictive type of anorexia nervosa, which have altered full-scale intelligence quotient and performance intelligence quotient, which may also be related to the experience of shame in people with eating disorders. In this sense, shame could be a link between pathological personality traits that are often observed in eating disorders and the final symptom, which can be overeating and vomiting, or food restriction. Gagliardini et al. [37] analyzed the latent profiles of patients with eating disorders and showed very interesting results and proposed the division of eating disorders into subtypes according to dimensions of mentalization related to affect imbalance, external imbalance, cognitive imbalance towards oneself and cognitive imbalance towards others. By understanding these latent profiles, we can assume that all patients with eating disorders are not the same and that mentalizing abilities can drastically differ over time, and thus their response to therapy. Furthermore, binge eating disorder is, for example, a disorder that is often neglected and patients usually start treatment quite late precisely because of the shame that underlies the disorder itself and sometimes the possible hiding of the disorder itself [38].

With greater understanding of mental states of one's self and others, the intensity of the symptoms should fade, and thus a better general state and a better quality of life. The goal in treatment of patients with eating disorders is to strengthen mentalizing capacities, which achieves better affect regulation and more fulfilling interpersonal relationships, which is shown through programs that used mentalization based treatment in the treatment of patients with eating disorders [39].

Psychometric Assessment of Shame

In the past decades, there has been an increasing trend in empirical research on shame and different assessments of shame (e.g., [40–42]). Various psychometric instruments have been developed to assess shame, and most of these scales have been created within clinical psychology and were constructed for individuals suffering from psychiatric symptoms and disorders [40, 43]. One of the most frequently used psychometric instrument is the Experience of Shame Scale (ESS) [40], a

25-item measure that assesses the frequency of shame experiences related to one's character ("Have you felt ashamed of the sort of person you are?"), behavior ("Have you tried to cover up or conceal things you felt ashamed of having done?"), and body ("Have you avoided looking at yourself in the mirror?"). Original prospective study carried out in university students has shown the ESS to have high internal consistency and test-retest reliability over 11 weeks and good discriminant and construct validity [40]. A recent study with psychiatric inpatients confirmed the initially proposed three-factor structure (i.e., characterological, behavioral, and bodily shame) and demonstrated adequate internal reliability and temporal stability for all three subscales [44]. Previous research has suggested meaningful relations with the measures of temperament traits, depression, eating pathology, and symptoms of borderline personality disorder (BPD) (e.g., [40, 44–46]). The authors of this chapter have previously used the ESS in relation to personality organization, pathological narcissism, depression and suicidality [47, 48].

However, the measurement of shame has encountered some conceptual and methodological issues. These obstacles are predominantly related to the lack of consensus among experts and scholars regarding the definition and various manifestations of shame. In addition, persons may deny their feelings of shame due to either lack of self-awareness or social desirability [49]. Also, when one feels ashamed, his or her tendency to self-isolate may reduce their study involvement thus increasing attrition rates [50]. Moreover, individuals may be less willing or almost unable to express themselves while shameful experiences are occurring [51]. On the other hand, there have been major methodological advances on shame measurement, with an emphasis on capturing both external and internal aspects of this self-conscious emotion. External shame relates to the experience of oneself as existing negatively in the minds of others, as having deficits, failures or flaws exposed; internal shame, on the other hand, is linked to the inner dynamics of the self and to how one judges oneself, being associated with global self-devaluations and feelings of inadequacy, inferiority, undesirability, emptiness, or isolation [52].

Several psychometric measures have been developed in order to assess these distinct aspects of shame: The Other As Shamer scale (OAS; [53]), a 18-item measure designed to assess external shame; and the Internalized Shame Scale (ISS; [54]), a 24-item questionnaire assessing internal shame. More recently, Ferreira et al. [52] have developed the External and Internal Shame Scale (EISS) to assess in a single measure these two dimensions. These authors have stated that the EISS allows the assessment of the specific dimensions of external and internal shame as well as a global sense of shame experience and may therefore be an important contribution for clinical work and research in human psychological functioning [52]. In recent years, however, some experts have argued for an even more nuanced conceptualization and assessment of shame that includes its maladaptive, but also adaptive dimensions of shame, as seen in a recently constructed dimensional instrument SHAME [42]. Here, existential shame was considered a maladaptive for of shame, and it was related to psychopathology, particularly in a clinical sample of patients with BPD.

Conclusions

Complete understanding of complex emotions such as shame is still ahead of us, but an overview of its historical roots as well as theoretical and empirical relationships with mental disorders should help facilitate a more systematic conceptual and quantitative approach to shame. Psychodynamic formulations of shame have given rise to numerous empirical investigations of this self-debilitating emotion, including its connection to related concepts such as narcissism and mentalizing capacities. Conceptually, shame is a transdiagnostic and multidimensional emotion that arises in healthy and clinical populations, but a large body of literature describes its role in personality and eating disorders. We encourage current and future mental health experts, particularly psychotherapists, to pay more attention to shameful experiences of their clients, as they are closely related to various psychopathological entites and negative therapeutic outcomes.

References

1. Nathanson DL. Affect theory and the compass of shame. In: Lansky MR, Morrison AP, editors. The widening scope of shame. New York, NY and London: Psychology Press; 2014a. p. 339–54.
2. Nathanson DL. Shame and the affect theory of Silvan Tomkins. In: Lansky MR, Morrison AP, editors. The widening scope of shame. New York, NY and London: Psychology Press; 2014b. p. 107–38.
3. Kohut H. Forms and transformations of narcissism. J Am Psychoanal Assoc. 1966;14:243–72.
4. Freud S. In: Strachey J, editor. New introductory lectures on psychoanalysis. New York, NY: Notron; 1933/1965.
5. Hare RD, Neumann CS. Psychopathy as a clinical and empirical construct. Annu Rev Clin Psychol. 2008;4:217–46.
6. Nastović I. Ego-psihologija psihopatije. Gornji Milanovac: Dečje Novine; 1989.
7. Yakeley J. Working with violence. A contemporary psychoanalytic approach. London: Palgrave; 2010.
8. Kernberg OF. Severe personality disorders. London: Yale University Press; 1984.
9. Freud S. On narcissism: an introduction. London: Hogarth Press; 1914/1957.
10. Kohut H. The analysis of the self. New York, NY: International Universities Press; 1971.
11. Winnicott DW. Ego distortion in terms of true and false self. In: Winnicott DW, editor. The maturational processes and the facilitating environment: studies in the theory of emotional development. London: Karnac Books; 1960.
12. Kernberg OF. Clinical dimensions of masochism. In: Glick RA, Meyers DI, editors. Masochism. New York, NY: The Analytic Press; 1988. p. 61–80.
13. Spillius EB, Milton J, Garvey P, Couve D, Steiner D. The new dictionary of Kleinian thought. London: Routledge; 2011.
14. Wurmser L. The shame about existing: a comment about the analysis of 'moral' masochism. In: Lansky MR, Morrison AP, editors. The widening scope of shame. New York, NY and London: Psychology Press; 2014. p. 367–82.
15. Katz J. Seductions to crime. New York: Basic Books; 1988.
16. Retzinger SM. Violent emotions: shame and rage in marital quarrels. Newbury Park, CA: Sage; 1991.

17. Lansky M. Shame and domestic violence. In: Nathanson D, editor. The many faces of shame. New York, NY: Guilford; 1987. p. 335–62.
18. Scheff TJ. The shame-rage spiral: a case study of an interminable quarrel. In: Lewis HB, editor. The role of shame in symptom formation. Hillsdale, NJ: Lawrence Erlbaum Associates; 1987. p. 109–49.
19. Kaufman G. The psychology of shame. London: Routledge; 1993.
20. Kaufman G. The meaning of shame: toward a self-affirming identity. J Couns Psychol. 1974;21(6):568–74.
21. English F. Shame and social control. Trans Anal J. 1975;5:24–8.
22. Brunt M. The use of transactional analysis in the treatment of eating disorders. Trans Anal J. 2005;35:3–253.
23. Erskine RG. Shame and self-righteousness: transactional analysis perspectives and clinical interventions. Trans Anal J. 1994;24(2):86–102.
24. Erskine RG. Script cure: behavioral, intrapsychic, and physiological. Trans Anal J. 1980;10(2):102–6.
25. Erskine RG. The schizoid process. In: International Transactional Analysis Association conference. San Francisco, CA; 1999.
26. Erskine RG. The fourth degree impasse. In: Moiso C, editor. TA in Europe: contributions to the European Association of Transactional Analysis Summer Conferences. Geneva: European Association for Transactional Analysis; 1977. p. 33–5.
27. Berne E. Transactional analysis in psychotherapy: a systematic individual and social psychiatry. New York, NY: Grove Press; 1961.
28. Adams SA. Using transactional analysis and mental imagery to help shame-based identity adults make peace with their past. Adultspan J. 2008;7(1):1–12.
29. Potter-Efron RT, Potter-Efron PS. Assessment of codependency with individuals from alcoholic and chemically dependent families. Alcohol Treat Q. 1989;16(4):37–57.
30. Corey G. Theory and practice of counseling and psychotherapy. 8th ed. Transactional analysis Web Tutor; 2009. p. 1–60.
31. Marques C, Simão M, Guiomar R, Castilho P. Self-disgust and urge to be thin in eating disorders: how can self-compassion help? Eat Weight Disord. 2021;26(7):2317–24.
32. Melo D, Oliveira S, Ferreira C. The link between external and internal shame and binge eating: the mediating role of body image-related shame and cognitive fusion. Eat Weight Disord. 2020;25(6):1703–10.
33. Marčinko D, Jakšić N, Šimunović Filipčić I, Mustač F. Contemporary psychological perspectives of personality disorders. Curr Opin Psychiatry. 2021;34(5):497–502.
34. Mustapic J, Marcinko D, Vargek P. Eating behaviours in adolescent girls: the role of body shame and body dissatisfaction. Eat Weight Disord. 2015;20:329–35.
35. Kupryjaniuk A. Cognitive dysfunction in people with eating disorders. Pol Merkur Lekarski. 2021;49(292):283–5.
36. Ogata K, Koyama KI, Fukumoto T, Kawazu S, Kawamoto M, Yamaguchi E, Fuku Y, Amitani M, Amitani H, Sagiyama KI, Inui A, Asakawa A. The relationship between premorbid intelligence and symptoms of severe anorexia nervosa restricting type. Int J Med Sci. 2021;18(7):1566–9.
37. Gagliardini G, Gullo S, Tinozzi V, Baiano M, Balestrieri M, Todisco P, Schirone T, Colli A. Mentalizing subtypes in eating disorders: a latent profile analysis. Front Psychol. 2020;11:564291.
38. Citrome L. Binge-eating disorder and comorbid conditions: differential diagnosis and implications for treatment. J Clin Psychiat. 2017;78(Suppl 1):9–13.
39. Zeeck A, Endorf K, Euler S, Schaefer L, Lau I, Flösser K, Geiger V, Meier AF, Walcher P, Lahmann C, Hartmann A. Implementation of mentalization-based treatment in a day hospital program for eating disorders - a pilot study. Eur Eat Disord Rev. 2021;29(5):783–801.
40. Andrews B, Qian M, Valentine JD. Predicting depressive symptoms with a new measure of shame: the experience of shame scale. Br J Clin Psychol. 2002;41(pt 1):29–42.

41. Gilbert P, McEwan K, Irons C, Bhundia R, Christie R, Broomhead C, Rockliff H. Self-harm in a mixed clinical population: the roles of self-criticism, shame, and social rank. Br J Clin Psychol. 2010;49:563–76.
42. Scheel CN, Eisenbarth H, Rentzsch K. Assessment of different dimensions of shame proneness: validation of the SHAME. Assessment. 2020;27(8):1699–717.
43. Garcia AF, Acosta M, Pirani S, Edwards D, Osman A. Factor structure, factorial invariance, and validity of the multidimensional shame-related response Inventory-21 (MSRI-21). J Couns Psychol. 2017;64:233–46.
44. Vizin G, Urbán R, Unoka Z. Shame, trauma, temperament and psychopathology: construct validity of the experience of shame scale. Psychiatry Res. 2016;246:62–9.
45. Harned MS, Korslund KE, Foa EB, Linehan MM. Treating PTSD in suicidal and self-injuring women with borderline personality disorder: development and preliminary evaluation of a dialectical behavior therapy prolonged exposure protocol. Behav Res Ther. 2021;50:381–6.
46. Kelly AC, Carter JC. Why self-critical patients present with more severe eating disorder pathology: the mediating role of shame. Br J Clin Psychol. 2013;52:148–61.
47. Jaksic N, Marcinko D, Skocic Hanzek M, Rebernjak B, Ogrodniczuk JS. Experience of shame mediates the relationship between pathological narcissism and suicidal ideation in psychiatric outpatients. J Clin Psychol. 2017;73:1670–81.
48. Jakšić N, Marčinko D, Bjedov S, Mustač F, Bilić V. Personality organization and depressive symptoms among psychiatric outpatients: the mediating role of shame. J Nerv Ment Dis. 2022;210(8):590–5.
49. Rizvi SL. Development and preliminary validation of a new measure to assess shame: the shame inventory. J Psychopathol Behav Assess. 2010;32:438–47.
50. Fischer KW, Tangney JP. Self-conscious emotions and the affect revolution: framework and overview. In: Tangney J, Fischer K, editors. Self-conscious emotions. New York, NY: Guilford Press; 1995. p. 3–24.
51. Babcock M, Sabini J. On differentiating embarrassment from shame. Eur J Soc Psychol. 1990;20:151–69.
52. Ferreira C, Moura-Ramos M, Matos M, Galhardo A. A new measure to assess external and internal shame: development, factor structure and psychometric properties of the external and internal shame scale. Curr Psychol. 2020;41:1892–901.
53. Goss K, Gilbert P, Allan S. An exploration of shame measures: I. the "other as Shamer scale". Personal Individ Differ. 1994;17:713–7.
54. Cook DR. Internalized shame scale: technical manual. North Tonawanda: Multi-Health Systems, Inc.; 1994/2001.

Chapter 7
Separation Anxiety Disorder: Is There a Justification for a Distinct Diagnostic Category?

Milan Latas, Stefan Jerotić, Danijela Tiosavljević, and Maja Lačković

Separation anxiety disorder is a diagnostic category included in the group of anxiety disorders within the DSM 5 classification system [1]. Namely, in the DSM-IV classification, this disorder was placed in the section of childhood disorders and could only be classified in adults when it first occurred before the age of 18 [2]. Within the current DSM 5 classification, this disorder can be classified in adults regardless of first appearance, i.e. even if its beginning is ascertained after the age of 18.

This disorder is characterized by intense anxiety, with its emotional, cognitive, and behavioural manifestations, occurring in situations of separation from significant figures, or loved ones, to whom the person is strongly emotionally attached. In adults, the objects of separation anxiety are, most often, emotional partners [3].

Within the ICD 10 classification, separation anxiety disorder is in the group of disorders that occur in childhood and it is not within the group of neurotic, stress-related and somatoform disorders (F40—F48), i.e. with other anxiety disorders [4]. The proposition of ICD 11 classifies separation anxiety disorder as a part of anxiety disorders, under the code 6B05, similarly to the DSM 5 [5].

Epidemiological Characteristics

Separation anxiety disorder is common in children and adolescents, but there are very few reliable studies that indicate the prevalence of this disorder in the adult population. Nevertheless, some of them indicate that the prevalence of this disorder in the adult population is as high as 6.6% [6]. As with other anxiety disorders,

M. Latas (✉) · S. Jerotić · D. Tiosavljević · M. Lačković
Faculty of Medicine, University of Belgrade, Belgrade, Serbia

Clinic for Psychiatry, University Clinical Center of Serbia, Belgrade, Serbia

© The Author(s), under exclusive license to Springer Nature Switzerland AG 2023
V. Demarin et al. (eds.), *Mind, Brain and Education*,
https://doi.org/10.1007/978-3-031-33013-1_7

separation anxiety disorder has a higher incidence in the female than in the male population [7].

It occurs much more often in young adulthood—up to the age of 30 and it also occurs frequently in very old population. Socio-epidemiological data further indicate that separation anxiety disorder is more common in: persons who are divorced, widowers and persons who have never married; persons with less education; persons who are not employed and do not work [6].

However, there is still little relevant data that would clearly and unambiguously determine specific epidemiological indicators for separation anxiety disorder. We hope that studies regarding these issues will be performed in the near future, thus evaluating the status of this disorder in the population with greater reliability.

Clinical Characteristics

Separation fears are quite common in children in the early stages of life. They most often occur in situations of separation from family members and those who are their principal caregivers—mother, father, grandmother, etc. Typical situations where the separation anxiety occurs are: going to kindergarten, going to a holiday with kindergarten/school, situations where parents are leaving to go to work, etc. In most children, with maturation, growth and personal development, such fears tend to disappear spontaneously. However, some children continue to experience fear in situations of separation from the people they are attached to. This phenomenon, if it becomes intense and if it persists, takes the form of a psychiatric disorder, and can cause various problems in life [8].

Separation anxiety in adults is most commonly caused by the progression of separation anxiety from childhood, but it may also occur as a *de novo* phenomenon in some individuals [6, 7].

As in children, separation anxiety in adults can be viewed through three basic components of anxiety [3]:

- *The cognitive component*. In adults, separation anxiety cognitively manifests as a strong belief that the person is abandoned, neglected and unable to take care of himself or herself in a situation when he/she is separated from the loved ones. Upon separation, a person might believe that some terrible consequences will befall them –they will become ill, they will not be able to manage everyday activities, they will be lost, etc. The irrational beliefs associated with separation anxiety in those situations, most often, are: "How will I manage alone?", "I will not be able to cope alone!", "Being left alone is awful!" ...
- *The emotional component*. The onset of separation anxiety is usually accompanied by somatic symptoms of anxiety—tachycardia, rapid respiration, tension headache, tension in the muscles and stomach. As in children, in addition to experiencing fear, there may be other emotional reactions, like anger and rage directed at the person who leaves them.

- ***The behavioural component***. Due to the consequence of separation anxiety, adults often avoid exposing themselves to situations where they might be separated from the person to whom they are attached. They are usually involved in symbiotic partnerships, and reluctant to leave their place of residence. All of the above results in a poorer lifestyle, stalling in academic and professional advancement, etc. [3]

In addition to the typical clinical presentation of separation anxiety, atypical presentation can occur in some individuals. Atypical presentations of separation anxiety are [3, 5]:

- **Experience and expression of jealousy** directed towards the person who is leaving. This phenomenon is different from pathological jealousy because it is fundamentally an expression of fear that he/she will be left alone, that he/she will not be able to cope with being alone, and that he/she will not be able to bear the loneliness. By showing jealousy, the person with a fear of separation can successfully control (i.e. command) his or her emotional partner and keep him/her close.
- **Overly controlling parenting**—sometimes referred to as "reverse separation anxiety". This phenomenon is characterized by hypercontrol of children (who may be adults) because of the fear that the child may leave home at some point, thus abandoning the parent. Therefore, various methods of deception and trickery are used by the parent, whether consciously or unconsciously, in order to keep the child at home for as long as possible, and in close connection with the parent.
- **Pathological relationships**. This phenomenon is characterized by remaining in detrimental, inappropriate partner relationships, even though the person might know that the relationship is harmful or damaging, he/she chooses to stay due to the fear of being alone.
- **In the elderly**. Although the incidence of separation anxiety disorder in the very old population is lower than in the younger population, there are a large number of older people with this problem. This population is, perhaps, the most exposed to separation from close persons—spouse, friends, family. Any such person might for various reasons, leave. For example, a child might leave home, a friend or spouse might leave for hospital stay or die, etc. Therefore, special attention should be paid to this disorder in the elderly in order to detect and provide adequate care for them.

Specific focus of the fear also varies in relation to age group of the patient. In children, the anxiety might be due to perceived unrealistic consequences of separation (e.g. fear of being kidnapped during the night), while adolescents might have rationalizations for their anxiety which are more realistic (e.g. parents or a significant other might be involved in a traffic accident upon separation). Moreover, behavioural manifestations of anger are more common in children. By comparison, adolescents and adults might manifest a certain degree of social isolation, by staying at home with parents of loved ones, instead of going "out into the world" and maintaining significant relationships with other people [5] .

Presentation of symptoms of anxiety can take many different forms. Personality and temperamental factors, as well as subtle avoidance behaviours may disguise typical symptoms of anxiety. As such, sometimes even prototypical anxiety disorders such as panic disorder may be difficult to diagnose. Lack of epidemiological studies on separation anxiety disorder may be due to some of these factors. Thus, to adequately recognize and diagnose separation anxiety disorder, clinicians must first have a clear understanding of this disorder in mind.

Diagnosis

Two short screening questions can be asked to quickly assess for the presence of separation anxiety:

1. Do you feel intense fear in situations where you are separating from your loved ones?
2. Do you avoid finding yourself in situations without your loved ones?

A positive answer to at least one question may indicate the existence of separation anxiety disorder, and attention needs to be paid to further diagnostic assessment [3].

ICD Classification

The diagnosis of separation anxiety disorder in adults does not exist in the tenth International Classification of Diseases. However, within the ICD 10, there is a diagnosis of "childhood separation anxiety disorder" - F93.0. According to the criteria of that diagnosis, separation anxiety disorder in childhood should be diagnosed when there is an intense fear of separation that appears during the early stages of life [5].

This has been ameliorated in the eleventh revision of International Classification of Diseases. ICD 11 classifies separation anxiety disorder under the code 6B05, and places it in the group of anxiety disorders. In order to make a diagnosis of separation anxiety disorder, the person should have the above-described symptomatology for several months, the symptoms of separation anxiety should be persistent and impact daily functioning in a significant manner [5].

DSM Classification

Within the DSM classification system, the diagnosis of separation anxiety disorder (in adults) first appeared in the 2013 Fifth Revision. Compared to the previous DSM-IV classification, a significant change in the diagnosis of this disorder has occurred. Namely, within the DSM-IV classification, the diagnosis of separation anxiety disorder could be made only in persons with whom it appeared before the age of 18. On the other hand, according to the DSM 5 classification, separation anxiety disorder is not a disorder that occurs exclusively in childhood. It could occur in any period of life—even in adulthood, and in this case, it should be diagnosed in the context of anxiety disorders in adults.

In order to diagnose this disorder according to the DSM 5 classification system, it is necessary to meet the following criteria: a person has to exhibit a strong and long-lasting anxiety, beyond developmental age, in situations of separation from those whom he or she is attached to—such as the carer, family members, etc., and this anxiety is manifested by the presence of some of the symptoms: (1) stress in periods of separation, (2) worry about the people they are attached to, (3) avoiding separation and autonomy, (4) complaints on somatic symptoms in the periods of separation [2].

Assessment instruments can be used to assist in the diagnostic process and evaluation of separation anxiety disorder. To the best of our knowledge, up until now, only one questionnaire has been specifically designed for this issue: the Separation Anxiety Questionnaire for Adults [8]. This instrument assesses the separation anxiety of adults. It consists of 27 statements related to the symptoms of separation anxiety and situations in which such symptoms might be triggered. This instrument has been used in several research studies, but similar instruments are expected to be developed in the future [7, 9, 10].

Aetiology

A little is known yet about the specific aetiology of separation anxiety disorder. It is assumed that the aetiology of this disorder, as in other anxiety disorders, is related to biological factors, hereditary factors, psychological factors that affect the growth and development of personality and many other factors that could be responsible for the manifestation of this psychiatric disorder. Only those factors for which there is a "reasonable doubt and a significant degree of evidence" that there may be aetiological factors in the onset of separation anxiety disorder will be listed [3, 11].

- **Heredity**. The research studies indicate that two-thirds of patients with separation anxiety disorder have at least one parent who has separation anxiety problems. This raises the question of the relative risk of transmission of separation anxiety disorder from a parent to children. Unfortunately, there are no published studies that have systematically examined this area in the adult population. The research on child population samples shows that there is a significant risk that separation anxiety will occur within families. However, it is not yet known whether this phenomenon is a specific consequence of genetic inheritance, exposure to particular behaviours in the family, or interplay.
- **Personality**. In addition to anxiety disorders and depression, dependent and avoidant personality disorders are often encountered in samples of patients with separation anxiety disorder [12]. However, little is known about whether the per-

sonality structure is primarily responsible for the occurrence of separation anxiety disorder, or if the separation anxiety disorder causes the development of (pathological/anxious) personality, since it occurs in the early formative period. Moreover, research has shown high association SAD with Dependent Personality Disorder, but in people with comorbid conditions such as addictions, eating disorders and obsessive compulsive disorder. Previous research has not shown a specific relationship between separation anxiety disorder and dependent personality disorder, but certainly suggest that separation anxiety disorder can predispose to the emergence of dependent personality disorder [12].

Separation anxiety is also present as a symptom within borderline personality disorder, bearing in mind the presence of similar symptoms such as intense fear of abandonment, hypersensitivity to rejection, insecurity in interpersonal relationships, anxious attachment style to others. Experiencing a current or upcoming broke of the relationship accompanied by anxiety, uncertainty, fear of abandonment and being hurt, regardless of whether their desires and needs can be met by another person is, among other things, characteristic of separation anxiety disorder and borderline personality disorder. It is especially important to emphasize that these two disorders have a partially common etiopathogenic model and that emphasizing separation anxiety is important for potentially new differential diagnostic goals and specific psychotherapeutic and pharmacotherapeutic approaches to the treatment of borderline personality disorder.

- **Information processing**. Information processing in patients with anxiety disorders exhibits certain specifics—mostly in the situation where a stimulus is unclear and information are confusing. Studies have shown that, in children with separation anxiety, compared with children who are not anxious, there are certain specificities in the cognitive processes. These children seem to have a greater propensity for fearful interpretations, a greater propensity to make plans for avoiding risky situations, and more frequent beliefs that they are less competent to cope with dangerous situations [11]. It is possible that these findings might be applicable to the separation anxiety in adults as well. However, there are still no data for the adult population to confirm this.

Thus, little is known about the aetiology of separation anxiety disorder in adults. We believe that future research in the fields of genetics, biomarker discovery, cognitive processes analyses and other areas will specify the underlying aetiological factors of the disorder, and that specific therapeutic techniques will be developed to assist patients with the disorder, in line with the results of these studies.

Treatment

An important point in the treatment of patients with separation anxiety disorder is early detection and early initiation of treatment. In this way, the problem of separation anxiety can be efficiently managed, especially in children and young people. If the problem is not detected and not treated in a timely manner, it can be expected that the disorder will progress and eventually enter a chronic phase, resistant to therapeutic interventions.

So far, no studies that adequately addressed the problem of effective treatment of adult patients with separation anxiety disorder have been published. It is therefore very difficult to make evidence-based therapeutic recommendations [3, 11].

Pharmacotherapy

To date, no randomized, placebo-controlled pharmacological studies have been published on a sample of adult patients with separation anxiety disorder. Therefore, no firm recommendation can be made for the pharmacological treatment of these patients. Nevertheless, based on the analogy with other anxiety disorders and pharmacological studies on samples of a paediatric population of patients with separation anxiety disorder, it can be hypothesized that SSRI and SNRI antidepressants could be successful in the treatment of this disorder.

Cognitive-Behavioural Therapy

For now, there is research suggesting that cognitive-behavioural therapy is a viable choice for young patients with this disorder [13]. The problem is that there are no differential treatment protocols, yet that indicate successful techniques for overcoming this problem in the adult population. However, if the general guidelines for cognitive-behavioural treatment of anxiety are applied, and if the problem of separation anxiety disorder is viewed through a model of cognitive-behavioural therapy, it can be assumed that this therapy could be successful in the adult patient population.

Therefore, standard cognitive-behavioural therapy techniques can be expected to be successful with patients with separation anxiety disorder:

Cognitive restructuring consists of discovering irrational (or rigid, i.e. non-flexible) beliefs - "I will not be able to do it alone", "He/she must not leave me", "I will not endure without her / him", which are related to the disorder and their replacement with adaptive, rational (or flexible) beliefs - "It will be difficult but I'll try, "" Independence can bring me new success" etc.

Exposure is a basic technique that should be applied in solving the problem of avoiding separation situations. It implies that the patient is exposed to separation from the person whom he or she is attached to for the purpose of learning to function independently.

Modelling is a technique that can be very useful for patients with separation anxiety because anxiety can be overcome by adapting a new behaviour, the behaviour of a role model—a therapist or another symbolic living model.

However, before providing recommendations for treatment of patients with separation anxiety disorder, these techniques need to be verified through adequate methodological studies.

Overall, it is expected that the future studies will provide specific therapeutic guidelines for the treatment of patients with separation anxiety disorder. Until then, the basic principles for treating all anxiety disorders should be followed.

Summary

Although studies of the prevalence and most significant psychopathological phenomena that characterize separation anxiety disorder in adults date back to two or more decades ago [14], there is still a distinctive lack of knowledge about this disorder. Some of the outstanding issues are:

1. its prevalence in the general and clinical population,
2. if there are any specific etiologically factors for the occurrence of this disorder,
3. its relation with other psychiatric disorders, especially anxiety disorders,
4. its specific psychopathological features.

In addition, little (or nothing) is known about effective therapeutic approaches to treating adults with this disorder.

Some outstanding questions remain and there are no clear answers so far regarding the differentiation of "normality" and pathology related to the separation anxiety disorder:

- Is a strong relationship between family members the consequence of specific cultural and social patterns or a pathological pattern that indicates separation anxiety?
- Are the pain and suffering resulting from separation from loved ones after wars, natural or other disasters "normal"/common/expected or are they psychopathological manifestations of separation anxiety?
- What should be "normal"/usual/expected reactions that occur as a result of the loss of a loved one after death or divorce, without the psychopathological manifestations of separation anxiety?

In order to clearly identify separation anxiety disorder as an unambiguous psychopathological category, it is necessary to answer these questions in detail.

Due to all of the above, substantial research is needed/expected in the forthcoming period. We hope that the research will confirm or challenge the justification for the existence of this diagnostic category. In any case, changes to the diagnostic systems are underway. Specifically, this disorder is included in the ICD 11 classification which is expected to be used in 2022 [5].

References

1. American Psychiatric Association. DSM. 2013:5.
2. American Psychiatric Association. (1994). DSM IV-TR.
3. Latas M. Anksiozni poremećaji: Teorija i praksa. Belgrade: Službeni glasnik; 2021.
4. World Health Organization. (1993). The ICD-10 classification of mental and behavioural disorders.
5. World Health Organization. (2019). The ICD-11 classification of mental and behavioral disorders.
6. Shear K, Jin R, Ruscio AM, Walters EE, Kessler RC. Prevalence and correlates of estimated DSM-IV child and adult separation anxiety disorder in the National Comorbidity Survey Replication. Am J Psychiatr. 2006;163(6):1074–83.
7. Pini S, Abelli M, Shear KM, Cardini A, Lari L, Gesi C, et al. Frequency and clinical correlates of adult separation anxiety in a sample of 508 outpatients with mood and anxiety disorders. Acta Psychiatr Scand. 2010;122(1):40–6.
8. Wijeratne C, Manicavasagar V. Separation anxiety in the elderly. J Anxiety Disord. 2003;17(6):695–702.
9. Eapen V, Dadds M, Barnett B, Kohlhoff J, Khan F, Radom N, Silove DM. Separation anxiety, attachment and inter-personal representations: disentangling the role of oxytocin in the perinatal period. PLoS One. 2014;9(9):e107745.
10. Lewinsohn PM, Holm-Denoma JM, Small JW, Seeley JR, Joiner TE Jr. Separation anxiety disorder in childhood as a risk factor for future mental illness. J Am Acad Child Adolesc Psychiatry. 2008;47(5):548–55.
11. Bögels SM, Knappe S, Clark LA. Adult separation anxiety disorder in DSM-5. Clin Psychol Rev. 2013;33(5):663–74.
12. Silove D, Marnane C, Wagner R, Manicavasagar V. Brief report--associations of personality disorder with early separation anxiety in patients with adult separation anxiety disorder. J Personal Disord. 2011;25(1):128–33.
13. Schneider S, Blatter-Meunier J, Herren C, Adornetto C, In-Albon T, Lavallee K. Disorder-specific cognitive-behavioral therapy for separation anxiety disorder in young children: a randomized waiting-list-controlled trial. Psychother Psychosom. 2011;80(4):206–15.
14. Lewinsohn PM, Holm-Denoma JM, Small JW, Seeley JR, Joiner TE Jr. Separation anxiety disorder in childhood as a risk factor for future mental illness. J Am Acad Child Adolesc Psychiatry. 2008;47(5):548–55. https://doi.org/10.1097/CHI.0b013e31816765e7. PMID: 18356763; PMCID: PMC2732357

Chapter 8
The Emerging Concept of and Treatment Approaches for Autoimmune Psychosis

Karl Bechter and Dominique Endres

Introduction

Emerging evidence indicates that a subgroup of severe mental illnesses (SMIs), especially those on the broadly defined schizophrenic and affective spectra, may be caused by an underlying neuroinflammatory/autoimmune process [1–4]. Such a scenario has been, in principle, predicted by the mild encephalitis (ME) hypothesis: ME is assumed to be elicited by various etiologies, including immunopathology/autoimmunity, infections, toxicity (exogenous and endogenous), trauma, and possibly unknown causes [5–7]. Most importantly, some patients with SMIs who were widely treatment resistant to conventional treatments rapidly improved when treated with immunotherapies [6, 8–11].

K. Bechter (✉)
Clinic for Psychiatry and Psychotherapy II, Ulm University, Bezirkskrankenhaus Günzburg, Günzburg, Germany
e-mail: Karl.Bechter@bkh-guenzburg.de

D. Endres
Department of Psychiatry and Psychotherapy, Medical Center-University of Freiburg, Faculty of Medicine, University of Freiburg, Freiburg, Germany
e-mail: dominique.endres@uniklinik-freiburg.de

V. Demarin et al. (eds.), *Mind, Brain and Education*,
https://doi.org/10.1007/978-3-031-33013-1_8

The New Diagnoses, Autoimmune Encephalitis and Autoimmune Psychosis, and Their Relation to the Mild Encephalitis Hypothesis

With the discovery of N-methyl-d-aspartate receptor (NMDA-R) antibodies, autoimmune encephalitis (AE)—which is often associated with tumors—was redescribed; now, AE constitutes a group of antibody-mediated encephalitides characterized by prominent neuropsychiatric symptoms that are associated with antibodies against neuronal cell-surface proteins, ion channels, or receptors [12]. Most AE cases have fulfilled established criteria for encephalitis (beyond the presence of central nervous system [CNS] antibodies), but an emerging number of cases have not. Therefore, a new and more inclusive but more arbitrary clinical approach to AE's diagnosis has been developed, and it includes a guide to major differential diagnoses that covers a small subgroup of SMI cases that were not previously diagnosed as AE [13]. From this valuable clinical consensus on AE, we intend to go one step further: The SMI cases now diagnosed as AE also would fulfill the proposed ME criteria [5–7]. Alternatively, for cases in which an autoimmune etiology of psychosis was plausible, the term autoimmune psychosis (AP) was introduced [1, 4, 14]. AP may be further differentiated into three AP subgroups [4]: (a) AP presenting with CNS antibodies, (b) AP accompanying systemic autoimmune diseases, and (c) AP presenting without CNS autoantibodies that is plausibly related to yet undefined and possibly innate immune pathology. The terminology AP focused on a certain psychiatric syndrome's assumed autoimmune pathogenesis without addressing the pathogenetic process's site and type. The term ME focused on the assumed pathogenetic process's type but did not name the assumed various etiological scenarios related to, for example, virus infections as an important model of ME. The now-known pathogenetic scenarios of AE, ranging from tumor-associated and infection-triggered to unknown causes of CNS autoimmunity, are relevant to discussions regarding ME. At the 14th Psychoimmunology Expert Meeting in late March 2018 (www.psychoimmunology-experts.de), various experts agreed that the term AP may be used to represent a special subgroup of ME but that other ME types, especially infectious ME, must be differentiated. They also agreed that renaming ME to ME-related psychiatric syndromes (MErPS) would also address the variant psychiatric syndromes that may be involved.

However, in addition to diagnostic issues, theoretical issues must be kept in mind. In the diagnosis of AP/MErPS, mild neuroinflammation must be assessed and distinguished from not only classical (severe) neuroinflammation but also systemic inflammation with possible consequent brain dysfunction (which is not necessarily identical with mild neuroinflammation). An ideal solution would categorize these various aspects and interactions of systemic and/or CNS-specific inflammation using site-specific grading of inflammatory responses and their respective short- and long-term consequences in the CNS. This type of categorization has been attempted with terms such as parainflammation, neuroprogression, microglia activation, and others, but they have remained in terminological and theoretical conflict with traded terms such as encephalitis and encephalopathy insofar as a refined approach to clinical categorization is missing. In this complicated situation, the terms AP and MErPS focus on a subgroup of SMI cases presenting with mild neuroinflammation of autoimmune or other (e.g., infectious) etiologies. Recent insights into AE have produced a new awareness in psychiatry and thus a red flag diagnostic approach focused on acute onset psychosis [2, 9, 14–16].

Autoimmune Psychosis: Present Status of Diagnosis and Treatment

Internationally, AP's consensus criteria were first published in 2020 [14]. Afterward, possible, probable, and definite AP could be distinguished. Clinical characteristics, such as acute onset psychosis with catatonia, disproportionate cognitive deficits, comorbid tumors, and neurological/autonomic symptoms, should lead to suspicion of possible AP. Furthermore, an extended diagnostic workup using electroencephalography, magnetic resonance imaging (MRI), neuronal antibody testing, and cerebrospinal fluid (CSF) analysis can justify the diagnosis of probable or definite AP under defined constellations, whereby the presence of well-characterized immunoglobulin G class neuronal antibodies in CSF (e.g., against the NMDA-R) is required for the diagnosis of definite AP [14].

The treatment of confirmed autoimmune psychosis includes immunotherapies such as high-dose steroids, immunoglobulins, plasmapheresis, and rituximab [13, 17]. First experiences suggest that many patients with confirmed AP can be successfully treated with immunotherapies [11, 14].

Continuing Challenges with a Diagnosis of AP/MErPS

Systemic Inflammation Evidence for signs of minor systemic inflammation in patients with SMI is broad, but its meaning remains difficult to evaluate, especially with a single-case approach. In patients with high-risk schizophrenia, the (minor) systemic inflammatory signature predicts brain volume reduction [18]. Anti-inflammatory add-on treatment has reduced positive symptoms in schizophrenia's acute stages [19]. In affective spectrum disorders, the immune–inflammatory signature has been used for immune stratification trials [20, 21]. However, a question remains: What does systemic inflammation mean for the brain and its function?

Postmortem Neuroinflammation Suicide subgroups of SMI patients presented microglial activation [22], and, in some, cellular and likely inflammatory abnormalities [23]. High cytokine signatures were found in 40% of schizophrenia cases [24, 25]. However, a challenge remains: How can mild neuroinflammation become assessed in vivo?

Neuroimaging Minor signs compatible with mild neuroinflammation are frequent but rarely definite in SMI-related (not identical to classical) neuroinflammation. Therefore, combined methods appear to become more specific. However, a challenge remains: Mild neuroinflammation may remain undetected or non-specific when using available neuroimaging methods, but which combined methods allow for specific conclusions?

CSF Examination Recent studies using advanced general CSF analysis (methods used in neurology) have consistently revealed frequent minor abnormalities in SMI subgroups: About 10% of psychotic patients had neuroinflammatory abnormalities (increased white blood cell counts, oligoclonal bands, intrathecal immunoglobulin production, and specific CNS antibodies); about 25% had increased albumin quotients; and approximately 30% had elevated neopterin. Moreover, cytokine changes and CSF cell activation have been detected in certain subgroups [26–31]. However, a question remains: What do minor CSF abnormalities, such as increased albumin quotients, indicate?

Immunogenetics The inflammatory response system's genes are risk genes for schizophrenia [32]. Genome editing analysis has produced the conclusion that inborn vasculopathy is a cardinal feature of schizophrenia [33]; mild vasculopathy could make individuals susceptible to mild neuroinflammation, potentially elicited by multiple events, such as infections, during a person's life and may be combined with acquired proneness to enhanced (neuro-)inflammatory responses from prenatal and perinatal hits, which seem to specifically induce lasting changes to inflammatory responses [34]. Therefore, the major site of neuropathological consequences may be the neurovascular unit [3]. In other words, subgroups of patients with SMI seem to suffer from an elevated risk of mild neuroinflammation, which may manifest in the form of major disorders with neuroinflammatory flares. However, a challenge remains: What can genetic findings contribute to individual treatment decisions?

Infection Associated with ME or AP Infections are known to trigger autoimmunity. However, although determining whether an infectious agent persists in the CNS (or elsewhere in the body) is not always an easy task (it may be relevant determining which treatment to use). Recently, it has been shown that after the acute phases of herpes simplex encephalitis, about 30% of patients develop NMDA-R antibodies and thus are treated as if they have induced AE [35]. An example similar to AP/MErPS is PANDAS syndrome, for which complicated diagnostic and treatment recommendations, including antibiotic and immune modulatory treatment options, have recently been worked out [36, 37]. However, the following questions remain: How can it be determined whether infectious agents persist or whether they may have triggered an autoimmune disorder? How can treatment be adapted accordingly?

Blood–Brain Interface (BBI) Neuroinflammation and systemic inflammation interact but can occur independently. Important anatomical sites of this interaction include the blood–brain barrier (BBB) and the blood–CSF barrier (BCSFB). Neuroinflammation is poorly reflected in the blood system, so its detection requires specific diagnostic methods. The diagnosis of acute meningoencephalitis, a high-grade neuroinflammatory disease, requires a CSF examination, although some cases show no (detectable) antibodies in the CSF [38]. The previous assumption of a strict BBB and the brain's immune privilege is now being replaced by a BBI model, which includes the BBB's multiregionality, a special role of the BCSFB, CSF outflow pathways, and bidirectional neural communication between the brain and the viscera [39, 40]. However, a challenge remains: How can the functions of the BBB, BCSFB, and BBI become optimally defined in individual cases?

The Continued Discovery of CNS Autoantibodies

The arsenal of known CNS antibodies in neurology has steadily increased in recent years [17]. In the psychiatric setting, novel CNS antibodies may still be undiscovered. A recent paper has described novel antibody patterns against CNS targets such as granule cells, vessels, myelin fibers, and Purkinje cells from CSF [41]. The exact target antigens remain to be discovered [42]. These developments indicate not only the field's research needs but also the possible clinical potential in patient subgroups. Similar processes may also play roles in mental disorders other than psychosis, such as obsessive–compulsive disorder [43]. The ongoing expansion in the spectrum of CNS antibodies and associated clinical syndromes makes neuroimmunology an innovative research topic.

References

1. Ellul P, Groc L, Tamouza R, Leboyer M. The clinical challenge of autoimmune psychosis: learning from anti-NMDA receptor autoantibodies. Front Psych. 2017;8 https://doi.org/10.3389/fpsyt.2017.00054.
2. Al-Diwani A, Pollak TA, Langford AE, Lennox BR. Synaptic and neuronal autoantibody-associated psychiatric syndromes: controversies and hypotheses. Front Psych. 2017;8 https://doi.org/10.3389/fpsyt.2017.00013.
3. Najjar S, Pahlajani S, Sanctis V de, Stern JNH, Najjar A, Chong D. Neurovascular unit dysfunction and blood–brain barrier hyperpermeability contribute to schizophrenia neurobiology: a theoretical integration of clinical and experimental evidence. Front Psych 2017. doi:https://doi.org/10.3389/fpsyt.2017.00083.
4. Najjar S, Steiner J, Najjar A, Bechter K. A clinical approach to new-onset psychosis associated with immune dysregulation: the concept of autoimmune psychosis. J Neuroinflammation. 2018;15(1):40. https://doi.org/10.1186/s12974-018-1067-y.

5. Bechter K. Mild encephalitis underlying psychiatric disorders-a reconsideration and hypothesis exemplified on Borna disease. Neurol Psychiatry Brain Res. 2001;9:55–70.
6. Bechter K. Die milde Enzephalitishypothese--neue Befunde und Studien. Psychiatr Prax 2004;31 Suppl 1:S41–S43. doi:https://doi.org/10.1055/s-2004-828428.
7. Bechter K. Updating the mild encephalitis hypothesis of schizophrenia. Prog Neuro-Psychopharmacol Biol Psychiatry. 2013;42:71–91. https://doi.org/10.1016/j.pnpbp.2012.06.019.
8. Pollak TA, Lennox BR. Time for a change of practice: the real-world value of testing for neuronal autoantibodies in acute first-episode psychosis. BJPsych Open. 2018;4(4):262–4. https://doi.org/10.1192/bjo.2018.27.
9. Herken J, Prüss H. Red flags: clinical signs for identifying autoimmune encephalitis in psychiatric patients. Front Psych. 2017;8:25. https://doi.org/10.3389/fpsyt.2017.00025. eCollection 2017
10. Mack A, Pfeiffer C, Schneider EM, Bechter K. Schizophrenia or atypical lupus erythematosus with predominant psychiatric manifestations over 25 years: case analysis and review. Front Psych. 2017;8:131. https://doi.org/10.3389/fpsyt.2017.00131. eCollection 2017
11. Endres D, Lüngen E, Hasan A, Kluge M, Fröhlich S, Lewerenz J, Bschor T, Haußleiter IS, Juckel G, Then Bergh F, Ettrich B, Kertzscher L, Oviedo-Salcedo T, Handreka R, Lauer M, Winter K, Zumdick N, Drews A, Obrocki J, Yalachkov Y, Bubl A, von Podewils F, Schneider U, Szabo K, Mattern M, Philipsen A, Domschke K, Wandinger KP, Neyazi A, Stich O, Prüss H, Leypoldt F, Tebartz van Elst L. Clinical manifestations and immunomodulatory treatment experiences in psychiatric patients with suspected autoimmune encephalitis: a case series of 91 patients from Germany. Mol Psychiatry. 2022;27(3):1479–89. https://doi.org/10.1038/s41380-021-01396-4. Epub 2022 Jan 19
12. Dalmau J. The case for autoimmune neurology. Neurol Neuroimmunol Neuroinflamm. 2017;4:e373. https://doi.org/10.1212/NXI.0000000000000373.
13. Graus F, Titulaer MJ, Balu R, Benseler S, Bien CG, Cellucci T, Cortese I, Dale RC, Gelfand JM, Geschwind M, Glaser CA, Honnorat J, Höftberger R, Iizuka T, Irani SR, Lancaster E, Leypoldt F, Prüss H, Rae-Grant A, Reindl M, Rosenfeld MR, Rostásy K, Saiz A, Venkatesan A, Vincent A, Wandinger KP, Waters P, Dalmau J. A clinical approach to diagnosis of autoimmune encephalitis. Lancet Neurol. 2016;15(4):391–404. https://doi.org/10.1016/S1474-4422(15)00401-9. Epub 2016 Feb 20
14. Pollak TA, Lennox BR, Müller S, Benros ME, Prüss H, Tebartz van Elst L, Klein H, Steiner J, Frodl T, Bogerts B, Tian L, Groc L, Hasan A, Baune BT, Endres D, Haroon E, Yolken R, Benedetti F, Halaris A, Meyer JH, Stassen H, Leboyer M, Fuchs D, Otto M, Brown DA, Vincent A, Najjar S, Bechter K. Autoimmune psychosis: an international consensus on an approach to the diagnosis and management of psychosis of suspected autoimmune origin. Lancet Psychiatry. 2020;7(1):93–108. https://doi.org/10.1016/S2215-0366(19)30290-1. Epub 2019 Oct 24
15. Lennox BR, Coles AJ, Vincent A. Antibody-mediated encephalitis: a treatable cause of schizophrenia. Br J Psychiatry. 2012;200(2):92–4. https://doi.org/10.1192/bjp.bp.111.095042.
16. Steiner J, Schiltz K, Bernstein H-G, Bogerts B. Antineuronal antibodies against neurotransmitter receptors and synaptic proteins in schizophrenia: current knowledge and clinical implications. CNS Drugs. 2015;29(3):197–206. https://doi.org/10.1007/s40263-015-0233-3.
17. Prüss H. Autoantibodies in neurological disease. Nat Rev Immunol. 2021;21(12):798–813. https://doi.org/10.1038/s41577-021-00543-w. Epub 2021 May 11
18. Cannon TD. Microglial activation and the onset of psychosis. Am J Psychiatry. 2016;173(1):3–4. https://doi.org/10.1176/appi.ajp.2015.15111377.
19. Müller N. Neuroprogression in schizophrenia and psychotic disorders: the possible role of inflammation. Mod Trends Pharmacopsychiatry. 2017;31:1–9. https://doi.org/10.1159/000470802.
20. Drexhage HA. Editorial: activation and de-activation of inflammatory pathways. The disequilibrium of immune-neuro-endocrine networks in psychiatric disorders. Brain Behav Immun. 2019;78:5–6. https://doi.org/10.1016/j.bbi.2019.01.007.

21. Drexhage RC, Hoogenboezem TH, Versnel MA, Berghout A, Nolen WA, Drexhage HA. The activation of monocyte and T cell networks in patients with bipolar disorder. Brain Behav Immun. 2011;25(6):1206–13. https://doi.org/10.1016/j.bbi.2011.03.013.
22. Steiner J, Gos T, Bogerts B, Bielau H, Drexhage H, Bernstein H-G. Possible impact of microglial cells and the monocyte-macrophage system on suicidal behavior. CNS Neurol Disord Drug Targets. 2013;12(7):971–9. https://doi.org/10.2174/18715273113129990099.
23. Black JE, Kodish IM, Grossman AW, Klintsova AY, Orlovskaya D, Vostrikov V, Uranova N, Greenough WT. Pathology of layer V pyramidal neurons in the prefrontal cortex of patients with schizophrenia. Am J Psychiatry. 2004;161(4):742–4. https://doi.org/10.1176/appi.ajp.161.4.742.
24. Cai HQ, Catts VS, Webster MJ, Galletly C, Liu D, O'Donnell M, Weickert TW, Weickert CS. Increased macrophages and changed brain endothelial cell gene expression in the frontal cortex of people with schizophrenia displaying inflammation. Mol Psychiatry. 2020;25(4):761–75. https://doi.org/10.1038/s41380-018-0235-x. Epub 2018 Sep 13
25. Laskaris L, Zalesky A, Weickert CS, Di Biase MA, Chana G, Baune BT, Bousman C, Nelson B, McGorry P, Everall I, Pantelis C, Cropley V. Investigation of peripheral complement factors across stages of psychosis. Schizophr Res. 2019;204:30–7. https://doi.org/10.1016/j.schres.2018.11.035. Epub 2018 Dec 6
26. Maxeiner H-G, Rojewski MT, Schmitt A, Tumani H, Bechter K, Schmitt M. Flow cytometric analysis of T cell subsets in paired samples of cerebrospinal fluid and peripheral blood from patients with neurological and psychiatric disorders. Brain Behav Immun. 2009;23(1):134–42. https://doi.org/10.1016/j.bbi.2008.08.003.
27. Kuehne LK, Reiber H, Bechter K, Hagberg L, Fuchs D. Cerebrospinal fluid neopterin is brain-derived and not associated with blood-CSF barrier dysfunction in non-inflammatory affective and schizophrenic spectrum disorders. J Psychiatr Res. 2013;47(10):1417–22. https://doi.org/10.1016/j.jpsychires.2013.05.027.
28. Endres D, Meixensberger S, Dersch R, Feige B, Stich O, Venhoff N, Matysik M, Maier SJ, Michel M, Runge K, Nickel K, Urbach H, Domschke K, Prüss H, Tebartz van Elst L. Cerebrospinal fluid, antineuronal autoantibody, EEG, and MRI findings from 992 patients with schizophreniform and affective psychosis. Transl. Psychiatry. 2020;10(1):279. https://doi.org/10.1038/s41398-020-00967-3.
29. Zetterberg H, Jakobsson J, Redsäter M, Andreasson U, Pålsson E, Ekman CJ, Sellgren C, Johansson AG, Blennow K, Landén M. Blood-cerebrospinal fluid barrier dysfunction in patients with bipolar disorder in relation to antipsychotic treatment. Psychiatry Res. 2014;217(3):143–6. https://doi.org/10.1016/j.psychres.2014.03.045. Epub 2014 Apr 5
30. Bechter K, Reiber H, Herzog S, Fuchs D, Tumani H, Maxeiner HG. Cerebrospinal fluid analysis in affective and schizophrenic spectrum disorders: identification of subgroups with immune responses and blood-CSF barrier dysfunction. J Psychiatr Res. 2010;44(5):321–30. https://doi.org/10.1016/j.jpsychires.2009.08.008.
31. Bechter K. CSF diagnostics in psychiatry – present status – future projects. Neurol Psychiatry Brain Res. 2016;22(2):69–74. https://doi.org/10.1016/j.npbr.2016.01.008.
32. Andreassen OA, Harbo HF, Wang Y, Thompson WK, Schork AJ, Mattingsdal M, Zuber V, Bettella F, Ripke S, Kelsoe JR, Kendler KS, O'Donovan MC, Sklar P, Psychiatric Genomics Consortium (PGC) Bipolar Disorder and Schizophrenia Work Groups; International Multiple Sclerosis Genetics Consortium (IMSGC), McEvoy LK, Desikan RS, Lie BA, Djurovic S, Dale AM. Genetic pleiotropy between multiple sclerosis and schizophrenia but not bipolar disorder: differential involvement of immune-related gene loci. Mol Psychiatry. 2015;20(2):207–14. https://doi.org/10.1038/mp.2013.195. Epub 2014 Jan 28
33. Moises HW, Wollschläger D, Binder H. Functional genomics indicate that schizophrenia may be an adult vascular-ischemic disorder. Transl Psychiatry. 2015;5(8):e616. https://doi.org/10.1038/tp.2015.103.
34. Meyer U. New serological evidence points toward an infectious route to bipolar disorder. Am J Psychiatry. 2014;171(5):485–8. https://doi.org/10.1176/appi.ajp.2014.14010095.

35. Prüss H. Postviral autoimmune encephalitis: manifestations in children and adults. Curr Opin Neurol. 2017;30(3):327–33. https://doi.org/10.1097/WCO.0000000000000445.
36. Thienemann M, Murphy T, Leckman J, Shaw R, Williams K, Kapphahn C, Frankovich J, Geller D, Bernstein G, Chang K, Elia J, Swedo S. Management of pediatric acute-onset neuropsychiatric syndrome: part I-psychiatric and behavioral interventions. J Child Adolesc Psychopharmacol. 2017;27(7):566–73. https://doi.org/10.1089/cap.2016.0145.
37. Spartz EJ, Freeman GM, Brown K, Farhadian B, Thienemann M, Frankovich J. Course of neuropsychiatric symptoms after introduction and removal of nonsteroidal anti-inflammatory drugs: a pediatric observational study. J Child Adolesc Psychopharmacol. 2017;27(7):652–9. https://doi.org/10.1089/cap.2016.0179.
38. Venkatesan A, Tunkel AR, Bloch KC, Lauring AS, Sejvar J, Bitnun A, et al. Case definitions, diagnostic algorithms, and priorities in encephalitis: consensus statement of the international encephalitis consortium. Clin Infect Dis. 2013;57(8):1114–28. https://doi.org/10.1093/cid/cit458.
39. Banks WA. The blood-brain barrier in neuroimmunology: Tales of separation and assimilation. Brain Behav Immun. 2015;44:1–8. https://doi.org/10.1016/j.bbi.2014.08.007.
40. Galea I, Perry VH. The blood-brain interface: a culture change. Brain Behav Immun. 2018;68:11–6. https://doi.org/10.1016/j.bbi.2017.10.014. Epub 2017 Oct 26
41. Endres D, von Zedtwitz K, Matteit I, Bünger I, Foverskov-Rasmussen H, Runge K, Feige B, Schlump A, Maier S, Nickel K, Berger B, Schiele MA, Cunningham JL, Domschke K, Prüss H, Tebartz van Elst L. Spectrum of novel anti-central nervous system autoantibodies in the cerebrospinal fluid of 119 patients with schizophreniform and affective disorders. Biol Psychiatry. 2022;92(4):261–74. https://doi.org/10.1016/j.biopsych.2022.02.010. Epub 2022 Feb 19
42. Duong SL, Prüss H. Molecular disease mechanisms of human antineuronal monoclonal autoantibodies. Trends Mol Med. 2022;S1471-4914(22):00259–3. https://doi.org/10.1016/j.molmed.2022.09.011.
43. Endres D, Pollak TA, Bechter K, Denzel D, Pitsch K, Nickel K, Runge K, Pankratz B, Klatzmann D, Tamouza R, Mallet L, Leboyer M, Prüss H, Voderholzer U, Cunningham JL, ECNP Network Immuno-NeuroPsychiatry, Domschke K, Tebartz van Elst L, Schiele MA. Immunological causes of obsessive-compulsive disorder: is it time for the concept of an "autoimmune OCD" subtype? Transl. Psychiatry. 2022;12(1):5. https://doi.org/10.1038/s41398-021-01700-4.

Chapter 9
"A Healthy Mind in a Healthy Body": An Overview on the Effects of Physical Activity on the Brain

Anna Maria Malagoni and Francesco Maffessanti

As the Roman poet Decimus Iunius Iuvenalis (Juvenal, late first and early second century AD) wrote *"Orandum est ut sit mens sana in corpore sano"* (Satire, X, 356), we should pray for a healthy mind in a healthy body. Such a statement has been interpreted over time as a reminder that a healthy body is fundamental to support a healthy mind and that being physically active plays a key role in sustaining the overall well-being of body and brain.

Moving from the narrative to the scientific literature, we can now reasonably state that current evidence supports the power of physical activity (PA) in modulating the brain plasticity, in maintaining and improving the cognitive and psychological functions throughout the life span [1, 2]. It is of interest to note that it has even been hypothesized that the brain size in the hominin was, at least in part, a correlated response to natural selection based on aerobic capacity [3]. PA related improvements in cognitive performance have been identified as early as the development phase in children of women who exercised regularly during pregnancy, as well as in subjects who were physically active during childhood and adolescence [4]. Moreover, a significant association between PA level in healthy pregnant women during first and second trimester and brain cortical thickness in newborns has been recently demonstrated [5]. Furthermore, this link can also be observed in other stages of life and numerous studies have reported a relationship between high levels

A. M. Malagoni (✉)
Maria Cecilia Hospital, GVM Care and Research, Cotignola, RA, Italy

UniCamillus-Saint Camillus International University of Medical and Health Sciences, Rome, Italy
e-mail: amalagoni@gvmnet.it

F. Maffessanti
Maria Cecilia Hospital, GVM Care and Research, Cotignola, RA, Italy
e-mail: fmaffessanti@gvmnet.it

V. Demarin et al. (eds.), *Mind, Brain and Education*,
https://doi.org/10.1007/978-3-031-33013-1_9

of physical activity, hippocampal size, and cognitive function in the elderly [6, 7]. Notably, a review of the 2018 Physical Activity Guidelines for Americans [2] concluded that "there is moderate-to-strong support that PA benefits cognitive functioning during early and late periods of the life span and in certain populations characterized by cognitive deficits".

According to Phillips et coll [6]. four main hypotheses might be proposed to explain the link between PA and brain health: the cardiovascular, immunologic, neuroendocrine, and neurotrophic signaling hypothesis. It is widely recognised that PA can have benefits on the cardiovascular system by preventing and lowering blood pressure, improving lipoprotein profile, managing body weight, increasing insulin sensitivity and exercise tolerance [8]. Besides, PA can induce improvements in cerebral blood flow, which is a physiological marker of cerebrovascular function [9, 10], and in small vessels condition [11]. All those effects are thought to contribute to sustaining brain health and function. A paradigm of such a statement is the association of high blood pressure with blood brain barrier alterations and functional and structural damage to the brain, suggesting that PA is important in preventing and reducing cognitive dysfunctions caused by hypertension [12]. Similarly, as the immunological and anti-inflammatory effects of PA are recognised [13], these benefits may also apply to cognitive impairments and neurodegenerative disorders where chronic inflammation has been etiologically linked, including Alzheimer's disease [14], Parkinson's disease [15] and multiple sclerosis [16].

On the molecular and cellular level, PA appears to be capable of inducing the expression of several neurotransmitters and neurotrophins, such as nerve growth factor, brain-derived neurotrophic factor (BDNF), neurotrophin-3 and, neurotrophin- 4/5 [7] which in turn influence synaptic plasticity and cell proliferation. Of all the neurotrophins, a key role seems to be played by BDNF as it is involved in all the most important aspects of neuroplasticity [2]. Of note, BDNF looks to be the most susceptible to regulation by PA, particularly by strength training and combined aerobic/strength training in older adults [17]. In addition, the aerobic energy expense would appear to be the major driver for the exercise impact on BDNF gene expression, namely the higher the volume of exercise, the higher the increase in BDNF concentration [18].

Intriguingly, new studies reveal that exercise has an impact on the epigenetic regulation of gene expression that can build an "epigenetic memory" affecting long-term brain function and behaviour, brain plasticity and neurogenesis [19]. In this very complex process, particular attention has been recently focused on microRNAs (miRNAs), single stranded RNA belonging to the class of non-coding RNAs, which appear to be the link between periphery and brain, and evidence, both in human and animals, has demonstrated that exercise can modulate blood levels of numerous miRNAs [19].

With regard to specific neurodegenerative disorders, several studies in the last few decades have shown that regular PA can help to improve symptoms, restore functions and improve quality of life in patients affected by multiple sclerosis [16, 20, 21], Alzheimer's disease [14], Parkinson's disease [15], Huntington's disease [22] and stroke survivors [23–25]. Many research studies on animal models and humans have been trying to explore and explain the molecular and cellular interplay between exercise and the brain in those diseases [1, 11]. Moreover, it also emerged that in patients affected by multiple sclerosis, regular physical exercise can mitigate mental and psychological disorders associated with COVID-19 infection along with the capacity to limit the effects of the cytokine storm [26].

Finally, it is well known that PA can improve the mental status, by lessening anxiety, inducing analgesia, and generating a general perception of well-being. PA seems able to induce those effects by favouring the production of some molecular mediators, such as dopamine, serotonin, nitric oxide, catecholamines, endogenous opioid and endocannabinoids [1, 27, 28].

In conclusion, there is growing evidence that PA is an effective tool to preserve, support and enhance brain functions and to prevent their decline. The role of PA on the brain is a fascinating and emerging research area certainly worthy of being further investigated.

References

1. Di Liegro CM, Schiera G, Proia P, Di Liegro I. Physical activity and brain health. Genes (Basel). 2019;10(9):720.
2. Erickson KI, Hillman C, Stillman CM, Ballard RM, Bloodgood B, Conroy DE, Macko R, Marquez DX, Petruzzello SJ, Powell KE, FOR 2018 PHYSICAL ACTIVITY GUIDELINES ADVISORY COMMITTEE*. Physical activity, cognition, and brain outcomes: a review of the 2018 physical activity Guidelines. Med Sci Sports Exerc. 2019;51(6):1242–51.
3. Hill T, Polk JD. BDNF, endurance activity, and mechanisms underlying the evolution of hominin brains. Am J Phys Anthropol. 2019;168(Suppl 67):47–62.
4. Gomes da Silva S, Arida RM. Physical activity and brain development. Expert Rev Neurother. 2015;15(9):1041–51.
5. Na X, Raja R, Phelan NE, Tadros MR, Moore A, Wu Z, Wang L, Li G, Glasier CM, Ramakrishnaiah RR, Andres A, Ou X. Mother's physical activity during pregnancy and newborn's brain cortical development. Front Hum Neurosci. 2022;6(16):943341.
6. Phillips C, Baktir MA, Srivatsan M, Salehi A. Neuroprotective effects of physical activity on the brain: a closer look at trophic factor signaling. Front Cell Neurosci. 2014;8:170.
7. Ji L, Steffens DC, Wang L. Effects of physical exercise on the aging brain across imaging modalities: a meta-analysis of neuroimaging studies in randomized controlled trials. Int J Geriatr Psychiatry. 2021;36(8):1148–57.
8. Garber CE, Blissmer B, Deschenes MR, Franklin BA, Lamonte MJ, Lee IM, Nieman DC, Swain DP. American College of Sports Medicine position stand. Quantity and quality of exercise for developing and maintaining cardiorespiratory, musculoskeletal, and neuromotor fitness in apparently healthy adults: guidance for prescribing exercise. Med Sci Sports Exerc. 2011;43(7):1334–59.
9. Swain RA, Harris AB, Wiener EC, Dutka MV, Morris HD, Theien BE, Konda S, Engberg K, Lauterbur PC, Greenough WT. Prolonged exercise induces angiogenesis and increases cerebral blood volume in primary motor cortex of the rat. Neuroscience. 2003;117(4):1037–46.
10. Kleinloog JPD, Nijssen KMR, Mensink RP, Joris PJ. Effects of physical exercise training on cerebral blood flow measurements: a systematic review of human intervention studies. Int J Sport Nutr Exerc Metab. 2022;33:1–13.

11. Schmidt W, Endres M, Dimeo F, Jungehulsing GJ. Train the vessel, gain the brain: physical activity and vessel function and the impact on stroke prevention and outcome in cerebrovascular disease. Cerebrovasc Dis. 2013;35(4):303–12.
12. Rêgo ML, Cabral DA, Costa EC, Fontes EB. Physical exercise for individuals with hypertension: it is time to emphasize its benefits on the brain and cognition. Clin Med Insights Cardiol. 2019;13:1179546819839411.
13. Gleeson M, Bishop NC, Stensel DJ, Lindley MR, Mastana SS, Nimmo MA. The anti-inflammatory effects of exercise: mechanisms and implications for the prevention and treatment of disease. Nat Rev Immunol. 2011;11(9):607–15.
14. Radić B, Blažeković A, Duraković D, Jurišić-Kvesić A, Bilić E, Borovečki F. Could mental and physical exercise alleviate Alzheimer's Disease? Psychiatr Danub. 2021;33(Suppl 4):1267–73.
15. Machado S, Teixeira D, Monteiro D, Imperatori C, Murillo-Rodriguez E, da Silva Rocha FP, Yamamoto T, Amatriain-Fernández S, Budde H, Carta MG, Caixeta L, de Sá Filho AS. Clinical applications of exercise in Parkinson's disease: what we need to know? Expert Rev Neurother. 2022;22:771–80. https://doi.org/10.1080/14737175.2022.2128768.
16. White LJ, Castellano V. Exercise and brain health--implications for multiple sclerosis: Part II--immune factors and stress hormones. Sports Med. 2008;38(3):179–86.
17. Marinus N, Hansen D, Feys P, Meesen R, Timmermans A, Spildooren J. The impact of different types of exercise training on peripheral blood brain-derived neurotrophic factor concentrations in older adults: a meta-analysis. Sports Med. 2019;49(10):1529–46.
18. De Assis GG, Gasanov EV, de Sousa MBC, Kozacz A, Murawska-Cialowicz E. Brain derived neutrophic factor, a link of aerobic metabolism to neuroplasticity. J Physiol Pharmacol. 2018;69(3):10.26402/jpp.2018.3.12.
19. Fernandes J, Arida RM, Gomez-Pinilla F. Physical exercise as an epigenetic modulator of brain plasticity and cognition. Neurosci Biobehav Rev. 2017;80:443–56.
20. Dalgas U, Langeskov-Christensen M, Stenager E, Riemenschneider M, Hvid LG. Exercise as medicine in multiple sclerosis-time for a paradigm shift: preventive, symptomatic, and disease-modifying aspects and perspectives. Curr Neurol Neurosci Rep. 2019;19(11):88.
21. White LJ, Castellano V. Exercise and brain health--implications for multiple sclerosis: Part 1--neuronal growth factors. Sports Med. 2008;38(2):91–100.
22. Trovato B, Magrì B, Castorina A, Maugeri G, D'Agata V, Musumeci G. Effects of exercise on skeletal muscle pathophysiology in Huntington's disease. J Funct Morphol Kinesiol. 2022;7(2):40.
23. Malagoni AM, Cavazza S, Ferraresi G, Grassi G, Felisatti M, Lamberti N, Basaglia N, Manfredini F. Effects of a "test in-train out" walking program versus supervised standard rehabilitation in chronic stroke patients: a feasibility and pilot randomized study. Eur J Phys Rehabil Med. 2016;52(3):279–87.
24. Lamberti N, Straudi S, Malagoni AM, Argirò M, Felisatti M, Nardini E, Zambon C, Basaglia N, Manfredini F. Effects of low-intensity endurance and resistance training on mobility in chronic stroke survivors: a pilot randomized controlled study. Eur J Phys Rehabil Med. 2017;53(2):228–39.
25. Moncion K, Allison EY, Al-Khazraji BK, MacDonald MJ, Roig M, Tang A. What are the effects of acute exercise and exercise training on cerebrovascular hemodynamics following stroke? A systematic review and meta-analysis. J Appl Physiol (1985). 2022;132:1379–93.
26. Razi O, Tartibian B, Laher I, Govindasamy K, Zamani N, Rocha-Rodrigues S, Suzuki K, Zouhal H. Multimodal benefits of exercise in patients with multiple sclerosis and COVID-19. Front Physiol. 2022 Apr;14(13):783251.
27. Da Silva SR, Galdino G. Endogenous systems involved in exercise-induced analgesia. J Physiol Pharmacol. 2018;69(1):3–13.
28. Chen C, Beaunoyer E, Guitton MJ, Wang J. Physical activity as a clinical tool against depression: opportunities and challenges. J Integr Neurosci. 2022;21(5):132.

Chapter 10
Next Generation Technologies in Functional Neurosurgery

Marina Raguž, Darko Orešković, Filip Derke, and Darko Chudy

Introduction

Neurosurgery has traditionally been a leader in advanced technologies, successfully adopting and adapting new techniques and devices in an effort to increase the safety and effectiveness of brain and spine surgery. Especially Deep Brain Stimulation (DBS) has evolved during last decades and become a leading stereotactic technique for the treatment of the Parkinson disease, essential tremor and dystonia, different psychiatric disorders such as obsessive compulsive disorder and major depression, refractory epilepsy, and chronic pain syndromes. In addition, the potential DBS role for disorders of consciousness, Alzheimer's disease, anorexia, obesity, addiction, traumatic brain injury etc. is still researched. As a procedure, DBS provides targeted circuit-based neuromodulation. Nowadays, DBS system includes intracranial electrode, an extension wire and a pulse generator. Software and hardware advances, as

M. Raguž (✉)
Department of Neurosurgery, University Hospital Dubrava, Zagreb, Croatia

Catholic University of Croatia, School of Medicine, Zagreb, Croatia

D. Orešković
Department of Neurosurgery, University Hospital Dubrava, Zagreb, Croatia

F. Derke
Department of Neurology, Dubrava University Hospital, Zagreb, Croatia

D. Chudy
Department of Neurosurgery, Dubrava University Hospital, Zagreb, Croatia

Department of Surgery, University of Zagreb, School of Medicine, Zagreb, Croatia

V. Demarin et al. (eds.), *Mind, Brain and Education*,
https://doi.org/10.1007/978-3-031-33013-1_10

well as targeting advances occurring daily in functional neurosurgery enables improve accuracy of implanted electrodes, individualized therapy to patient and less surgical intervention, improving overall the quality of patients life.

By introduction of a stereotactic apparatus by Spiegel and Wycis in 1947 started the development of modern DBS technology [1]. Next years following, a number of neuroscientist started experiments with implanted electrodes in animals and humans applying chronicle stimulation to various brain regions [2]. The real breakthrough for DBS occurred in the late 1980's by Benabid and colleagues, who reported successful chronic electrode implantation in the ventral intermediate nucleus of the thalamus for treatment of tremor in essential tremor and Parkinsons disease [3]. Good results in reported case series has proven that patients underwent DBS had less side effect compared with ablation techniques [2, 4].

Due to proven good outcome and technological development, a number of research has been conducted [5, 6] resulting in several approved targets for treating mentioned neurological and neuropsychiatric disorders, such as subthalamic nucleus (STN), the globus pallidus internus (GPi), ventral intermediate nucleus (VIM), anterior nucleus and centromedian-parafascicle nucleus of the thalamus, anterior limb of the internal capsule, fornix etc. [3, 6–11].

DBS Technology

As mentioned previously, DBS system consist of the lead and electrodes implanted intracranially, the extensions and the implantable pulse generator (IPG), usually placed in the subcutaneous chest or abdominal pocket, depending on the patient. In addition, external components include the clinician and the patient programmer, as well as recharger for rechargeable IPG, developed and widely used in a last few years. Rapid evolution of DBS is mostly dependent on several DBS system manufactures.

Electrodes, usually composed of platinum-iridium with nickel alloy connectors, play crucial role in the stimulation [12]. Besides being made of material which provide proper electrical properties for continuous stimulation, the electrode design with four ring contacts placed from 0.5 to 1.5 mm apart contributes adequate activation of targeted brain structure [12, 13]. Model to estimate the stimulated area has been described as the volume of tissue activated (VTA) enabling to presume which brain area will be stimulated. Still, VTA depends on number of contacts, polarities, chose stimulation parameters and brain structure itself i.e. cell type, fiber type, fiber orientation etc. [14]. The aim to shape the VTA to cover the therapeutic area and exclude or minimize the stimulation of close area or fibers led to segmental and directional leads providing improved precision and selectivity to targeted areas. A number of research's done recently provided information regarding better stimulation and outcome in patients; directional leads activating targeted and individualized VTA would eventually lead to preserved battery of IPG [15]. Still, it is to keep in mind that although segmented and directional leads provides good patient outcome, it increase significantly the complexity of clinician programming. Therefore, further technical development should be followed with proper targeted learning from clinicians involved.

Extension, composed of similar material as electrodes, connects electrodes and IPG. As mentioned previously, IPG is placed in the subcutaneous chest or abdominal pocket, depending on the patient. In last decade IPG decreased in weight and size, although there are preclinical research showing that IPG could be even smaller, and therefore more comfortable for patient. Also, IPGs nowadays has dual-lead channel capability i.e. one IPG is used for bilateral DBS electrodes [15]. One of the most important thing for patient regarding IPG is its longevity. As known, it is dependent on used stimulation parameters; double monopolar stimulation consume more IPG power than monopolar stimulation leading to reduced battery life, as well as higher frequency, amplitude, and pulse [16], while bipolar stimulation may lead to extended IPG life [17]. Furthermore, numerous patients reported sudden appearance of neurologic or psychiatric symptoms sometimes leading to life threatening conditions when IPG reaches the end of battery life, although the IPG should provide consistent power regardless to battery status [12]. All of the above has led to the necessity of delivering stimulation without affecting the battery life and development and clinical use of the rechargeable IPG systems. Rechargeable IPG systems are estimated to last approximately 15 years before neurosurgical replacement, leading to both cost-savings (less surgeries) and patient satisfaction [18], with only minor objections - need to recharge on a regular basis (depending on the stimulation parameters) and possible technical issues with external charging device.

DBS Targeting

DBS system properly implanted represents a precondition for a positive clinical outcome. Targeting initially relies of finding the proper anatomical structure-target, including stereotaxy and neuroimaging.

Stereotaxy is used to determine position of targeted structure in a 3-D coordinate system [19]. A stereotactic head frame is rigidly mounted to the patient's skull; afterwards the patients is scanned using a computed tomography (CT) scan in order to establish a frame-based coordinate system. Number of available commercially available softwares can then be used to merge a previously obtained magnetic resonance imaging (MRI) with head frame CT. Thus, trajectory planning performed on the preoperative MRI is translated to the frame-based CT scan providing a mechanism for accurate DBS lead implantation [20]. Frameless systems using fiducial markers that are merged or co-registered to preoperative MRI scans, obtain the same mechanism to establish a coordinate system.

On the other hand, many institutions use microelectrode recording (MER) in order to identify the physiologic target. Number of studies showed improved localization of DBS lead placement by using intraoperative MER and its benefits: safe identification of neural structures and borders, approximation of the best lead location, understanding disease pathophysiology [21]. Still, it is to mention intracranial hemorrhage in approximately 5% of patients, with hemorrhage risk increasement with the number of microelectrode passes used to determine the correct position within the anatomical structure [22, 23]. Regarding total procedure time, several studies showed that MER does extend the length of the DBS procedure [24]. Although MER will likely continue to be an important technique, improved neuroimaging and atlas techniques call into question the role of MER [25]. Since data in literature are still insufficient, larger, randomized multicenter international clinical studies are needed to establish which technique, image vs. MER, will be chosen.

Along with imaging advances, atlas-based direct targeting is used widely, where an anatomic atlas is overlaid onto a patient's MRI, providing a detailed estimation of nuclei borders. Classical stereotactic atlases, Talairach and Tournoux [26], Schaltenbrand et al. [27], and Schaltenbrand and Bailey [28] atlases, have been digitized and are still used for targeting. Additionally, number of new atlases for deep brain structured have been published, based on histology [29], connectivity [30], and 7 T MRI [31]. The majority of atlases have been developed based on data from healthy controls, population-specific atlases, such as the PD25 atlas for Parkinson disease have been published [32]. In order to improve atlas-patient registration a combination of automated algorithms and manual refinement is used for ensuring accuracy of the DBS lead trajectory and maximize the VTA in the targeted zone.

Many new MRI protocols have been developed to improve visualization of targeted structures, such as Fast Gray matter Acquisition T1 Inversion Recovery (FGATIR) sequence [33], Quantitative susceptibility mapping (QSM) and susceptibility-weighted imaging (SWI) [34]. Although 7 T MRI has several advantages compared to 1.5 T or 3 T, it is unavailable in the majority of centers. To understand better the pathophysiology, connectomes became a part of neurosurgical targeting as two main components: a model of the effect of stimulation on surrounding structures, VTA and neuroimaging based connectivity. Neuroimaging modalities for connectivity measures include diffusion tensor imaging (DTI) for structural connectivity and functional MRI (fMRI) for functional connectivity. Combining mentioned models provides comparison of the modulated brain networks [35].

DBS Patient Stimulation

To shape a VTA and control the patients stimulation, the amplitude, pulse width, frequency and the polarity of the electrode contacts is used to deliver therapeutic effects. Stimulation parameters are mostly based on structure-effect relationships learned empirically or using computational modeling. Various and numerous combination of mentioned parameters can be set when programing a patient.

Additionally to three main stimulation parameters, the contacts polarity plays an important role in shaping the VTA. Usually, monopolar configurations consist of the IPG assigned as the anode and a single contact assigned as the cathode. Recently, bipolar and multipolar electrode configuration is enabled, resulting in minimizing stimulation in areas that are more prone to side effects [36]. The amplitude the first parameter adjusted in initial programming visits. A higher amplitude leads to a larger VTA, resulting in both therapeutic benefit and increased probability of stimulation-induced side effects. Studies have shown highest correlation between amplitude and motor improvement in patients with Parkinsons disease after STN-DBS [37].

The pulse width is set between 60 and 90 µs. Recently, shortening the pulse width in patients with STN-DBS for Parkinson's disease [38, 39] and VIM-DBS for essential tremor [40] has been shown to widen the therapeutic window. It seems that lower pulse width may focus the stimulation on smaller diameter myelinated axons near the electrode as opposed to larger diameter axons located farther away, making the stimulation area more precise and potentially saving battery life [38]. On the other hand, dystonia typically require higher pulse widths, as high as 450 µs, to achieve good clinical outcome [41].

The rate of stimulation was usually delivered at 130 Hz. In certain patient populations, adjusting the frequency of stimulation may be an important programming strategy to improve therapeutic benefit or to reduce stimulation-induced side effects. Low of high frequency stimulation is described previously to affects better some of the Parkinson's disease symptoms [42, 43]. In VIM-DBS for essential tremor studies showed maximal tremor benefit to be around 100–130 Hz [44], while studies evaluating GPi-DBS for dystonia showed that higher frequencies in the range of 180-250 Hz led to significant clinical improvement [45].

In addition to conventional stimulation parameters, several advances that enable new stimulation approaches has been described, such as interleaving, cycling, biphasic, and current fractionation. Interleaving enables rapid alternation between two contacts with different amplitudes and pulse widths but the same frequencies, while cycling alternates between an active stimulation phase and an off phase in order to reduce or avoid stimulation-induced side effects. Variable frequency stimulation provides combination of multiple frequencies, on the same electrode contact, patterned in blocks to provide better symptoms management [46], while theta burst stimulation is form of alternative therapy that uses a similar concept, delivering a bursts of stimulation that cycles on and off at a 5 Hz rate, and showing benefits for patients with Parkinsons disease and dystonia [47, 48].

DBS Softwares

Recently, several software advances that have improved upon the clinical programming strategies. Although telemedicine system has been used in the medicine in the last decades, only recent has been used for DBS programming [49], and became especially relevant in the setting of the Covid-19 pandemic. By using wearable sensors for objective symptom assessment and advanced video recognition software, clinicians are able to gain objective measurements and insight into symptom severity prior to the telemedicine visit [50, 51]. Although benefits of remote DBS programming are numerous: reductions in patients travel and cost, improved access for patients in rural locations, frequent DBS programming visits for specific cases that require frequent titrations, the limitations include difficulty targeting symptoms that are challenging to assess virtually, patient difficulty using novel technology for remote DBS access, prevention of potential security breaches [52].

Automated programming is a novel area with the potential to reduce the burdens and shortens the time for the clinician and patient. The primary focus is the use of objective symptom assessment paired with computer-controlled therapy updates, like wearable sensors for tremor. Still, the primary challenge is the capability of therapy adjustment in real-time available to the computer running symptom assessment.

One of the recent update for the IPG systems is the sensing technology that has the potential to expand closed-loop stimulation to a broad patient population. The Medtronic Percept PC is currently the only DBS IPG system capable of sensing in vivo brain activity [53]. Local field potentials from targeted structure are being recorded in a natural setting which help understand the underlying neurophysiology of the disease and the mechanisms of DBS by identifying biomarkers of neural dysfunction; if potentials correspond to clinical symptoms closed-loop technologies can be developed to stimulation parameters in real time.

Closed-loop DBS can be described as adaptive DBS (aDBS) or responsive DBS (rDBS). aDBS adjusts the stimulation amplitude based on the detection of symptomatic events, like the use of subthalamic beta oscillations as a biomarker for the presence of symptoms [54]. rDBS is a commonly used for the treatment of epilepsy [55] and movement disorders [56]. The main difference between aDBS and rDBS is the duration of stimulation after an event is observed: aDBS turns off stimulation when the detector identifies the disappearance of the event, while rDBS turns off stimulation after a fixed duration.

Conclusion—Next Generation Technologies

DBS technology advances in have led to new implications for the clinical treatment of patients with different disorders. Advances in atlases, imaging techniques, and connectomes improved DBS targeting strategies, novel lead design is used for directional stimulation, while improvement of the IPG systems have led to smaller, MRI compatible devices with increased flexibility for programming strategies and rechargeable capabilities. In combination with directional leads, huge potential is offered for patients stimulation. Additionally, closed-loop forms of stimulation are going to emerge in clinical practice very soon, since case report good clinical outcomes.

Still, it is crucial assess all technological developments and determine is it possible to use them for clinical improvements of patients. Further, extensive research is needed to determine which patients will benefit most from these types of technology, as well as systematic approach to programming.

Comparisons between targeting strategies mentioned previously is needed, as well as comparison studies between awake and asleep DBS, including short-term and long-term follow-up. Privacy and security of patients data is one more factor that should be analyzed, especially nowadays, when every device is somehow connected, which is especially relevant when telemedicine has become important way of communicating in the healthcare world. International, multicentric studies and registers may be one solution for overbridged institutional variability and enable practitioners to share protocols and recommendations and targeting strategies.

References

1. Spiegel EA, Wycis HT, Marks M, Lee AJ. Stereotaxic apparatus for operations on the human brain. Science. 1947;106:349–50. https://doi.org/10.1126/science.106.2754.349.
2. Hariz MI, Blomstedt P, Zrinzo L. Deep brain stimulation between 1947 and 1987: the untold story. Neurosurg Focus. 2010;29:E1. https://doi.org/10.3171/2010.4.FOCUS10106.
3. Benabid AL, Pollak P, Louveau A, Henry S, de Rougemont J. Combined (thalamotomy and stimulation) stereotactic surgery of the VIM thalamic nucleus for bilateral Parkinson disease. Appl Neurophysiol. 1987;50:344–6. https://doi.org/10.1159/000100803.
4. Sironi VA. Origin and evolution of deep brain stimulation. Front Integr Neurosci. 2011;5:42. https://doi.org/10.3389/fnint.2011.00042.
5. DeLong MR. Primate models of movement disorders of basal ganglia origin. Trends Neurosci. 1990;13:281–5. https://doi.org/10.1016/0166-2236(90)90110-V.
6. Sarem-Aslani A, Mullett K. Industrial perspective on deep brain stimulation: history, current state, and future developments. Front Integr Neurosci. 2011;5:46. https://doi.org/10.3389/fnint.2011.00046.
7. Herrington TM, Cheng JJ, Eskandar EN. Mechanisms of deep brain stimulation. J Neurophysiol. 2016;115:19–38. https://doi.org/10.1152/jn.00281.2015
8. Deeb W, Salvato B, Almeida L, Foote KD, Amaral R, Germann J, et al. Fornix region deep brain stimulation-induced memory flashbacks in Alzheimer's disease. N Engl J Med. 2019;381:783–5. https://doi.org/10.1056/NEJMc1905240.
9. Cagnan H, Denison T, McIntyre C, Brown P. Emerging technologies for improved deep brain stimulation. Nat Biotechnol. 2019;37:1024–33. https://doi.org/10.1038/s41587-019-0244-6.
10. Chudy D, Deletis V, Almahariq F, Marčinković P, Škrlin J, Paradžik V. Deep brain stimulation for the early treatment of the minimally conscious state and vegetative state: experience in 14 patients. J Neurosurg. 2018;128(4):1189–98.
11. Chudy D, Raguž M, Deletis V. Deep brain stimulation for treatment patients in vegetative state and minimally conscious state. In: Deletis V, Shils JL, Sala F, Seidel K, editors. Neurophysiology in neurosurgery: a modern approach. Elsevier; 2020.
12. Krauss JK, Lipsman N, Aziz T, Boutet A, Brown P, Chang JW, et al. Technology of deep brain stimulation: current status and future directions. Nat Rev Neurol. 2021;17:75–87. https://doi.org/10.1038/s41582-020-00426-z.
13. Vissani M, Isaias IU, Mazzoni A. Deep brain stimulation: a review of the open neural engineering challenges. J Neural Eng. 2020;17:051002. https://doi.org/10.1088/1741-2552/abb581.
14. Zhang S, Tagliati M, Pouratian N, Cheeran B, Ross E, Pereira E. Steering the volume of tissue activated with a directional deep brain stimulation lead in the globus pallidus pars interna: a modeling study with heterogeneous tissue properties. Front Comput Neurosci. 2020;14:561180. https://doi.org/10.3389/fncom.2020.561180.

15. Paff M, Loh A, Sarica C, Lozano AM, Fasano A. Update on current technologies for deep brain stimulation in Parkinson's disease. J Mov Disord. 2020;13:185–98. https://doi.org/10.14802/jmd.20052.

16. van Riesen C, Tsironis G, Gruber D, Klostermann F, Krause P, Schneider GH, et al. Disease-specific longevity of impulse generators in deep brain stimulation and review of the literature. J Neural Transm. 2016;123:621–30. https://doi.org/10.1007/s00702-016-1562-1.

17. Almeida L, Rawal PV, Ditty B, Smelser BL, Huang H, Okun MS, et al. Deep brain stimulation battery longevity: comparison of monopolar versus bipolar stimulation modes. Mov Disord Clin Pract. 2016;3:359–66. https://doi.org/10.1002/mdc3.12285.

18. Hitti FL, Vaughan KA, Ramayya AG, McShane BJ, Baltuch GH. Reduced long-term cost and increased patient satisfaction with rechargeable implantable pulse generators for deep brain stimulation. J Neurosurg. 2018;131:799–806. https://doi.org/10.3171/2018.4.JNS172995.

19. Schiefer TK, Matsumoto JY, Lee KH. Moving forward: advances in the treatment of movement disorders with deep brain stimulation. Front Integr Neurosci. 2011;5:69. https://doi.org/10.3389/fnint.2011.00069.

20. Almahariq F, Sedmak G, Vuletić V, Dlaka D, Orešković D, Marčinković P, Raguž M, Chudy D. The accuracy of direct targeting using fusion of MR and CT imaging for deep brain stimulation of the subthalamic nucleus in patients with Parkinson's disease. J Neurol Surg A Cent Eur Neurosurg. 2021;82:518–25. https://doi.org/10.1055/s-0040-1715826.

21. Koirala N, Serrano L, Paschen S, Falk D, Anwar AR, Kuravi P, et al. Mapping of subthalamic nucleus using microelectrode recordings during deep brain stimulation. Sci Rep. 2020;10:19241. https://doi.org/10.1038/s41598-020-74196-5.

22. Obeso JA, Olanow CW, Rodriguez-Oroz MC, Krack P, Kumar R, Lang AE. Deep-brain stimulation for Parkinson's disease study group. Deep-brain stimulation of the subthalamic nucleus or the pars interna of the globus pallidus in Parkinson's disease. N Engl J Med. 2001;345(13):956–63.

23. Zrinzo L, Foltynie T, Limousin P, Hariz MI. Reducing hemorrhagic complications in functional neurosurgery: a large case series and systematic literature review. J Neurosurg. 2012;116(01):84–94.

24. Kocabicak E, Alptekin O, Aygun D, Yildiz O, Temel Y. Microelectrode recording for deep brain stimulation of the subthalamic nucleus in patients with advanced Parkinson's disease: advantage or loss of time? Turk Neurosurg. 2019;29:677–82. https://doi.org/10.5137/1019-5149.JTN.23307-18.3.

25. Pastor J, Vega-Zelaya L. Can we put aside microelectrode recordings in deep brain stimulation surgery? Brain Sci. 2020;10:571. https://doi.org/10.3390/brainsci10090571.

26. Talairach J, Tournoux P. Co-planar stereotaxic atlas of the human brain. New York, NY: Thieme Medical Publishers; 1988.

27. Schaltenbrand G, Hassler R, Wahren W. Atlas for Stereotaxy of the human brain. MO: Maryland Heights; 1977.

28. Schaltenbrand G, Bailey P. Einführung in die stereotaktischen Operationen, mit einem Atlas des menschlichen Gehirns. In: Introduction to Stereotaxis, with an atlas of the human brain. Stuttgart: Thieme; 1959.

29. Morel A. Stereotactic atlas of the human thalamus and basal ganglia. Boca Raton, FL: CRC Press; 2007.

30. Akram H, Dayal V, Mahlknecht P, Georgiev D, Hyam J, Foltynie T, et al. Connectivity derived thalamic segmentation in deep brain stimulation for tremor. NeuroImage Clin. 2018;18:130–42. https://doi.org/10.1016/j.nicl.2018.01.008.

31. Su JH, Thomas FT, Kasoff WS TT, Choi EY, Rutt BK, et al. Thalamus optimized multi atlas segmentation (THOMAS): fast, fully automated segmentation of thalamic nuclei from structural MRI. NeuroImage. 2019;194:272–82. https://doi.org/10.1016/j.neuroimage.2019.03.021.

32. Xiao Y, Lau JC, Anderson T, DeKraker J, Collins DL, Peters T, et al. An accurate registration of the BigBrain dataset with the MNI PD25 and ICBM152 atlases. Sci Data. 2019;6:210. https://doi.org/10.1038/s41597-019-0217-0.

33. Sudhyadhom A, Haq IU, Foote KD, Okun MS, Bova FJ. A high resolution and high contrast MRI for differentiation of subcortical structures for DBS targeting: the fast gray matter acquisition T1 inversion recovery (FGATIR). NeuroImage. 2009;47(Suppl. 2):T44–52. https://doi.org/10.1016/j.neuroimage.2009.04.018.
34. Vertinsky AT, Coenen VA, Lang DJ, Kolind S, Honey CR, Li D, et al. Localization of the subthalamic nucleus: optimization with susceptibility weighted phase MR imaging. AJNR Am J Neuroradiol. 2009;30:1717–24. https://doi.org/10.3174/ajnr.A1669.
35. Gunalan K, Howell B, McIntyre CC. Quantifying axonal responses in patient-specific models of subthalamic deep brain stimulation. NeuroImage. 2018;172:263–77. https://doi.org/10.1016/j.neuroimage.2018.01.015.
36. Amon A, Alesch F. Systems for deep brain stimulation: review of technical features. J Neural Transm. 2017;124:1083–91. https://doi.org/10.1007/s00702-017-1751-6.
37. Moro E, Esselink RJ, Xie J, Hommel M, Benabid AL, Pollak P. The impact on Parkinson's disease of electrical parameter settings in STN stimulation. Neurology. 2002;59:706–13. https://doi.org/10.1212/WNL.59.5.706.
38. Reich MM, Steigerwald F, Sawalhe AD, Reese R, Gunalan K, Johannes S, et al. Short pulse width widens the therapeutic window of subthalamic neurostimulation. Ann Clin Transl Neurol. 2015;2:427–32. https://doi.org/10.1002/acn3.168.
39. Dayal V, Grover T, Limousin P, Akram H, Cappon D, Candelario J, et al. The effect of short pulse width settings on the therapeutic window in subthalamic nucleus deep brain stimulation for Parkinson's disease. J Parkinsons Dis. 2018;8:273–9. https://doi.org/10.3233/JPD-171272.
40. Moldovan AS, Hartmann CJ, Trenado C, Meumertzheim N, Slotty PJ, Vesper J, et al. Less is more - pulse width dependent therapeutic window in deep brain stimulation for essential tremor. Brain Stimul. 2018;11:1132–9. https://doi.org/10.1016/j.brs.2018.04.019.
41. Picillo M, Lozano AM, Kou N, Munhoz RP, Fasano A. Programming deep brain stimulation for tremor and dystonia: the Toronto western hospital algorithms. Brain Stimul. 2016;9:438–52. https://doi.org/10.1016/j.brs.2016.02.003.
42. Stegemöller EL, Vallabhajosula S, Haq I, Hwynn N, Hass CJ, Okun MS. Selective use of low frequency stimulation in Parkinson's disease based on absence of tremor. NeuroRehabilitation. 2013;33:305–12. https://doi.org/10.3233/NRE-130960.
43. Huang H, Watts RL, Montgomery EB. Effects of deep brain stimulation frequency on bradykinesia of Parkinson's disease. Mov Disord. 2014;29:203–6. https://doi.org/10.1002/mds.25773.
44. Ushe M, Mink JW, Tabbal SD, Hong M, Schneider Gibson P, Rich KM, et al. Postural tremor suppression is dependent on thalamic stimulation frequency. Mov Disord. 2006;21:1290–2. https://doi.org/10.1002/mds.20926.
45. Moro E, Piboolnurak P, Arenovich T, Hung SW, Poon YY, Lozano AM. Pallidal stimulation in cervical dystonia: clinical implications of acute changes in stimulation parameters. Eur J Neurol. 2009;16:506–12. https://doi.org/10.1111/j.1468-1331.2008.02520.x.
46. Jia F, Hu W, Zhang J, Wagle Shukla A, Almeida L, Meng FG, et al. Variable frequency stimulation of subthalamic nucleus in Parkinson's disease: rationale and hypothesis. Parkinsonism Relat Disord. 2017;39:27–30. https://doi.org/10.1016/j.parkreldis.2017.03.015.
47. Bologna M, Di Biasio F, Conte A, Iezzi E, Modugno N, Berardelli A. Effects of cerebellar continuous theta burst stimulation on resting tremor in Parkinson's disease. Parkinsonism Relat Disord. 2015;21:1061–6. https://doi.org/10.1016/j.parkreldis.2015.06.015.
48. Bologna M, Paparella G, Fabbrini A, Leodori G, Rocchi L, Hallett M, et al. Effects of cerebellar theta-burst stimulation on arm and neck movement kinematics in patients with focal dystonia. Clin Neurophysiol. 2016;127:3472–9. https://doi.org/10.1016/j.clinph.2016.09.008.
49. Chirra M, Marsili L, Wattley L, Sokol LL, Keeling E, Maule S, et al. Telemedicine in neurological disorders: opportunities and challenges. Telemed J E Health. 2019;25:541–50. https://doi.org/10.1089/tmj.2018.0101.
50. Heldman DA, Harris DA, Felong T, Andrzejewski KL, Dorsey ER, Giuffrida JP, et al. Telehealth management of parkinson's disease using wearable sensors: an exploratory study. Digit Biomark. 2017;1:43–51. https://doi.org/10.1159/000475801.

51. Wong DC, Relton SD, Fang H, Qhawaji R, Graham CD, Alty J, et al. Supervised classification of bradykinesia for Parkinson's disease diagnosis from smartphone videos. IEEE; 2019. p. 32–7. https://doi.org/10.1109/CBMS.2019.00017.

52. Sharma VD, Safarpour D, Mehta SH, Vanegas-Arroyave N, Weiss D, Cooney JW, et al. Telemedicine and deep brain stimulation – current practices and recommendations. Parkinsonism Relat Disord. 2021;89:199–205. https://doi.org/10.1016/j.parkreldis.2021.07.001.

53. Jimenez-Shahed J. Device profile of the percept PC deep brain stimulation system for the treatment of Parkinson's disease and related disorders. Expert Rev Med Devices. 2021;18:319–32. https://doi.org/10.1080/17434440.2021.1909471.

54. Little S, Beudel M, Zrinzo L, Foltynie T, Limousin P, Hariz M, et al. Bilateral adaptive deep brain stimulation is effective in Parkinson's disease. J Neurol Neurosurg Psychiatry. 2016;87:717–21. https://doi.org/10.1136/jnnp-2015-310972.

55. Sohal VS, Sun FT. Responsive neurostimulation suppresses synchronized cortical rhythms in patients with epilepsy. Neurosurg Clin N Am. 2011;22:481–8, vi. https://doi.org/10.1016/j.nec.2011.07.007.

56. Molina R, Okun MS, Shute JB, Opri E, Rossi PJ, Martinez-Ramirez D, et al. Report of a patient undergoing chronic responsive deep brain stimulation for Tourette syndrome: proof of concept. J Neurosurg. 2018;129:308–14. https://doi.org/10.3171/2017.6.JNS17626.

Chapter 11
Successful Maintenance of Brain Sharpness

Vida Demarin and Filip Derke

The human brain, the key to individual and social human behaviour is certainly the most complicated system on the earth and enormous investigations tried and are still trying to resolve the secret of its functioning. Results of numerous investigations during the Decade of the Brain, by sophisticated diagnostic methods, point out the importance of the brain's neuroplasticity, the mechanism that was described already at the end of the nineteenth century but at that time, still without scientific proof. This mechanism shows that the brain is not a static organ, but on the contrary, by the development of the new connections between cells and new pathways, its functions could be restored as well as preserved even in older age.

Spanish neuroscientist, and Nobel Prize winner Santiago Ramon y Cajal set the roots of neuroplasticity in his book Textura del Sistema nervioso del hombre y de los vertebrados, published in Madrid 1904 [1]. and his famous sentence: "Every man can, if he so desires, become a sculptor of his own brain" can serve as a source for numerous future research [2]. Several scientists have devoted their investigation to the importance of neuroplasticity in maintaining a healthy brain. Bruce McEven published his important work on changing the structure of the brain during stress in Nature 1968 [3]. Further research followed by Paul Bach y Rita with his sensory substitution [4], Pascal Leone with imagining and mental practice [5], and Beatriz Calvo Merino with neurocognitive mechanisms involved in action observation, expertise and dance [6]. Another Nobel prize winner Eric Kandel investigated the molecular biology of memory storage as a dialogue between genes and synapses

V. Demarin (✉)
International Institute for Brain Health, Zagreb, Croatia

F. Derke
Department of Neurology, University Hospital Dubrava, Zagreb, Croatia

Department of Psychiatry and Neurology, Faculty of Dental Medicine and Health,
JJ Strossmayer University of Osijek, Osijek, Croatia
e-mail: filip@mozak.hr

V. Demarin et al. (eds.), *Mind, Brain and Education*,
https://doi.org/10.1007/978-3-031-33013-1_11

[7]. Michael Merzenich's research is devoted to the role of neuroplasticity in preserving the brain's health [8]. And Norman Doidge in his groundbreaking book „The Brain that Changes itself "explains the brain's ability to change its structure based on neuroplasticity [9], together with the same approach in his second book „The Brain's Way of Healing." [10].

Neuroplasticity, the capacity of neurons and neural networks in the brain to change their connections and behaviour in response to new information, sensory stimulation, development, damage, or dysfunction is considered generally to be a complex, multifaceted, fundamental property of the brain.

As with many medical and health-related fields, during the lifespan, health and disease interventions targeting mechanisms of plasticity should follow an individualized approach by harnessing individual differences to best utilize the brain's innate capacity to change.

Mirror neurons are one of the most important discoveries in the last decade of neuroscience. These are a variety of visuospatial neurons which indicate fundamentally human social interaction. Essentially, mirror neurons respond to actions that we observe in others. The interesting part is that mirror neurons fire in the same way when we recreate that action ourselves. Apart from imitation, they are responsible for the myriad of other sophisticated human behaviour and thought processes [11, 12].

Another important approach reflects the role of epigenetics. We are witnessing a recent burst of research efforts, discussions, and debates on epigenetics. The enthusiasm in this area reflects the realization that these processes form a fundamental biological basis for the interplay among environmental signals, the genome and heritability. Scientists are working on addressing many additional issues, such as what controls the propagation of epigenetic information throughout developmental stages and how changes in the epigenome are inherited, as well as the interplay of environment and human behaviour in health and disease. Continued research will hopefully yield more discoveries shortly. Revealing how epigenetic marks work and what they do will surely open important new chapters in genetics and human health [13].

Until recently, there was a general belief that over the years brain function weakens and cognitive decline is inevitable, being a kind of a threat and a reason to worry about. But the results of numerous research and investigations led to a completely new approach, identifying risk factors that could be modified by certain lifestyle changes, thus preventing cognitive impairment and even dementia.

Be Ambitious About Prevention!

In July 2017. The Lancet Commission on Dementia Prevention, Intervention and Care presented a life-course model showing potentially modifiable and non-modifiable risk factors for dementia [14, 15]. According to this model, it is estimated that 35% of dementia cases could be prevented if we eliminate risk factors. A key recommendation is to focus on interventions to build up resilience and brain reserve, activate neuroplasticity, detect and treat risk factors and live a healthier lifestyle. This life course model shows us the need for preventative actions from early childhood, or even, from birth. As we already know, the most common form of dementia, Alzheimer's disease, is in large part modulated by genetics. Genetics is one non-modifiable risk factor, meaning that with birth we either have already inherited the ApoE4 allele, or not. Along with other genetic material, this is among the only non-modifiable risk factors. All other risk factors less education, hypertension, hearing loss, obesity, smoking, depression, physical inactivity, diabetes, and social isolation, belong to potentially modifiable risk factors and they account for about 35%.

The same group of authors, members of The Lancet Commission on Dementia Prevention, Intervention and Care, published their new report in July 2020, adding three more factors for dementia, with newer, convincing evidence. These factors are excessive alcohol consumption, traumatic brain injury (TBI) and air pollution. They completed new reviews and meta-analyses and incorporated these into an updated 12 risk factors life-course model of dementia prevention. Together the 12 modifiable risk factors account for around 40% of worldwide dementias, which consequently could theoretically be prevented or delayed. The potential for prevention is high and might be higher in low-income and middle-income countries (LMIC) where more dementias occur [16].

In early childhood, we need to start taking care of an important risk factor for dementia, and that is education. Studies have shown that a lower grade of education brings a higher risk for dementia, pointing to the conclusion that education protects against the onset of dementia. Education also influences the course and the outcome of the disease in terms of a pattern of cognitive decline and underlying brain pathology. Study shows that adult life work complexity, social network and complex leisure activities also reduce the occurrence of dementia [14, 17, 18].

The modifiable risk factors for dementia during midlife are hearing loss, hypertension and obesity. These three factors attack people at the age of 40 or 45 (45 is officially the beginning of middle age), and if they are present for the rest of the middle age or longer, they cause an increased risk for developing dementia [14]. So, keeping fit, taking care of the extra weight, as well as early recognition and treatment of hypertension, will not only guard the body against disease but also the brain [19, 20].

The potential public health impact of hearing loss in the context of dementia is substantial, given the high worldwide prevalence of hearing loss in older adults. What we urgently need is an interdisciplinary effort to bring together hearing and mental health and to investigate further early hearing loss in the context of brain and cognitive ageing [21, 22]. Later on in life, it is necessary to take care of smoking, depression, physical inactivity, social isolation, and diabetes [17, 22–24].

Change from a sedentary lifestyle to moderate physical activity has beneficial effects on cognitive functioning, and preliminary evidence suggests that such change may reduce the incidence of dementia [25–27].

Dance is a very useful complex activity and dance training is superior to repetitive physical exercise in inducing brain plasticity in the elderly [28].

Targeting Modifiable Risk Factors

The National Academy of Sciences in 2017 reported that there are no specific interventions that have sufficient evidence to warrant a public health campaign for the prevention of dementia except cognitive training, blood pressure management in people with hypertension, and increased physical activity [29–31]. In 2017, the presidential advisory from the American Heart Association/American Stroke Association, tried to decide on a definition of initial optimal brain health in adults [17]. The working group identified seven metrics to define optimal brain health in adults, and these originated from the well-known Life's Simple 7 [32], identified by Ralph Sacco in 2011. He then identified four ideal health behaviours; non-smoking, physical activity, a healthy diet, and a body mass index under 25 kg/m2, and three ideal health factors such as untreated blood pressure under 120/80 mmHg, untreated total cholesterol under 200 mg/dL and fasting blood glucose less than 100 mg/dL. Along with these recommendations to maintain cognitive health, it is advised to incorporate control of cardiovascular risks and suggest social engagement and other related strategies. There is always an opportunity to improve brain health through adult prevention and other interventions.

Overall, white matter fibre-tracking on MRI evidenced an early signature of damage in hypertensive patients when otherwise undetectable by conventional neuroimaging. In perspective, this approach could allow identifying those patients that are in the initial stages of brain damage and could benefit from therapies aimed at limiting the transition to dementia and neurodegeneration [33]. In adults with high baseline blood pressure, those using any blood-pressure lowering drug, regardless of drug class, had a reduced risk for developing all-cause dementia and Alzheimer's disease compared with those not using blood-pressure medication. It is also interesting, that this meta-analysis looked not only at dementia but also Alzheimer's disease specifically and found a benefit of blood pressure lowering. This suggests that the onset of Alzheimer's disease may be slowed through the treatment of high blood pressure [33].

SPRINT Memory and Cognition in Decreased Hypertension (MIND), a substudy of the SPRINT MIND Study [34] evaluated the effect of intensive systolic blood pressure lowering on mild cognitive impairment and probable dementia, with a subset of participants completing MRI. Intensive lowering of blood pressure did not result in a significant reduction in the incidence of probable dementia, compared with standard management of blood pressure (the primary outcomes in SPRINT MIND). However, mild cognitive impairment and the composite of mild cognitive impairment and probable dementia (the secondary outcomes in SPRINT MIND) were significantly reduced in the intensive lowering group compared with the standard treatment group [35].

Intensive lowering of systolic blood pressure was also associated with a significantly smaller increase in cerebral white matter change but a greater decrease in total brain volume compared with standard treatment, although the changes were small [35].

Several studies showed that increased arterial stiffness has greater value in predicting cognitive decline in healthy subjects, than blood pressure [36]. It is superior to blood pressure in predicting cognitive decline in all domains and in explaining the hypertension-executive function association. Arterial stiffness, especially in hypertension, may be a target in the prevention of cognitive decline [37–39].

An increasing number of studies confirm the positive correlation of obesity and inflammation with cognitive impairment [40, 41]. There is sufficiently strong evidence, from a population-based perspective, to conclude that regular physical activity [42] and management of cardiovascular risk factors (diabetes, obesity, smoking, and hypertension) [43] reduce the risk of cognitive decline and may reduce the risk of dementia. Also, there is sufficiently strong evidence to conclude that a healthy diet and lifelong learning/cognitive training may also reduce the risk of cognitive decline, thus enhancing the inborn mechanism of neuroplasticity [44–47].

Findings indicate that older men with a history of depression are at increased risk of developing dementia, although depression in later life is more likely to be a marker of incipient dementia than a truly modifiable risk factor. Older people with depression may be better viewed as potential targets of indicated prevention strategies, rather than people with mild cognitive impairment [48].

The window of opportunity for beneficial effects of physical activity seems to be broad and may extend to people who become active later in life. However, beyond already available general recommendations for health promotion, it is very challenging to draw specific practical recommendations from the current evidence regarding the type, frequency, intensity, and duration of physical activity that could protect against AD. Physical activities that have additional social and cognitive stimulation components may likely be most effective. The multi-domain approach to dementia prevention also seems more promising compared with the traditional, single-domain approach [14, 49, 50].

Loneliness predicted greater dementia risk, whereas being married and having many close relationships with friends and family were related to a lower risk of dementia. Further epidemiological research is needed to understand the possible causal nature of these associations, including the likely underlying mechanisms [51].

The Finnish Geriatric Intervention Study to Prevent Cognitive Impairment and Disability (FINGER) is the first multi-domain lifestyle intervention that has shown that a combination of lifestyle interventions can prevent or slow down a cognitive decline [50]. Lots of evidence from epidemiological studies indicate that these different modifiable lifestyle factors are related to dementia and Alzheimer's disease. The intervention areas were diet (Nordic diet), exercise, cognitive training (individualized) and vascular risk monitoring. The results showed a reduction of cognitive decline by 30%. There are now 3 multi-domain trials going on globally. The FINGER study was taken as a model. The FINGER study included 1109 participants in the analysis: 362 APOE ε4 allele carriers (173 interventions and 189 controls) and 747 non-carriers (380 interventions and 367 controls). The difference between the intervention and control groups in the annual neuropsychological test battery total score change was 0.037 (95% CI, 0.001 to 0.073) among carriers and 0.014 (95% CI, −0.011 to 0.039) among non-carriers. The intervention effect was not significantly different between carriers and non-carriers (0.023; 95% CI, −0.021 to 0.067). Healthy lifestyle changes may be beneficial for cognition in older at-risk individuals even in the presence of APOE-related genetic susceptibility to dementia.

In another study, it was found that lifestyle factors, such as physical activity, sleep, and social activity appear to be associated with cognitive function among older people. Physical activity and appropriate durations of sleep and conversation are important for cognitive function [52].

To quantify the impact of a healthy lifestyle on the risk of Alzheimer's dementia, using data from the Chicago Health and Aging Project (CHAP; n = 1845) and the Rush Memory and Aging Project (MAP; n = 920), a healthy lifestyle score was defined based on nonsmoking, ≥150 min/wk. moderate/vigorous-intensity physical activity, light to moderate alcohol consumption, high-quality Mediterranean-DASH Diet Intervention for Neurodegenerative Delay diet (upper 40%), and engagement in late-life cognitive activities (upper 40%) giving an overall score ranging from 0 to 5. The results suggest that a healthy lifestyle as a composite score is associated with a substantially lower risk of Alzheimer's dementia [53].

Assuming a causal relation and intervention at the correct age for prevention, relative reductions of 10 or 20% per decade in the prevalence of each of the 7 risk factors would potentially reduce the prevalence of AD in 2050 by between 8 and 15% - between 8,8 and 16,two million cases worldwide. After accounting for non-independence between risk factors, around a third of Alzheimer's disease cases worldwide can be attributed to potentially modifiable risk factors. The incidence of Alzheimer's disease can be reduced through improved access to education, reduction of vascular risk factors (through the use of effective methods such as physical activity, non-smoking, diagnosing and treating midlife hypertension, obesity, and diabetes) and depression [54, 55].

Three More Things for Brain Health

As already mentioned, the Lancet Commission on Dementia continued its work and published a new report in September 2020. adding three more risk factors to their model of preventing cognitive decline and dementia [16].

TBI is usually caused by car, motorcycle, and bicycle injuries; military exposures; boxing, horse riding, and other recreational sports; firearms; and falls [56]. A nationwide Danish cohort study of nearly three million people aged 50 years or older, followed for a mean of 10 years, found increased dementia and Alzheimer's disease risk [57]. Dementia risk was highest in the 6 months after TBI and increased with several injuries in people with TBI to reduce reverse causation bias [57].

Similarly, a Swedish cohort of over three million people aged 50 years or older, found TBI increased 1-year dementia risk); and risk remained elevated, albeit attenuated over 30 years [58].

The term chronic traumatic encephalopathy describes sports head injury, which is not yet fully characterized and covers a broad range of neuropathologies and outcomes, with current views largely conjecture [59].

Heavy drinking is associated with brain changes, cognitive impairment, and dementia, a risk known for centuries [60]. An increasing body of evidence is emerging on alcohol's complex relationship with cognition and dementia outcomes from a variety of sources including detailed cohorts and large-scale record-based studies. Alcohol is strongly associated with cultural patterns and other socio-cultural and health-related factors, making it particularly challenging to understand the evidence base. Several studies are analyzing the role of alcohol consumption in cognitive impairments, each of them with varying quantities of alcohol and based upon these results the Lancet Commission proposed the quantity of more than 21 units of alcohol per week as a risk factor [16].

As previously mentioned, air pollution and particulate pollutants are associated with poor health outcomes, including those related to non-communicable diseases, with special attention to their potential effect on the brain [61, 62]. A systematic review of studies until 2018 including 13 longitudinal studies with 1–15 years of follow-up of air pollutants exposure and incident dementia found exposure to particulate matter ≤ 2.5 µ (PM2.5), nitrogen dioxide (NO_2), nitrous oxides (NOx), carbon monoxide (CO), and ozone were all associated with increased dementia risk [63].

Lifestyle changes for modifying 12 risk factors might prevent or delay up to 40% of dementias. We should be ambitious about prevention. Contributions to the risk and mitigation of dementia begin early and continue throughout life, so it is never too early or too late to act. These actions require both public health programs and individually tailored interventions [16, 49].

Mitochondria: The Important Player

Mitochondria are crucial for our survival—we literally can't live without them. However, how well a person lives, how much energy she/he has, how well their cells communicate, and how well the organs function can vary depending on how well the mitochondria are functioning. So, much of the longevity research is deeply intertwined with functions that rely on mitochondrial health: sirtuin activation, mTOR inhibition, activation of AMPK, telomere length, healthy inflammation and immune function, and the list goes on. These microscopic organelles hold the key to healthy living, it is necessary to take good care of them. Several studies analysed current evidence that signposts a role for mitochondria as a key hub that also supports and integrates activity across four domains: circadian clocks, metabolic pathways, the intestinal microbiota, and the immune system, coordinating their integration and crosstalk [64–66].

Gut Microbiota and Cognitive Protection

A consequence of the progressively ageing global population is the increasing prevalence of worldwide age-related cognitive decline and dementia. In the absence of effective therapeutic interventions, identifying risk factors associated with cognitive decline becomes increasingly vital. Novel perspectives suggest that a dynamic bidirectional communication system between the gut, its microbiome, and the central nervous system, commonly referred to as the microbiota-gut-brain axis, may be a contributing factor to cognitive health and disease. However, the exact mechanisms remain undefined. Microbial-derived metabolites produced in the gut can cross the intestinal epithelial barrier, enter systemic circulation, and trigger physiological responses both directly and indirectly affecting the central nervous system and its functions. Dysregulation of this system (i.e., dysbiosis) can modulate cytotoxic metabolite production, promote neuroinflammation and negatively impact cognition. It is important to explore critical connections between microbial-derived metabolites (secondary bile acids, trimethylamine-N-oxide (TMAO), tryptophan derivatives and others) and their influence on cognitive function and neurodegenerative disorders, with a particular interest in their less-explored role as risk factors of cognitive decline [67, 68].

Let us summarize the effect of the human intestinal microbiome on cognitive impairments and focus primarily on the impact of diet and eating habits on learning processes. A better understanding of the microbiome could revolutionize the possibilities of therapy for many diseases. The digestive tube is populated by billions of living microorganisms including viruses, bacteria, protozoa, helminths, and microscopic fungi. In adulthood, under physiological conditions, the intestinal microbiome appears to be relatively steady. However, it can be influenced, both in the positive sense and in the negative one. The basic pillars that maintain a steady microbiome are genetics, lifestyle, diet and eating habits, geography, and age. The gastrointestinal tract and the brain communicate with each other through several pathways considered the gut-brain axis. New evidence is published every year about the association between intestinal dysbiosis and neurological/psychiatric diseases. On the other hand, specific diets and eating habits can have a positive effect on balanced microbiota composition and thus contribute to harmonious cognitive functions [69].

Several mechanisms in which gut microbiota may modulate changes in cognitive function with age are proposed: issues related to the measurement of cognition in the elderly and in particular a standardized model of cognition that could be utilised to better understand cognitive outcomes related to gut microbiota and cognition in the elderly, biological processes such as oxidative stress and inflammation which are related to cognitive changes with age and which are also influenced by gut microbiota [70].

Give Priority to Circadian Rhythm

Good sleep and mood are important for health and for keeping active. Numerous studies have suggested that the incidence of insomnia and depressive disorder are linked to biological rhythms, immune function, and nutrient metabolism, but the exact mechanism is not yet clear. There is considerable evidence showing that the gut microbiome not only affects the digestive, metabolic, and immune functions of the person but also regulates his sleep and mental states through the microbiome-gut-brain axis. Preliminary evidence indicates that microorganisms and circadian genes can interact with each other. The characteristics of the gastrointestinal microbiome and metabolism are related to the person's sleep and circadian rhythm. Moreover, emotion and physiological stress can also affect the composition of the gut microorganisms. The gut microbiome and inflammation may be linked to sleep loss, circadian misalignment, affective disorders, and metabolic diseases. Exploring the effects of the gut microbiome on insomnia and depression will help further our understanding of the pathogenesis of mental disorders as well as potential cognitive impairments. It is therefore important to regulate and maintain a normal gastrointestinal micro-ecological environment [71].

These preliminary data suggest that circadian misalignment and sleep loss in humans are associated with dysbiosis and that the resulting microbiota configurations may contribute to the occurrence of metabolic imbalances. The reduction of an individual's sleep time or the disruptions of a person's circadian rhythms will lead to physiological stress response and change the normal intestinal microbiota. Furthermore, these changes will cause certain inflammatory reactions, metabolic disorders and impaired immune function. This process will also change the metabolism of neurotransmitters and cause nervous system dysfunction. The person will then experience sleep problems or psychiatric or cognitive symptoms, which ultimately initiate a vicious cycle [72].

Circadian rhythm disruption, causing among other symptoms, sleep problems is an important risk for brain health, especially cognitive impairments. The link between sleep and cognition has been well established in laboratory studies. In brief, when sleep is shortened or disrupted, cognitive performance on a range of tasks suffers. For instance, total sleep deprivation leads to poorer short-term memory, attention, and processing speed, as shown by a recent meta-analysis. Similarly, several nights of experimentally induced chronic sleep restriction (e.g. 5 h of sleep per night) can affect vigilance as severely as a single night of total sleep deprivation.

By the same token, extending sleep past habitual levels, which is achieved by allowing individuals to sleep more than usual, improves cognition relative to baseline. Taken together, these results suggest there is a linear relationship between sleep and cognitive functioning: more sleep leads to better cognitive performance and vice versa [73].

Disruption of a circadian rhythm with sleep disturbances and cognitive impairment is common in older adults. Mounting evidence points to a potential connection between sleep and cognitive function. Findings from observational studies support the role of sleep disturbances (particularly for sleep duration, sleep fragmentation and sleep-disordered breathing) in the development of cognitive impairment. Less consistent evidence exists for associations of insomnia and circadian rhythm dysfunction with cognition. These findings suggest that the sleep-wake cycle plays a crucial part in brain ageing, pointing to a potential avenue for improvement of cognitive outcomes in people at risk of cognitive decline and dementia. Several biological mechanisms might underlie the association between sleep and cognition [31].

Trouble falling or staying asleep, poor sleep quality, and short or long sleep duration are gaining attention as potential risk factors for cognitive decline and dementia, including Alzheimer's disease (AD). Sleep-disordered breathing (SDB) has also been linked to these outcomes. Recent studies with the objective and self-report sleep measures support sleep disturbance as a risk factor for cognitive decline and Alzheimer's disease in older adults. Observational studies in humans are beginning to identify associations between sleep parameters and Alzheimer's disease biomarkers, even among individuals without clinical dementia. Experimental studies suggest that sleep loss and hypoxia (a consequence of sleep-disordered breathing) affect Alzheimer's disease biomarkers. Maintenance of healthy sleep may be an important avenue for the prevention of dementia [74].

How to Boost Mitochondria

Continuous investigations on the importance of mitochondria, a sub-cellular organelle with its genome, producing the energy required for life and generating signals that enable stress adaptation to add an emerging concept proposing that mitochondria sense, integrate, and transduce psychosocial and behavioural factors into cellular and molecular modifications. Mitochondrial signalling might in turn contribute to the biological embedding of psychological states [75, 76].

Considering ageing there are well-documented reductions of tissue CoQ10 in senescence. It is still not completely clear if low CoQ10 is an effect of ageing, perhaps matching the fall in mitochondrial electron transport function or a contributing cause to the ageing process.

There is accumulating evidence that some diseases of ageing may benefit from supplemental ubiquinol or CoQ10 treatment. Studies to date have supported the safety and the potential of CoQ10 in reducing oxidative stress biomarkers [77].

Early supplementation with CoQ10 in various primary deficiencies can improve mitochondrial functions. Supplementation with CoQ10 may also confer antioxidant protection to organs and tissues affected by various pathophysiological conditions. The ability of CoQ10 to protect against the release of proinflammatory markers provides an attractive anti-inflammatory therapeutic for the treatment of some human diseases in ageing [78].

Hidden Elements for Fine-Tuning Brain Sharpness

The new scientific discipline psychoneuroendocrinoimmunology (PNEI) promote the new concept of the role of chronic stress, weakening the immune system and thus causing low-grade inflammation in various parts of the organism. The research in this particular field and its translation to clinical medicine contributes to a more comprehensive approach and communication to the patient in preserving health [79, 80].

Recent research has shown that the immune system plays an integral role in many serious disease conditions and that psychosocial factors can modulate immune system function. In one meta-analysis, the authors found that psychosocial interventions were associated with improvements in immune system function over time—in particular, with decreased proinflammatory cytokines or markers and increased immune cell counts—and that these associations were most consistent for interventions that incorporate cognitive behavioural therapy (CBT) or multiple interventions. Given the effectiveness and relative affordability of psychosocial interventions for treating chronic disease, they suggest that psychosocial interventions may represent a viable strategy for reducing disease burden and improving human health [81].

Eating Smart for Clear Thoughts

The importance of nutrition in preserving brain health is a subject of investigation for many years, pointing out the role of polyunsaturated fatty acids from fish consumption, an abundance of fruits and vegetables, whole grains, olive oil and red wine, which are all main ingredients of the Mediterranean Diet. Adherence to this diet leads to improved endothelial function, increased plasma antioxidant capacity and reduction of insulin resistance, which contributes to the prevention of stroke, neurodegenerative disorders, metabolic syndrome etc. [47]

Greater intake of fruits and vegetables, antioxidants, folate and flavonoids, polyphenols, resveratrol and vitamins have been associated with lower levels of inflammation markers (CRP, IL-6, TNF, PGF2) as well as oxidative stress (prostaglandin F2, isoprostane) in adults and adolescence. Anti-inflammatory mechanisms (antioxidants) can slow down atherosclerosis and have a positive effect on cardiovascular as well as cognitive status.

Adherence to a Mediterranean diet has been associated with a lower risk of various age-related diseases including dementia. Higher adherence to the Mediterranean diet was associated with better cognitive function, lower rates of cognitive decline, and reduced risk of Alzheimer's disease in nine out of 12 studies, whereas results for mild cognitive impairment were inconsistent. Published studies suggest that greater adherence to the Mediterranean diet is associated with slower cognitive decline and a lower risk of developing Alzheimer's disease. Several studies were investigating the role of the Mediterranean diet in the prevention of stroke, showing at the same time, positive effects in the prevention of cognitive decline as well. Further studies would be useful to clarify the association between mild cognitive impairment and vascular dementia. Long-term randomized controlled trials promoting a Mediterranean diet may help establish whether improved adherence helps to prevent or delay the onset of Alzheimer's disease and dementia [44, 82–85].

In older Spanish adults with overweight or obesity, higher adherence to the MedDiet may help mitigate the risk of cognitive decline, specifically as it relates to general and executive cognitive functioning, even over a short (2-year) period. [86]The findings indicate that adherence to MedDiet has a positive effect on both cognitively impaired and unimpaired older populations, especially on their memory, both in the short and long run. The results show that the higher adherence to MedDiet proves to have a better effect on the global cognitive performance of older people. In addition, adherence to MedDiet offers other benefits to older people, such as reduction of depressive symptoms, lowered frailty, as well as reduced length of hospital stays [87].

The MIND diet, a hybrid of the Mediterranean diet and the Dietary Approaches to Stop Hypertension diet, is associated with a slower cognitive decline and lower risk of Alzheimer's disease (AD) dementia in older adults. The MIND diet is associated with better cognitive functioning independently of common brain pathology, suggesting that the MIND diet may contribute to cognitive resilience in the elderly [88].

The benefit of coffee, tea and chocolate have been investigated in several studies. One review suggests that caffeine consumption, especially moderate quantities consumed through coffee or green tea in women, may reduce the risk of dementia and cognitive decline, and may ameliorate cognitive decline in cognitively impaired individuals [89–91].

Keep Fit for Cognitive Benefit

Numerous published studies investigate the role of exercise and physical activity in general in preserving brain health, especially cognitive functioning. And this research goes on. The rate of age-associated cognitive decline varies considerably between individuals. It is important, both on a societal and individual level to investigate factors that underlie these differences to identify those which might realistically slow cognitive decline. Physical activity is one such factor with substantial support in the literature. Regular exercise can positively influence cognitive ability, reduce the rate of cognitive ageing, and even reduce the risk of Alzheimer's disease (AD) and other dementias. However, while there is substantial evidence in the extant literature for the effect of exercise on cognition, the processes that mediate this relationship are less clear. This review examines cardiovascular health, production of brain-derived neurotrophic factor (BDNF), insulin sensitivity, stress, and inflammation as potential pathways, via which exercise may maintain or improve cognitive functioning, and may be particularly pertinent in the context of the ageing brain [92].

Exercise has been shown to improve executive function acutely in adults of all ages. Dance routines or other exercise regimens requiring some cognitive input may confer additional benefits to cognitive function. Exercise has a moderate effect on the ability of people with dementia to perform activities of daily living and may improve cognitive function. Midlife exercise may also have an impact on later cognitive function [93].

Forty-six trials involving 5099 participants were included in this review. Meta-analysis of the data estimated that aerobic exercise reduced the decline in global cognition, with a standardized mean difference (SMD) of 0.44, 95% CI 0.27 to 0.61, I2 = 69%. For individual cognitive functions, meta-analysis estimated that exercise lessened working memory decline (SMD 0.28, 95% CI 0.04 to 0.52, I2 = 40%). The estimated mean effect on reducing the decline in language function was favourable (SMD 0.17). Physical exercise can reduce global cognitive decline and lessen behavioural problems in people with MCI or dementia. Its benefits on cognitive function can be primarily attributed to its effects on working memory. Aerobic exercise at moderate intensity or above and a total training duration of 24 hours can lead to a more pronounced effect on global cognition [94].

Nonetheless, positive outcomes from BDNF and executive functions (EF) variables were displayed by all the populations exposed to exercise across studies. Aerobic exercise was shown to be a major source of the enhancement of BDNF-dependent executive functioning when compared to cognitive stimulation. Moreover, the effect of exercise-dependent BDNF on domains of executive functioning appears to occur in a dose-dependent manner for ageing individuals, independently of cognitive condition [95].

Several longitudinal studies found an inverse relationship between levels of physical activity and cognitive decline, dementia, and/or Alzheimer's disease (AD). The physical activity index (PAI) was assessed in the Framingham Study Original and Offspring cohorts, aged 60 years or older. The association between PAI and risk of incident all-cause dementia and AD in participants of both cohorts who were cognitively intact and had available PAI was examined. The association between PAI and brain MRI in the Offspring cohort was also examined (n = 1987). Low physical activity was associated with a higher risk for dementia in older individuals, suggesting that a reduced risk of dementia and higher brain volumes may be additional health benefits of maintaining physical activity into old age [96]. If provided with the most potent modalities, older adults can get clinical meaningful benefits with lower doses than the WHO guidelines [97].

Considering all that has been said, today we know that there are many preventable factors of cognitive decline and even dementia, and new ones are also being discovered. We also know that prevention is not only based on activities and habits, but also takes place at the level of organelles such as mitochondria, and even molecules such as coenzyme Q10. The new studies, in the idea of preventing cognitive decline, increasingly take into account stress as an important factor. There is also the new scientific discipline, psychoneuroendocrinoimmunology which studies and explores it. Our citizens are living longer, and old age is not a disease. Future research will provide more data on the mechanisms of protection and sharpening of our brains.

References

1. Cajal SRy. Textura del Sistema Nervioso del Hombre y de los Vertebrados, 1899–1904. In: Stripped Bare. Princeton: Princeton University Press; 2018. p. 216–9.
2. Demarin V. Zdrav mozak danas - za sutra: Medicinska naklada; 2017.
3. McEwen BS, Weiss JM, Schwartz LS. Selective retention of corticosterone by limbic structures in rat brain. Nature. 1968;220(5170):911–2. https://doi.org/10.1038/220911a0.
4. Bach-y-Rita P. Brain plasticity as a basis of sensory substitution. J Neurol Rehabil. 1987;1(2):67–71. https://doi.org/10.1177/136140968700100202.
5. Pascual-Leone A, Nguyet D, Cohen LG, Brasil-Neto JP, Cammarota A, Hallett M. Modulation of muscle responses evoked by transcranial magnetic stimulation during the acquisition of new fine motor skills. J Neurophysiol. 1995;74(3):1037–45. https://doi.org/10.1152/jn.1995.74.3.1037.
6. Calvo-Merino B, Glaser DE, Grèzes J, Passingham RE, Haggard P. Action observation and acquired motor skills: an fMRI study with expert dancers. Cereb Cortex. 2004;15(8):1243–9. https://doi.org/10.1093/cercor/bhi007.
7. Kandel ER, Dudai Y, Mayford MR. The molecular and systems biology of memory. Cell. 2014;157(1):163–86. https://doi.org/10.1016/j.cell.2014.03.001.
8. Merzenich MM, Van Vleet TM, Nahum M. Brain plasticity-based therapeutics. Front Hum Neurosci. 2014;8:8. https://doi.org/10.3389/fnhum.2014.00385.
9. Doidge N. The brain that changes itself : stories of personal triumph from the frontiers of brain science: Revised edition. Carlton North, Vic: Scribe Publications; 2010.
10. Doidge N. The Brain's way of healing: remarkable discoveries and recoveries from the frontiers of neuroplasticity. Penguin Publishing Group; 2015.
11. Rizzolatti G, Fadiga L, Gallese V, Fogassi L. Premotor cortex and the recognition of motor actions. Brain Res Cogn Brain Res. 1996;3(2):131–41. https://doi.org/10.1016/0926-6410(95)00038-0.
12. Kilner JM, Lemon RN. What we know currently about Mirror neurons. Curr Biol. 2013;23(23):R1057–R62. https://doi.org/10.1016/j.cub.2013.10.051.
13. Mansuy IM, Mohanna S. Epigenetics and the human brain: where nurture meets nature. Cerebrum. 2011;2011:8. Epub 20110525. PubMed PMID: 23447777; PubMed Central PMCID: PMC3574773
14. Livingston G, Sommerlad A, Orgeta V, Costafreda SG, Huntley J, Ames D, et al. Dementia prevention, intervention, and care. Lancet. 2017;390(10113):2673–734. https://doi.org/10.1016/S0140-6736(17)31363-6.
15. Frankish H, Horton R. Prevention and management of dementia: a priority for public health. Lancet. 2017;390(10113):2614–5. Epub 20170720. https://doi.org/10.1016/s0140-6736(17)31756-7.

16. Livingston G, Huntley J, Sommerlad A, Ames D, Ballard C, Banerjee S, et al. Dementia prevention, intervention, and care: 2020 report of the <em>lancet</em> commission. Lancet. 2020;396(10248):413–46. https://doi.org/10.1016/S0140-6736(20)30367-6.
17. Gorelick PB, Furie KL, Iadecola C, Smith EE, Waddy SP, Lloyd-Jones DM, et al. Defining optimal brain health in adults: a presidential advisory from the american heart association/american stroke association. Stroke. 2017;48(10):e284–303. https://doi.org/10.1161/str.0000000000000148. Epub 20170907. PubMed PMID: 28883125; PubMed Central PMCID: PMC5654545.
18. Morovic S, Budincevic H, Govori V, Demarin V. Possibilities of dementia prevention - it is never too early to start. J Med Life. 2019;12(4):332–7. https://doi.org/10.25122/jml-2019-0088. PubMed PMID: 32025250; PubMed Central PMCID: PMC6993301
19. McMaster M, Kim S, Clare L, Torres SJ, Cherbuin N, D'Este C, et al. Lifestyle risk factors and cognitive outcomes from the multidomain dementia risk reduction randomized controlled trial, body brain life for cognitive decline (BBL-CD). J Am Geriatr Soc. 2020;68(11):2629–37. https://doi.org/10.1111/jgs.16762.
20. Klimova B, Valis M, Kuca K. Cognitive decline in normal aging and its prevention: a review on non-pharmacological lifestyle strategies. Clin Interv Aging. 2017;12:903–10. https://doi.org/10.2147/cia.S132963. Epub 20170525. PubMed PMID: 28579767; PubMed Central PMCID:PMC5448694
21. Uhlmann RF, Larson EB, Rees TS, Koepsell TD, Duckert LG. Relationship of hearing impairment to dementia and cognitive dysfunction in older adults. JAMA. 1989;261(13):1916–9.
22. Thomson RS, Auduong P, Miller AT, Gurgel RK. Hearing loss as a risk factor for dementia: A systematic review. Laryngoscope. Investig Otolaryngol. 2017;2(2):69–79. https://doi.org/10.1002/lio2.65. Epub 20170316. PubMed PMID: 28894825; PubMed Central PMCID: PMC5527366
23. Almeida OP, Hankey GJ, Yeap BB, Golledge J, Flicker L. Depression as a modifiable factor to decrease the risk of dementia. Transl. Psychiatry. 2017;7(5):e1117. https://doi.org/10.1038/tp.2017.90. Epub 20170502. PubMed PMID: 28463236; PubMed Central PMCID: PMC5534958
24. Morović S, Demarin V. Role of physical activity on human brain functions. Period Biol. 2014;116(2):219–21.
25. Lövdén M, Xu W, Wang HX. Lifestyle change and the prevention of cognitive decline and dementia: what is the evidence? Curr Opin Psychiatry. 2013;26(3):239–43. https://doi.org/10.1097/YCO.0b013e32835f4135.
26. Hewston P, Kennedy CC, Borhan S, Merom D, Santaguida P, Ioannidis G, et al. Effects of dance on cognitive function in older adults: a systematic review and meta-analysis. Age Ageing. 2021;50(4):1084–92. https://doi.org/10.1093/ageing/afaa270.
27. Salinas-Rodríguez A, Manrique-Espinoza B, Palazuelos-González R, Rivera-Almaraz A, Jáuregui A. Physical activity and sedentary behavior trajectories and their associations with quality of life, disability, and all-cause mortality. Eur Rev Aging Phys Act. 2022;19(1):13. https://doi.org/10.1186/s11556-022-00291-3. Epub 20220429. PubMed PMID: 35488197; PubMed Central PMCID: PMC9052456
28. Rehfeld K, Lüders A, Hökelmann A, Lessmann V, Kaufmann J, Brigadski T, et al. Dance training is superior to repetitive physical exercise in inducing brain plasticity in the elderly. PLoS One. 2018;13(7):e0196636. https://doi.org/10.1371/journal.pone.0196636. Epub 20180711. PubMed PMID: 29995884; PubMed Central PMCID: PMC6040685
29. Stephen R, Hongisto K, Solomon A, Lönnroos E. Physical activity and Alzheimer's disease: a systematic review. J Gerontol A Biol Sci Med Sci. 2017;72(6):733–9. https://doi.org/10.1093/gerona/glw251.
30. Rebok GW, Ball K, Guey LT, Jones RN, Kim H-Y, King JW, et al. Ten-year effects of the advanced cognitive training for independent and vital elderly cognitive training trial on cognition and everyday functioning in older adults. J Am Geriatr Soc. 2014;62(1):16–24. https://doi.org/10.1111/jgs.12607.

31. Yaffe K, Falvey CM, Hoang T. Connections between sleep and cognition in older adults. Lancet Neurol. 2014;13(10):1017–28. https://doi.org/10.1016/s1474-4422(14)70172-3.
32. Sacco RL. Achieving ideal cardiovascular and brain health: opportunity amid crisis: presidential address at the American Heart Association 2010 scientific sessions. Circulation. 2011;123(22):2653–7. https://doi.org/10.1161/CIR.0b013e318220dec1.
33. Carnevali L, Koenig J, Sgoifo A, Ottaviani C. Autonomic and brain morphological predictors of stress resilience. Front Neurosci. 2018;12:12. https://doi.org/10.3389/fnins.2018.00228.
34. Williamson JD, Pajewski NM, Auchus AP, Bryan RN, Chelune G, Cheung AK, et al. Effect of intensive vs standard blood pressure control on probable dementia: a randomized clinical trial. JAMA. 2019;321(6):553–61. https://doi.org/10.1001/jama.2018.21442. PubMed PMID: 30688979; PubMed Central PMCID: PMC6439590
35. Gorelick PB, Sorond F. Cognitive function in SPRINT: where do we go next? Lancet Neurol. 2020;19(11):880–1. https://doi.org/10.1016/s1474-4422(20)30356-2.
36. Marfella R, Paolisso G. Increased Arterial Stiffness Trumps on Blood Pressure in Predicting Cognitive Decline in Low-Risk Populations. Hypertension. 2016;67(1):30–1. https://doi.org/10.1161/hypertensionaha.115.06450.
37. Hajjar I, Goldstein FC, Martin GS, Quyyumi AA. Roles of arterial stiffness and blood pressure in hypertension-associated cognitive decline in healthy adults. Hypertension. 2016;67(1):171–5. https://doi.org/10.1161/hypertensionaha.115.06277. Epub 20151102. PubMed PMID: 26527049; PubMed Central PMCID: PMC4715367
38. Morović S, Jurašić M-J, Martinić Popović I, Šerić V, Lisak M, Demarin V. Vascular characteristics of patients with dementia. J Neurol Sci. 2009;283(1):41–3. https://doi.org/10.1016/j.jns.2009.02.330.
39. Demarin V, Morovic S. Ultrasound subclinical markers in assessing vascular changes in cognitive decline and dementia. J Alzheimers Dis. 2014;42(Suppl 3):S259–66. https://doi.org/10.3233/jad-132507.
40. OBESITY.
41. Spyridaki EC, Avgoustinaki PD, Margioris AN. Obesity, inflammation and cognition. Curr Opin Behav Sci. 2016;9:169–75. https://doi.org/10.1016/j.cobeha.2016.05.004.
42. Dominguez LJ, Veronese N, Vernuccio L, Catanese G, Inzerillo F, Salemi G, et al. Nutrition, physical activity, and other lifestyle factors in the prevention of cognitive decline and dementia. Nutrients. 2021;13(11) https://doi.org/10.3390/nu13114080. PubMed PMID: 34836334; PubMed Central PMCID: PMC8624903. Epub 20211115
43. Medic N, Ziauddeen H, Ersche KD, Farooqi IS, Bullmore ET, Nathan PJ, et al. Increased body mass index is associated with specific regional alterations in brain structure. Int J Obes (Lond). 2016;40(7):1177–82. https://doi.org/10.1038/ijo.2016.42. Epub 20160322. PubMed PMID: 27089992; PubMed Central PMCID: PMC4936515
44. Scarmeas N, Anastasiou CA, Yannakoulia M. Nutrition and prevention of cognitive impairment. Lancet Neurol. 2018;17(11):1006–15. Epub 20180921. doi: 10.1016/s1474-4422(18)30338-7
45. Demarin V, Morović S. Neuroplasticity. Periodicum biologorum [Internet]. 2014;116(2):209–11. Dostupno na: https://hrcak.srce.hr/126369. [pristupljeno 22.09.2022.]
46. Demarin V, Lisak M, Morović S. Mediterranean diet in healthy lifestyle and prevention of stroke. Acta Clin Croat. 2011;50(1):67–77.
47. Demarin V, Derke F. Healthy lifestyle and prevention of stroke and other brain impairments: Croatian Academy of Sciences and arts. Department of Medical Sciences; 2019.
48. Tetsuka S. Depression and dementia in older adults: a neuropsychological review. Aging Dis. 2021;12(8):1920–34. https://doi.org/10.14336/ad.2021.0526. PubMed PMID: 34881077; PubMed Central PMCID: PMC8612610. Epub 20211201
49. Salinas J, Beiser AS, Samra JK, O'Donnell A, DeCarli CS, Gonzales MM, et al. Association of Loneliness with 10-year dementia risk and early markers of vulnerability for neurocognitive decline. Neurology. 2022;98(13):e1337–e48. https://doi.org/10.1212/wnl.0000000000200039.
50. Ngandu T, Lehtisalo J, Solomon A, Levälahti E, Ahtiluoto S, Antikainen R, et al. A 2 year multidomain intervention of diet, exercise, cognitive training, and vascular risk monitoring versus

control to prevent cognitive decline in at-risk elderly people (FINGER): a randomised controlled trial. Lancet. 2015;385(9984):2255–63. https://doi.org/10.1016/s0140-6736(15)60461-5. PubMed PMID: 25771249. Epub 20150312

51. Rafnsson SB, Orrell M, d'Orsi E, Hogervorst E, Steptoe A. Loneliness, social integration, and incident dementia over 6 years: prospective findings from the English longitudinal study of ageing. J Gerontol B Psychol Sci Soc Sci. 2020;75(1):114–24. https://doi.org/10.1093/geronb/gbx087. PubMed PMID: 28658937; PubMed Central PMCID: PMC6909434

52. Kimura A, Sugimoto T, Kitamori K, Saji N, Niida S, Toba K, et al. Malnutrition is associated with behavioral and psychiatric symptoms of dementia in older women with mild cognitive impairment and early-stage alzheimer's disease. Nutrients. 2019;11(8) https://doi.org/10.3390/nu11081951. Epub 20190820. PubMed PMID: 31434232; PubMed Central PMCID: PMC6723872

53. Dhana K, Evans DA, Rajan KB, Bennett DA, Morris MC. Healthy lifestyle and the risk of Alzheimer dementia: Findings from 2 longitudinal studies. Neurology. 2020;95(4):e374–e83. https://doi.org/10.1212/wnl.0000000000009816. Epub 20200617. PubMed PMID: 32554763; PubMed Central PMCID: PMC7455318

54. Fallahpour M, Fritz H, Thunborg C, Akenine U, Kivipelto M. Experiences of active everyday life among persons with prodromal alzheimer's disease: a qualitative study. J Multidiscip Healthc. 2022;15:1921–32. https://doi.org/10.2147/jmdh.S369878. PubMed PMID: 36068878; PubMed Central PMCID: PMC9441142

55. Norton S, Matthews FE, Barnes DE, Yaffe K, Brayne C. Potential for primary prevention of Alzheimer's disease: an analysis of population-based data. Lancet Neurol. 2014;13(8):788–94. https://doi.org/10.1016/s1474-4422(14)70136-x.

56. Bruns J Jr, Hauser WA. The epidemiology of traumatic brain injury: a review. Epilepsia. 2003;44(s10):2–10. https://doi.org/10.1046/j.1528-1157.44.s10.3.x.

57. Fann JR, Ribe AR, Pedersen HS, Fenger-Grøn M, Christensen J, Benros ME, et al. Long-term risk of dementia among people with traumatic brain injury in Denmark: a population-based observational cohort study. Lancet Psychiatry. 2018;5(5):424–31. https://doi.org/10.1016/s2215-0366(18)30065-8. Epub 20180410. PubMed PMID: 29653873

58. Nordström A, Nordström P. Traumatic brain injury and the risk of dementia diagnosis: a nationwide cohort study. PLoS Med. 2018;15(1):e1002496. https://doi.org/10.1371/journal.pmed.1002496. Epub 20180130. PubMed PMID: 29381704; PubMed Central PMCID: PMC5790223

59. Smith DH, Johnson VE, Trojanowski JQ, Stewart W. Chronic traumatic encephalopathy - confusion and controversies. Nat Rev Neurol. 2019;15(3):179–83. https://doi.org/10.1038/s41582-018-0114-8. PubMed PMID: 30664683; PubMed Central PMCID: PMC6532781

60. Rehm J, Manthey J, Lange S, Badaras R, Zurlyte I, Passmore J, et al. Alcohol control policy and changes in alcohol-related traffic harm. Addiction. 2020;115(4):655–65. https://doi.org/10.1111/add.14796.

61. Demarin V, Morović S, Derke F. Air pollution: a new risk factor for developing stroke. RAD CASA - Med Sci. 2019;540(48–49):51–7. https://doi.org/10.21857/mzvkptz3n9.

62. Béjot Y, Reis J, Giroud M, Feigin V. A review of epidemiological research on stroke and dementia and exposure to air pollution. Int J Stroke. 2018;13(7):687–95. https://doi.org/10.1177/1747493018772800. PubMed PMID: 29699457. Epub 20180427

63. Peters R, Ee N, Peters J, Booth A, Mudway I, Anstey KJ. Air pollution and dementia: a systematic review. J Alzheimers Dis. 2019;70(s1):S145–s63. https://doi.org/10.3233/jad-180631. PubMed PMID: 30775976; PubMed Central PMCID: PMC6700631

64. Aguilar-López BA, Moreno-Altamirano MMB, Dockrell HM, Duchen MR, Sánchez-García FJ. Mitochondria: an integrative hub coordinating circadian rhythms, metabolism, the microbiome, and immunity. Front Cell Dev Biol. 2020;8:51. https://doi.org/10.3389/fcell.2020.00051. Epub 20200207. PubMed PMID: 32117978; PubMed Central PMCID: PMC7025554

65. Zefferino R, Di Gioia S, Conese M. Molecular links between endocrine, nervous and immune system during chronic stress. Brain Behav. 2021;11(2):e01960. https://doi.org/10.1002/brb3.1960.

66. Pourcet B, Duez H. Circadian control of inflammasome pathways: implications for circadian medicine. Front Immunol. 2020;11:1630. https://doi.org/10.3389/fimmu.2020.01630. Epub 20200731. PubMed PMID: 32849554; PubMed Central PMCID: PMC7410924
67. Connell E, Le Gall G, Pontifex MG, Sami S, Cryan JF, Clarke G, et al. Microbial-derived metabolites as a risk factor of age-related cognitive decline and dementia. Mol Neurodegener. 2022;17(1):43. https://doi.org/10.1186/s13024-022-00548-6.
68. Meyer K, Lulla A, Debroy K, Shikany JM, Yaffe K, Meirelles O, et al. Association of the gut Microbiota with Cognitive Function in midlife. JAMA Netw Open. 2022;5(2):e2143941. https://doi.org/10.1001/jamanetworkopen.2021.43941. Epub 20220201. PubMed PMID: 35133436; PubMed Central PMCID: PMC8826173
69. Novotný M, Klimova B, Valis M. Microbiome and cognitive impairment: can any diets influence learning processes in a positive way? Front Aging Neurosci. 2019;11:170. https://doi.org/10.3389/fnagi.2019.00170. Epub 20190628. PubMed PMID: 31316375; PubMed Central PMCID: PMC6609888
70. Komanduri M, Gondalia S, Scholey A, Stough C. The microbiome and cognitive aging: a review of mechanisms. Psychopharmacology (Berl). 2019;236(5):1559–71. https://doi.org/10.1007/s00213-019-05231-1. PubMed PMID: 31055629. Epub 20190504
71. Li Y, Hao Y, Fan F, Zhang B. The role of microbiome in insomnia, circadian disturbance and depression. Front Psychiatry. 2018;9:669. https://doi.org/10.3389/fpsyt.2018.00669. Epub 20181205. PubMed PMID: 30568608; PubMed Central PMCID: PMC6290721
72. Logan RW, McClung CA. Rhythms of life: circadian disruption and brain disorders across the lifespan. Nat Rev Neurosci. 2019;20(1):49–65. https://doi.org/10.1038/s41583-018-0088-y.
73. Mantua J, Simonelli G. Sleep duration and cognition: is there an ideal amount? Sleep. 2019;42(3) https://doi.org/10.1093/sleep/zsz010.
74. Wennberg AMV, Wu MN, Rosenberg PB, Spira AP. Sleep disturbance, cognitive decline, and dementia: a review. Semin Neurol. 2017;37(4):395–406. https://doi.org/10.1055/s-0037-1604351. Epub 20170824. PubMed PMID: 28837986; PubMed Central PMCID: PMC5910033
75. Picard M, McEwen BS. Psychological stress and mitochondria: a conceptual framework. Psychosom Med. 2018;80(2):126–40. https://doi.org/10.1097/psy.0000000000000544. PubMed PMID: 29389735; PubMed Central PMCID: PMC5901651
76. Picard M, Sandi C. The social nature of mitochondria: implications for human health. Neurosci Biobehav Rev. 2021;120:595–610. https://doi.org/10.1016/j.neubiorev.2020.04.017.
77. Barcelos IP, Haas RH. CoQ10 and Aging. Biology. 2019;8(2):28. https://doi.org/10.3390/biology8020028.
78. Aaseth J, Alexander J, Alehagen U. Coenzyme Q(10) supplementation - In ageing and disease. Mech Ageing Dev. 2021;197:111521. https://doi.org/10.1016/j.mad.2021.111521. Epub 20210612. PubMed PMID: 34129891
79. Furman D, Campisi J, Verdin E, Carrera-Bastos P, Targ S, Franceschi C, et al. Chronic inflammation in the etiology of disease across the life span. Nat Med. 2019;25(12):1822–32. https://doi.org/10.1038/s41591-019-0675-0. Epub 20191205. PubMed PMID: 31806905; PubMed Central PMCID: PMC7147972
80. Demarin V, Toljan S. Klinička psihoneuroendokrinoimunologija. Zagreb: Croatian Academy of Sciences and Arts; 2020.
81. Shields GS, Spahr CM, Slavich GM. Psychosocial interventions and immune system function: a systematic review and meta-analysis of randomized clinical trials. JAMA Psychiat. 2020;77(10):1031–43. https://doi.org/10.1001/jamapsychiatry.2020.0431. PubMed PMID: 32492090; PubMed Central PMCID: PMC7272116
82. Lourida I, Soni M, Thompson-Coon J, Purandare N, Lang IA, Ukoumunne OC, et al. Mediterranean diet, cognitive function, and dementia: a systematic review. Epidemiology. 2013;24(4):479–89. https://doi.org/10.1097/EDE.0b013e3182944410.

83. Estruch R, Ros E, Salas-Salvadó J, Covas M-I, Corella D, Arós F, et al. Primary prevention of cardiovascular disease with a Mediterranean diet supplemented with extra-virgin olive oil or nuts. N Engl J Med. 2018;378(25):e34. https://doi.org/10.1056/NEJMoa1800389.
84. van den Brink AC, Brouwer-Brolsma EM, Berendsen AAM, van de Rest O. The Mediterranean, dietary approaches to stop hypertension (DASH), and Mediterranean-DASH intervention for neurodegenerative delay (MIND) diets are associated with less cognitive decline and a lower risk of Alzheimer's disease-a review. Adv Nutr. 2019;10(6):1040–65. https://doi.org/10.1093/advances/nmz054. PubMed PMID: 31209456; PubMed Central PMCID: PMC6855954
85. Fu J, Tan LJ, Lee JE, Shin S. Association between the mediterranean diet and cognitive health among healthy adults: A systematic review and meta-analysis. Front Nutr. 2022;9:946361. https://doi.org/10.3389/fnut.2022.946361. Epub 20220728. PubMed PMID: 35967772; PubMed Central PMCID: PMC9372716
86. Nishi SK, Babio N, Gómez-Martínez C, Martínez-González MÁ, Ros E, Corella D, et al. Mediterranean, DASH, and MIND dietary patterns and cognitive function: the 2-year longitudinal changes in an older Spanish cohort. Frontiers in aging. Neuroscience. 2021;13:13. https://doi.org/10.3389/fnagi.2021.782067.
87. Klimova B, Novotny M, Schlegel P, Valis M. The effect of Mediterranean diet on cognitive functions in the elderly population. Nutrients. 2021;13(6) Epub 20210616
88. Dhana K, James BD, Agarwal P, Aggarwal NT, Cherian LJ, Leurgans SE, et al. MIND diet, common brain pathologies, and cognition in community-dwelling older adults. J Alzheimers Dis. 2021;83(2):683–92. https://doi.org/10.3233/jad-210107. PubMed PMID: 34334393; PubMed Central PMCID: PMC8480203
89. Chen JQA, Scheltens P, Groot C, Ossenkoppele R. Associations between caffeine consumption, cognitive decline, and dementia: a systematic review. J Alzheimers Dis. 2020;78(4):1519–46. https://doi.org/10.3233/jad-201069. PubMed PMID: 33185612; PubMed Central PMCID: PMC7836063
90. Moreira A, Diógenes MJ, de Mendonça A, Lunet N, Barros H. Chocolate consumption is associated with a lower risk of cognitive decline. J Alzheimers Dis. 2016;53(1):85–93. https://doi.org/10.3233/jad-160142.
91. Zhang Y, Yang H, Li S, Li W-d, Wang Y. Consumption of coffee and tea and risk of developing stroke, dementia, and poststroke dementia: a cohort study in the UK biobank. PLoS Med. 2021;18(11):e1003830. https://doi.org/10.1371/journal.pmed.1003830.
92. Kennedy G, Hardman RJ, Macpherson H, Scholey AB, Pipingas A. How does exercise reduce the rate of age-associated cognitive decline? A review of potential mechanisms. J Alzheimers Dis. 2017;55(1):1–18. https://doi.org/10.3233/jad-160665.
93. Behrman S, Ebmeier KP. Can exercise prevent cognitive decline? Practitioner. 2014;258(1767):17–21.
94. Law CK, Lam FM, Chung RC, Pang MY. Physical exercise attenuates cognitive decline and reduces behavioural problems in people with mild cognitive impairment and dementia: a systematic review. J Physiother. 2020;66(1):9–18. https://doi.org/10.1016/j.jphys.2019.11.014. Epub 20191213
95. de Assis GG, Almondes KM. Exercise-dependent BDNF as a modulatory factor for the executive processing of individuals in course of cognitive decline. A systematic review. Front Psychol. 2017;8:8. https://doi.org/10.3389/fpsyg.2017.00584.
96. Tan ZS, Spartano NL, Beiser AS, DeCarli C, Auerbach SH, Vasan RS, et al. Physical activity, brain volume, and dementia risk: the Framingham study. J Gerontol A Biol Sci Med Sci. 2017;72(6):789–95. https://doi.org/10.1093/gerona/glw130. PubMed PMID: 27422439; PubMed Central PMCID: PMC6075525
97. Gallardo-Gómez D, Del Pozo-Cruz J, Noetel M, Álvarez-Barbosa F, Alfonso-Rosa RM, Del Pozo Cruz B. Optimal dose and type of exercise to improve cognitive function in older adults: A systematic review and bayesian model-based network meta-analysis of RCTs. Ageing Res Rev. 2022;76:101591. https://doi.org/10.1016/j.arr.2022.101591. Epub 20220217

Chapter 12
Dietary Habits of the Patients with Epilepsy

Filip Derke and Vida Demarin

Epilepsy is one of the oldest known diseases. It's been like this for 2000 years. ex. Kr. in the ancient Mesopotamian Code of Hammurabi we find the mentioned "sacred disease" (lat. "Morbus sacer"). However, the Greek physician Hippocrates already indicated the natural cause of the disease, considering that its cause is in the brain. In 1873, the English neurologist John Hughlings Jackson defined epilepsy as a disorder caused by sudden, excessive and rapid bursting of neurons (brain nerve cells) [1, 2]. Epilepsy is now defined as a paroxysmal disorder of the central nervous system, which is stereotyped and tends to repeat itself. At the same time, the pathologically changed cells of the cerebral cortex become overly excitable and react with synchronous bursts of electrical impulses, which is clinically manifested by epileptic seizures. Many famous people in history suffered from epilepsy: Napoleon Bonaparte, Julius Caesar, George Gershwin, Alfred Nobel, Vincent Van Gogh, Alexander the Great, etc. [3–8] How common is epilepsy? And more than we think. Epilepsy affects 1% of the population and is therefore one of the most common neurological diseases, and due to its characteristics, it represents a serious medical and social problem. About 50 million people in the world suffer from epilepsy and six million in Europe. It is also important to say that an isolated epileptic attack, i.e. one attack based on which the diagnosis of epilepsy cannot yet be established, occurs in about 20 per 100,000 people every year. The risk of developing epilepsy

F. Derke (✉)
Department of Neurology, Referral Centre for Preoperative Assessment of Patients with Pharmacoresistant epilepsy, University Hospital Dubrava, Zagreb, Croatia

Department of Psychiatry and Neurology, Faculty of Dental Medicine and Health, JJ Strossmayer University of Osijek, Osijek, Croatia
e-mail: filip@mozak.hr

V. Demarin
International Institute for Brain Health, Zagreb, Croatia

V. Demarin et al. (eds.), *Mind, Brain and Education*,
https://doi.org/10.1007/978-3-031-33013-1_12

during life is between 3 and 5%, with the highest incidence in infancy, the period around puberty, and the generative age, and it increases significantly again in the elderly population, especially those over 65 years of age [9–11].

How Much Do We Know About Epilepsy?

So far, numerous types of research have been conducted in the field of epilepsy, hundreds of clinical studies and dozens of larger (demographic-) epidemiological studies on patients treated for epilepsy have been published. The aforementioned research was conducted using national health databases and databases by patient associations. Such research was carried out as the first step towards creating a register of patients suffering from epilepsy. The studies mentioned above collected only demographic and basic clinical parameters (diagnosis, time of placement).

The clinical characteristics and lifestyle habits of patients have not been investigated on a significant sample of subjects. There are some sporadic studies with a smaller number of respondents, most often related to a specific workplace (hospital, neurological clinic, city, or county). In recent decades, thanks to advances not only in diagnostic procedures but also in therapeutic options, epilepsy has become a disease that can be well treated and no longer represents the stigma in society that it once did. Due to the development of the pharmaceutical industry, there are more quality antiepileptic drugs available today, and the disease can be satisfactorily controlled in about 65–70% of patients. The rest, the so-called pharmacoresistant patients are candidates for neurosurgical treatment, which has also improved significantly in recent years and shows a high percentage of success [12–15].

Eating Habits and Epilepsy

Nutrients and dietary habits in the context of disease control have been mostly investigated in patients with a pharmacoresistant form of epilepsy. In this sense, at one time it was recommended to starve, then in children's forms of pharmacoresistant epilepsy where a ketogenic diet was recommended as one of the possible methods of treatment. Recently, especially in the last 10 years or so, a healthy lifestyle has been emphasized more and more as a prevention of many diseases, thanks to new gadgets that we can carry with us and that makes it easier for us to monitor calorie consumption, calorie intake and create recommendations based on these results, it all started to analyze more with which nutrients the patient has fewer epileptic attacks, that is, what patients eat who do not have epileptic attacks. In this connection, a completely new theory of thinking about the so-called functional foods has been established. Studies are still being conducted for the mentioned theories, and the ones available so far cannot give us a clear cause-and-effect relationship between the consumption of certain foods and the number of epileptic attacks, that is, between the way of eating and the number of epileptic attacks [16–18].

Ketogenic Diet

The ketogenic diet is a term that refers to any diet therapy in which dietary composition would be expected to result in a ketogenic state of human metabolism. A ketogenic diet is generally defined as a high-fat, low-carbohydrate, moderate protein diet that aims to force the body to break down fat instead of glucose, both of which provide adenosine triphosphate synthesis, essentially mimicking the metabolic state of starvation or fasting. Ketogenic diets do not induce starvation; instead, they are precisely calculated to maintain adequate nutrient intake to prevent malnutrition associated with starvation, therefore, ensuring healthy growth and development. Calculations for classic KDs to this day remain like those first proposed by the Mayo Clinic group — approximately 1 g protein/kg of body weight, 10–15 g carbohydrate/day, and the remaining calories from fat [19, 20].

Functional Nutrients

Various nutritional interventions have been performed in patients with epilepsy to investigate the clinical significance of functional nutrients in seizure control. Despite abundant experimental evidence demonstrating the therapeutic potential of various nutrients, clinical data evaluating the anticonvulsant effects of dietary supplements are scarce. So far, only certain treatments have been investigated in detail, which are useful in other chronic non-communicable diseases, and potential benefits have been sporadically proven in patients suffering from epilepsy [21].

Omega-3 Polyunsaturated Fatty Acids

Omega-3 polyunsaturated fatty acids are essential fatty acids that cannot be synthesized in sufficient quantities in the body. Therefore, they must be obtained from the diet or nutritional supplements. Marine fish (salmon, tuna, and mackerel), nuts and seeds (flax seeds, chia seeds, almonds, and walnuts) are rich in omega-3 fatty acids. The three main omega-3 polyunsaturated fatty acids are docosahexaenoic acid and eicosapentaenoic acid, which are mainly found in marine fish oils, and alpha-linolenic acid, which is a major component of vegetable oils [22, 23]. Among these three, docosahexaenoic acid is the primary omega-3 fatty acid in the brain and makes up 10–20% of the total fatty acids, while others make up less than 1%.

Docosahexaenoic acid can serve as a structural component of neuronal membranes, modulating membrane biophysical properties, ion channel functions, and neurotransmitter signalling [21, 23].

Therapeutic benefits of omega-3 fatty acids for epilepsy were expected since they have a proven effect on cardiac arrhythmia, which involves hyperexcitability of heart cells. The first clinical study of the therapeutic potential of fish oil in patients with epilepsy revealed an anticonvulsant effect associated with omega-3 fatty acids [21]. However, later clinical studies produced more controversial results. For example, while some randomized clinical trials have found that consumption of approximately 0.6–2 g of fish oil reduces the frequency and duration of seizures, other, non-randomized studies have shown no seizure-regulatory efficacy. Despite several studies suggesting that omega-3 fatty acids suppress seizures, it remains unclear whether they affect seizure control [21, 24–26].

Vitamin D3 (Cholecalciferol)

Vitamin D is a fat-soluble vitamin that comes in several forms, including ergocalciferol (vitamin D2) and cholecalciferol (vitamin D3). Among them is vitamin D3, which is naturally present in animals and can be synthesized by exposure to sunlight. However, vitamin D can also be obtained from food sources such as fatty fish (cod, swordfish, salmon, tuna and sardines), dairy products (milk, cheese and yoghurt), meat (beef liver) and mushrooms. Vitamin D functions in the central nervous system include neurotransmitter regulation, neuronal differentiation, axonal growth, voltage-gated calcium channels, and neurotrophic factors, which can affect proper neuronal function [21, 27]. Several studies have reported that both young and adult patients with epilepsy show vitamin D deficiency in the blood, providing a rationale for administering vitamin D to people with epilepsy. To date, three clinical studies have investigated the effects of vitamin D on seizure control. The first study involved 23 patients with epilepsy who were divided into two experimental groups, one of which were given 4000 international units (IU) of vitamin D3 per day for 28 days and 16,000 IU of vitamin D3 for the next 28 days, followed by an initial observation period, and the other was given placebo first and then 8000 IU of vitamin D3 for each of the 28 days periods [28]. In patients receiving vitamin D3, the mean seizure frequency was significantly reduced to 67–71% of baseline, indicating an anticonvulsant effect. A similar observation was reported in another clinical study in which correction of vitamin D3 deficiency reduced seizure frequency by up to 40% for drug-resistant epilepsy. However, a recent cross-sectional cohort study evaluating 160 patients with epilepsy reported conflicting results showing no therapeutic efficacy of vitamin D3 in seizure control [29–31]. More carefully designed studies with larger sample sizes will be needed to resolve the anticonvulsant effects of vitamin D3 on epilepsy.

Vitamin E

Vitamin E refers to a group of fat-soluble antioxidants. Vitamin E consists of eight vitamin chemicals with similar molecular structures and functions. Among them, the one called α-tocopherol is the most biologically active in humans. Vitamin E is found in nuts, seeds, vegetable oils and green leafy vegetables such as spinach and broccoli. Well known for its antioxidant role, vitamin E is associated with versatile health promotion and additive effects in the treatment of many diseases, including cardiovascular, liver and Alzheimer's diseases. For patients with epilepsy, 400 IU daily vitamin E supplementation for 3 months was shown to reduce the frequency of seizures by approximately 60% compared to a placebo group that showed no difference in the number of seizures. Two other clinical studies confirmed the beneficial effects of chronic administration of vitamin E in patients with refractory epilepsy, supporting the use of vitamin E as an additional therapeutic option for epilepsy. However, the authors of another randomized, double-blind clinical trial reported that vitamin E supplementation for 3 months did not affect the occurrence of seizures regardless of the different types of epilepsy. As with the nutrients, further research is needed to answer the possible effects of vitamin E in the regulation of epileptic seizures [32–34].

Vitamin B6

Vitamin B6 is a water-soluble vitamin that plays a vital role in the development and maintenance of the central nervous system. Dietary vitamin B6 is present in fish (tuna and salmon), meat (beef liver, pork, lamb and chicken), dairy products (cheese and yoghurt), cereals (yeast bread and biscuits), vegetables (carrots, onions and tomatoes), and fruit (berries, apple, watermelon and banana). It contains six vitaminic chemicals: pyridoxine, pyridoxamine, pyridoxal, pyridoxine 5′-phosphate, pyridoxamine 5′-phosphate and pyridoxal 5′-phosphate. In general, vitamin B6 derived from meat consists of phosphorylated pyridoxal and pyridoxamine, while vitamin B6 derived from plants contains phosphorylated pyridoxine. In neurons, pyridoxal is regenerated by pyridoxal kinases, which means that it can act as an active cofactor. It may play a key role in the synthesis of multiple neurotransmitters, including GABA, glycine, dopamine, serotonin, and histamine. Pyridoxal can convert glutamate to GABA by facilitating the activation of glutamate decarboxylase, suggesting that vitamin B6 deficiency may increase network excitability in the brain. Pyridoxine deficiency can cause unprovoked seizures that are classified as a type of symptomatic epilepsy. On that postulate all future studies have been based and the thesis that vitamin B6 supplementation is beneficial in some patients with epilepsy. The researchers found that some patients who received vitamin B6 showed significant clinical improvement, although this was not observed in all patients. Later trials was conducted in which 30 or 50 mg of pyridoxine infusion was able to resolve recurrent seizures. However, other clinical studies involving different types of epilepsy have failed to illustrate anticonvulsant effects after treatment with vitamin B6 [35–38]. Since all clinical studies evaluating the therapeutic efficacy of vitamin B6 are small studies interpreting mixed results in different epilepsies, more detailed studies will be needed to provide definitive information on the role of vitamin B6 in epilepsy.

Conclusion

Although epilepsy, according to its clinical manifestation, is one of the first described diagnoses back in the Old Testament, and the search for a cure and control of the disease has lasted for thousands of years, we still do not have answers to many questions. Previously, in history, the treatment of epilepsy was based on the modest experiences of individual doctors, while today knowledge is spreading much faster. On the other hand, with more and more effective drugs, primarily with the appearance of phenobarbital in 1912, and until today, all the magical manifestations of the disease disappear. Patients suffering from epilepsy mostly, and in the modern world completely, are successfully treated with antiepileptic drug therapy. With the progress of medicine, and at the same time we emphasize the development of new antiepileptic drugs, the interest in researching eating habits in the context of disease treatment is declining. In the 2000s, interest began to grow again, primarily due to the patient's awareness of the importance of a healthy diet. When we compare the number of research on the influence of eating habits on epilepsy with the number of research on other neurological diseases, we are rather surprised by how little research is done on this topic, especially in comparison with cerebrovascular and neurodegenerative diseases. For now, we cannot safely make recommendations about dietary habits that could reduce epileptic seizures. What we can say is that a healthy, calorically balanced diet, based on the Mediterranean diet, cannot be harmful, that is, there is no evidence that it is harmful to patients with epilepsy.

References

1. Petelin Gadže Ž, Poljaković Z, Nanković S, Šulentić V. Epilepsija – dijagnostički i terapijski pristup. Zagreb: Medicinska naklada; 2019. p. Str.1.
2. Petelin GŽ. Klinička semiologija žarišnih epileptičkih napadaja. In: Petelin Gadže Ž, Poljaković Z, Nanković S, Šulentić V, editors. Epilepsija – dijagnostički i terapijski pristup. Zagreb: Medicinska naklada; 2019. Str. 20.
3. Allendorfer JB, Arida RM. Role of physical activity and exercise in alleviating cognitive impairment in people with epilepsy. Clin Ther. 2018;40(1):26–34. https://doi.org/10.1016/j.clinthera.2017.12.004.

4. Carrizosa-Moog J, Ladino LD, Benjumea-Cuartas V, Orozco-Hernández JP, Castrillón-Velilla DM, Rizvi S, i sur. Epilepsy, physical activity and sports: a narrative review. Can J Neurol Sci. 2018;45(6):624–32. https://doi.org/10.1017/cjn.2018.340.

5. Canova C, Danieli S, Barbiellini Amidei C, Simonato L, Di Domenicantonio R, Cappai G, i sur. A systematic review of case-identification algorithms based on Italian healthcare administrative databases for three relevant diseases of the nervous system: Parkinson's disease, multiple sclerosis, and epilepsy. Inferenze scarl. 2019;43(4S2):62–74. https://doi.org/10.19191/EP19.4.S2.P062.093.

6. Kostev K, Osina G, Rider F, Guekht A. Prevalence of valproate prescriptions in women of childbearing age in certain regions of Russia. E&B. 2019;101:106584. https://doi.org/10.1016/j.yebeh.2019.106584.

7. Orozco-Hernández JP, Quintero-Moreno JF, Marín-Medina DS, Castaño-Montoya JP, Hernández-Coral P, Pineda M, i sur. Perfil clínico y sociodemográfico de la epilepsia en adultos de un centro de referencia de Colombia. Neurologia. 2019;34(7):437–44. https://doi.org/10.1016/j.nrl.2017.02.013.

8. Buono V, Giussani G, Bianchi E, D'Ambrosio V, Coppola A, Bilo L, i sur. Epidemiology and familial clustering of pediatric epilepsy in the geographic isolate of Ischia. Epilepsy Res. 2019;154:86–9. https://doi.org/10.1016/j.eplepsyres.2019.05.004.

9. Kim DW, Lee SY, Chung SE, Cheong HK, Jung KY. Clinical characteristics of patients with treated epilepsy in Korea: a nationwide epidemiologic study. Epilepsia. 2013;55(1):67–75. https://doi.org/10.1111/epi.12469.

10. Forsgren L, Beghi E, Oun A, Sillanpaa M. The epidemiology of epilepsy in Europe – a systematic review. Eur J Neurol. 2005;12:245–53. https://doi.org/10.1111/j.1468-1331.2004.00992.x.

11. Banerjee PN, Filippi D, Allen HW. The descriptive epidemiology of epilepsy: a review. Epilepsy Res. 2009;85(1):31–45. https://doi.org/10.1016/j.eplepsyres.2009.03.003.

12. Wallace H, Shorvon S, Tallis R. Age-specific incidence and prevalence rates of treated epilepsy in an unselected population of 2 052 922 and age-specific fertility rates of women with epilepsy. Lancet. 1998;352(9145):1970–3. https://doi.org/10.1016/S0140-6736(98)04512-7.

13. Sillanpää M, Kälviäinen R, Klaukka T, Helenius H, Shinnar S. Temporal changes in the incidence of epilepsy in Finland: nationwide study. Epilepsy Res. 2006;71:206–15. https://doi.org/10.1016/j.eplepsyres.2006.06.017.

14. Kurth T, Lewis BE, Walker AM. Health care resource utilization in patients with active epilepsy. Epilepsia. 2010;51:874–82. https://doi.org/10.1111/j.1528-1167.2009.02404.x.

15. Christensen J, Vestergaard M, Pedersen MG, Pedersen CB, Olsen J, Sidenius P. Incidence and prevalence of epilepsy in Denmark. Epilepsy Res. 2007;76:60–5. https://doi.org/10.1016/j.eplepsyres.2007.06.012.

16. Cervenka MC, Kossoff EH. Dietary treatment of intractable epilepsy. Continuum (Minneap Minn). 2013;19(3 Epilepsy):756–66. https://doi.org/10.1212/01.Con.0000431396.23852.56.

17. Gudden J, Arias Vasquez A, Bloemendaal M. The effects of intermittent fasting on brain and cognitive function. Nutrients. 2021;13(9) https://doi.org/10.3390/nu13093166. Epub 20210910

18. Pfeifer HH, Thiele EA. Low-glycemic-index treatment: a liberalized ketogenic diet for treatment of intractable epilepsy. Neurology. 2005;65(11):1810–2. https://doi.org/10.1212/01.wnl.0000187071.24292.9e.

19. Wilder RM, editor. The effects of ketonemia on the course of epilepsy. Mayo Clin Proc; 1921.

20. Peterman MG. The ketogenic diet in epilepsy. JAMA. 1925;84(26):1979–83. https://doi.org/10.1001/jama.1925.02660520007003.

21. Kim JE, Cho KO. Functional Nutrients for Epilepsy. Nutrients. 2019;11(6) https://doi.org/10.3390/nu11061309. Epub 20190610

22. Cysneiros RM, Arida RM, Terra VC, Sonoda EY, Cavalheiro EA, Scorza FA. To sushi or not to sushi: can people with epilepsy have sushi from time to time? Epilepsy Behav. 2009;16(3):565–6. Epub 20090917. https://doi.org/10.1016/j.yebeh.2009.08.019.

23. Pourmasoumi M, Vosoughi N, Derakhshandeh-Rishehri SM, Assarroudi M, Heidari-Beni M. Association of Omega-3 fatty acid and epileptic seizure in epileptic patients: a systematic review. Int. J Prev Med. 2018;9:36. Epub 20180405. https://doi.org/10.4103/ijpvm.IJPVM_281_16.
24. Scorza CA, Cavalheiro EA, Calderazzo L, de Almeida AC, Scorza FA. Chew on this: sardines are still a healthy choice against SUDEP. Epilepsy Behav. 2014;41:21–2. Epub 20140929. https://doi.org/10.1016/j.yebeh.2014.08.019.
25. Taha AY, Burnham WM, Auvin S. Polyunsaturated fatty acids and epilepsy. Epilepsia. 2010;51(8):1348–58. Epub 20100701. https://doi.org/10.1111/j.1528-1167.2010.02654.x.
26. Yuen AWC, Flugel D, Poepel A, Bell GS, Peacock JL, Sander JW. Non-randomized open trial of eicosapentaenoic acid (EPA), an omega-3 fatty acid, in ten people with chronic epilepsy. Epilepsy Behav. 2012;23(3):370–2. https://doi.org/10.1016/j.yebeh.2011.11.030.
27. Eyles DW, Burne TH, McGrath JJ. Vitamin D, effects on brain development, adult brain function and the links between low levels of vitamin D and neuropsychiatric disease. Front Neuroendocrinol. 2013;34(1):47–64. Epub 20120711. https://doi.org/10.1016/j.yfrne.2012.07.001.
28. Christiansen C, Rødbro P, Sjö O. "Anticonvulsant action" of vitamin D in epileptic patients? A controlled pilot study. Br Med J. 1974;2(5913):258–9. https://doi.org/10.1136/bmj.2.5913.258.
29. Scorza FA, Albuquerque M, Arida RM, Terra VC, Machado HR, Cavalheiro EA. Benefits of sunlight: vitamin D deficiency might increase the risk of sudden unexpected death in epilepsy. Med Hypotheses. 2010;74(1):158–61. Epub 20090808. https://doi.org/10.1016/j.mehy.2009.07.009.
30. Mann MC, Exner DV, Hemmelgarn BR, Hanley DA, Turin TC, MacRae JM, et al. The VITAH trial-vitamin D supplementation and cardiac autonomic tone in patients with end-stage kidney disease on hemodialysis: a blinded, randomized controlled trial. Nutrients. 2016;8:10. https://doi.org/10.3390/nu8100608. Epub 20160928
31. Pendo K, DeGiorgio CM. Vitamin D3 for the Treatment of Epilepsy: Basic Mechanisms, Animal Models, and Clinical Trials. Front Neurol. 2016;7:218. Epub 20161208. https://doi.org/10.3389/fneur.2016.00218.
32. Ayyıldız M, Yıldırım M, Agar E. The effects of vitamin E on penicillin-induced epileptiform activity in rats. Exp Brain Res. 2006;174(1):109–13. https://doi.org/10.1007/s00221-006-0425-7.
33. Raju GB, Behari M, Prasad K, Ahuja GK. Randomized, double-blind, placebo-controlled, clinical trial of D-alpha-tocopherol (vitamin E) as add-on therapy in uncontrolled epilepsy. Epilepsia. 1994;35(2):368–72. https://doi.org/10.1111/j.1528-1157.1994.tb02446.x.
34. Sozen E, Demirel T, Ozer NK. Vitamin E: Regulatory role in the cardiovascular system. IUBMB Life. 2019;71(4):507–15. https://doi.org/10.1002/iub.2020. Epub 20190218
35. Fox JT, Tullidge G. Pyridoxine (vitamin B6) in epilepsy: a clinical trial. Lancet. 1946;248(6419):345. https://doi.org/10.1016/S0140-6736(46)90842-2.
36. Bowling FG. Pyridoxine supply in human development. Semin Cell Dev Biol. 2011;22(6):611–8. https://doi.org/10.1016/j.semcdb.2011.05.003.
37. Spector R, Greenwald LL. Transport and metabolism of vitamin B6 in rabbit brain and choroid plexus. J Biol Chem. 1978;253(7):2373–9.
38. di Salvo ML, Contestabile R, Safo MK. Vitamin B6 salvage enzymes: Mechanism, structure and regulation. Biochimica et Biophysica Acta (BBA) - Proteins and Proteomics. 2011;1814(11):1597–608. https://doi.org/10.1016/j.bbapap.2010.12.006.

Chapter 13
The Role of Amature Sports on Cognitive Functions

Sandra Morovic and Vida Demarin

Mens sana in corpore sano
  (Decimus Iunius Iuvenalis - Roman poet, 1.-2. century B. C.)

What man is happy? "He who has a healthy body, a resourceful mind and a docile nature."
  (Thales - Greek philosopher, 624 – 546 B. C.)

Cognition is the psychological activity or process of gaining information and comprehension through idea, experience, and the sense. As we are growing up, people always speak of physical changes and physical growth. "You've grown up to be tall" or "your hair is long", etc. Rarely we will hear someone mentioning psychological growth or cognitive maturation. Cognitive development represents an important domain of human growth. It consists of two parts, nonsocial and social cognition. Nonsocial cognition refers to the mental abilities of an individual, such as his or her attention span, processing speed, problem solving and reasoning skills, as well as working memory. The psychological processes involved in perception, encoding, storage, retrieval, and control of knowledge about oneself, and others are collectively labeled as social cognition [1]. A complex construct representing our capacity to achieve learning and problem-solving skills through the optimal usage of mental resources is referred to as cognitive efficiency [1].

It has long been known that exercise is beneficial to physical and mental health. The nineties were a turnover in our understanding of how exercise's effect on the human brain. Until then, most benefits of exercise on the human brain and overall health were attributed to better oxygenation of the brain and muscles. Little was

S. Morovic (✉)
Juraj Dobrila University of Pula, Zagreb, Croatia

Arsano Medical Group, Zagreb, Croatia
e-mail: sandra.morovic@poliklinika-aviva.hr

V. Demarin
International Institute for Brain Health, Zagreb, Croatia

V. Demarin et al. (eds.), *Mind, Brain and Education*,
https://doi.org/10.1007/978-3-031-33013-1_13

known about the neurobiological basis of exercise benefits. During that time, Rita Levi Montalcini and Stanly Cohen discovered neurotrophins, and for their discovery received a Nobel prize in 1986. This discovery forever changed the way we view the brain and respect exercise. Neurotrophins belong to a growth factor group called - neural growth factor (NGF). These proteins promote neuronal survival. Neurotropic factor is a growth factor with a special effect on neurons, it signals neurons to survive, differentiate or grow [2].

Soon after, the brain derived neurotrophic factor (BDNF) was discovered. This neurotrophic factor is important for long-term memory, acts on certain neurons in the central nervous system (CNS) and peripheral nervous system (PNS), helps the survival of existing neurons, and stimulates the growth and differentiation of new neurons and synapses. It was soon discovered that physical activity stimulates increased production and secretion of BDNF.

Characteristics of BDNF are antegrade and retrograde transport into the synapse, to stimulates transmission, participate in gene transcription, modify synaptic morphology, enhance neuronal elasticity, and the quantity of BDNF mRNA and protein levels are activity dependent. Released BDNF binds to its Tropomyosin receptor kinase B (TrkB) receptors. Presynaptically it regulates neurotransmitter secretion, postsynaptically it regulates postsynaptic sensitivity (i.e. by interaction with NMDA receptors) [3].

BDNF serves as our regulator of energy consumption. It participates in prevention of excessive body temperature loss during exposure to the cold, or during food deprivation. On the other hand, increased oxidative stress lowers BDNF levels.

Ability of physical activity to improve cognition includes the BDNF effect on metabolic processes. BDNF acts as a metabotrophin in the hippocampus. Therefore, we can conclude that exercise is beneficial to physical and mental health. It has been proven that regular exercise has a positive effect on functional changes in the brain, by not only improving brain volume but by also inducing angiogenesis and neurogenesis. Different sports show different areas of benefit. Studies show jogging can increase life expectancy; yoga makes us happy. Martial arts practice helps with difficult motor tasks and the cognition of the human brain. They alleviate stress, reduce frustration, increase self-esteem, reduce anger, boost focus, call for introspection, release the feel-good hormone. Martial arts improve the human body's executive functions, which is the most critical metric when it comes down to cognitive performance. It improves attention and alertness through attention state training, getting into a specific state of mind that allows a stronger focus. Their practice may increase selective attention of adolescents with attention deficit hyperactivity disorder (ADHD); therefore, they could be considered as an appropriate non-pharmacological therapeutic method to lessen the attentional impairment of individuals from a younger age. Studies also suggest that regular martial arts training may be effective for enhancing cognitive function and academic self-efficacy in growing children, as well as participation in regular martial arts training may be effective in improving the cognitive function of elderly individuals by increasing their levels of neurotrophic growth factor. Therefore, it is only fair to conclude that martial arts are not only beneficial for physical and mental health, but they can also enhance cognition [4].

During exercise, increase in nerve growth factors and neurotransmitters seem to play an important role in explaining the mechanisms of increasing nerve cell production. BDNF is known to promote neuronal cell formation and show an exercise-induced increase in expression [5]. It is involved in promoting the survival of progenitor cells that have the potential to differentiate into neurons or glia, as well as in differentiating these progenitor cells into neurons [6]. Neurotransmitters, such as serotonin, also show increased levels after exercise and have been reported to induce the generation of neurons [7]. In addition, various growth factors, such as vascular endothelial growth factor (VEGF) and insulin-like growth factor 1 (IGF-1), are known to promote the production of new cells [8].

Earlier investigations suggested that the main benefit of regular exercise on reduction of cerebrovascular risk and neurological diseases is exercise induced increased cerebral blood flow. A decade ago, a group of investigators reported that a 12-week aerobic exercise training program significantly increased middle cerebral artery blood velocity (MCAv) in young (mean age: 23 ± 5 years) and older (mean age: 63 ± 5 years) participants [9, 10].

Studies investigating the influence of regular marital arts practice can help improve body composition by improving aerobic capacity and flexibility as well as fat loss [11]. The effect of improving physical fitness and physical development of children has been reported for regular martial arts practice. But studies also suggest that martial arts practice could be effective in improving not only physical fitness, but also cognitive function [12], improving children's brain connectivity from the cerebellum to the parietal and frontal cortex [13, 14]. Evidence strongly supports the idea that regular martial arts exercise improves brain health, including cognitive functions, of children during their growth period. The mechanism underlying this effect could be explained by neuroplasticity-related growth factors in the blood and the changes in cerebral blood flow with corresponding changes in their cognitive function. These neuroplasticity-related neurotrophic factors and other growth factors [1, 15] are involved in neuronal proliferation, migration, survival, differentiation, and synaptic plasticity, and on cerebral blood flow velocity [16], which is associated in cognitive functions.

Significant changes in serum neuroplasticity-related growth factors (BDNF, vascular endothelial growth factor- VEGF, insulin like growth factor- IGF-1) were detected in martial arts training group.

Cerebral blood flow velocities in the training and control groups before and after the intervention revealed no significant differences in middle cerebral artery systolic blood flow velocity (MCAs), middle cerebral artery diastolic blood flow velocity (MCAd), middle cerebral artery mean blood flow velocity (MCAm), and pulsatility index (PI).

Regular physical activity and exercise through childhood helps the development of the physical constitution [17], and significantly participates in prevention of metabolic syndrome and similar disorders. Preventive effect is achieved through advancements in body composition and physical fitness [18]. Aerobic exercises increase the expression of growth factors such as BDNF, VEGF, and IGF-1, which promote the production of neurons.

BDNF regulates neuroplasticity, through the growth and survival of neurons [19]. IGF-1 is also an important factor in regulation of BDNF expression. VEGF contributes to the production of neurons in the hippocampus [1, 20]. Reports show that new neurons produced by exercise are mainly observed around the blood vessels [21].

Aerobic exercise and martial arts training induce an increase in neurotrophic and growth factors, suggesting that enhancement of aerobic fitness has a major role in this development. Increased aerobic fitness is highly correlated with increased IGF-1 levels. Specifically, resting serum IGF-1 levels are in positive correlation with aerobic fitness [22], and increased IGF-1 levels are related to increased expression of BDNF.

Martial arts (taekwondo) training interventions do not significantly influence the cerebral blood flow velocity of the MCA. The mechanism underlying exercise-induced cerebral blood flow regulation is understood to be related to the autoregulatory ability of brain vasculature to maintain constant blood flow, despite blood pressure changes and brain activation demands.

Therefore, we can conclude that martial arts training can help improve cognitive functions in children and increase the levels of factors such as BDNF.

We can categorize exercises into cardiovascular exercises, resistance training, martial arts, racquet sports, dancing and mind-body exercises. Considering the differences of these categories, some differences in cognitive functioning are present, while changing the type of physical activity performed. Different imaging modalities were used to prove this finding (viz. functional magnetic resonance imaging (fMRI), functional near-infrared spectroscopy (fNIRS) and electroencephalography (EEG).

Studies reveal that physical exercise plays an important role in enhancing cognitive function for all age groups. It elevates the levels of cerebral blood flow and growth factors (i.e., BDNF), and neurotransmitters (e.g., dopamine and norepinephrine) [23]. Furthermore, physical exercise enhances cognitive functioning and prevents cognitive decline in aged individuals [24]. Exercise is effective in making lifestyle changes in the elderly and lowering the risk for cognitive decline and other neurological disorders [25]. It boosts academic performance (this finding was more evident in females).

Exercise programs are helpful for improvements in cognition for elderly individuals with moderate cognitive impairment. Studies have shown improvements in cognition after a five-month exercise intervention [26]. A boost in cognitive function and brain activation in elderly, was noticed after a three-month intervention.

Clinical research has demonstrated an abundant hippocampal and basal ganglia volume, greater white matter integrity, efficient brain activity along with superior cognitive performance and academic achievements for physically active and fit children and preadolescents. Physical fitness and regular exercise positively impact the prefrontal cortex, functional brain connectivity and executive and memory functions [27].

What is the impact of certain types of exercise?

Twenty minutes of aerobic exercise demonstrated enhanced cognitive function among preadolescents and young adults. An increase in activation of the left dorsolateral prefrontal cortex and left orbital frontal cortex were seen. These benefits were not noted in the control group, which did not perform exercise [28]. Additional benefits in cognition were seen when aerobic training was combined with resistance training. Also, regular aerobic exercise shows reduction in neuronal decline in the elderly population.

Aerobic exercise is a method the restoration of normal functions occurring due to brain structure changes [29]. It is beneficial in enhancing cognition, especially driving memory, inhibition of redundant information and multitasking. These functions can be attributed to frontal lobe functions. The impacts of aerobic exercises on cognition and executive functioning are due to the influence of the temporal, frontal and parietal brain regions.

The frontal lobe performs a significant part in maintaining executive functioning in aging. Regular aerobic exercise impacts the hippocampal region [30].

Resistance Training is getting more and more popular among athletes other than bodybuilders (basketball, soccer players). Thirty minutes of resistance training shows results similar to aerobic exercise on condition. Positive correlation has been established between cortical hemodynamics of the prefrontal cortex and the handgrip strength of the individual. Addition of resistance training to regular aerobic exercise can enhance memory. Resistance training potentially elevates the level of insulin-like growth factor 1 (IGF-1). IGF-1 raises the synthesis of BDNF and vascular endothelial growth factor (VEGF), which improves cognitive functioning. A six-month intervention consisting of cognitive training and resistance training showed resistance training to be superior for improving global cognition and expanding gray matter volume in the posterior cingulate rather than cognitive training alone. Cognitive training was beneficial to memory due to greater functional communication between the hippocampus and superior frontal cortex [31].

Sports in general, martial arts and combat sports bring many benefits to cognitive health. Nowadays, martial arts are viewed as a sport and are taken responsible for many health benefits including strength, power, endurance, and balance [32]. Martial arts are also studied as means to develop cognitive functioning, given its predisposition to mind-body nature. The complex movements have a greater cortical requirement in comparison with aerobic exercises or resistance training. Certain elements of cognition, such as concentration and mindfulness, which may not be adequately developed in aerobic, or resistance training sessions are enhanced in martial arts sessions. An increased regional cerebral blood flow has also been observed following more complex moves, enhancing executive functioning, which serves as a pivotal component in cognition, while aerobic exercises could only improve attention and the processing speed [32].

Dance has a positive influence on overall health of an individual. Musical experience is only one aspect of influence. Additional benefits can be observed in the physical and social dimensions. Dance intervention to groups of elderly individuals showed improvements in brain volumes. Enhanced brain volumes from the 6-month intervention caused higher cognitive functions, such as working memory and attention, hence attenuating the age-related cognitive decline.

Yoga is a popular mind-body workout which consists of meditation, breathing exercises and posture control. It has physical and psychological health benefits, including conditions like arthritis, pain, and musculoskeletal conditions. Certain psychological conditions such as depression and anxiety can improve through yoga. Experienced yoga practitioners have higher gray matter volume in the left hippocampus. Yoga demonstrates greater activation of the prefrontal cortex, and higher executive functions independent of negative emotional stimuli. Performing yoga has been displayed as a successful strategy for enhancing mindfulness, leading to prevention of age-related decline in fluid intelligence along with enhanced brain functioning [33].

Exercise helps enhance functioning of the BDNF, enhances memory and changes in brain volume. Open skill sports such as martial arts and tennis demonstrate greater corticospinal excitability, motor cortex function, faster reaction times with better accuracy and better inhibitory control in comparison with performing aerobic exercises or resistance training. It could be understood that variations in physical activity have shown different effects on cognitive functioning.

References

1. Srinivas NS, Vimalan V, Padmanabhan P, Gulyás B. An overview on cognitive function enhancement through physical exercises. Brain Sci. 2021;11(10):1289. https://doi.org/10.3390/brainsci11101289.
2. Dinoff A, Herrmann N, Swardfager W, Liu CS, Sherman C, Chan S, Lanctôt KL. The effect of exercise training on resting concentrations of peripheral brain-derived neurotrophic factor (BDNF): a meta-analysis. PLoS One. 2016;11(9):e0163037. https://doi.org/10.1371/journal.pone.0163037. PMID: 27658238; PMCID: PMC5033477
3. Marie C, Pedard M, Quirié A, Tessier A, Garnier P, Totoson P, Demougeot C. Brain-derived neurotrophic factor secreted by the cerebral endothelium: a new actor of brain function? J Cereb Blood Flow Metab. 2018;38(6):935–49. https://doi.org/10.1177/0271678X18766772. Epub 2018 Mar 20. PMID: 29557702; PMCID: PMC5998997
4. Ho S-Y, So W-Y, Roh H-T. The effects of taekwondo training on peripheral neuroplasticity-related growth factors, cerebral blood flow velocity, and cognitive functions in healthy children: a randomized controlled trial. Int J Environ Res Public Health. 2017;14(5):454. https://doi.org/10.3390/ijerph14050454.
5. Cotman CW, Berchtold NC, Christie LA. Exercise builds brain health: key roles of growth factor cascades and inflammation. Trends Neurosci. 2007;30:464–72.
6. Shetty AK, Turner DA. In vitro survival and differentiation of neurons derived from epidermal growth factor-responsive postnatal hippocampal stem cells: inducing effects of brain-derived neurotrophic factor. J Neurobiol. 1998;35:395–425.
7. Yuan TF, Paes F, Arias-Carrión O, Ferreira Rocha NB, de Sá Filho AS, Machado S. Neural mechanisms of exercise: anti-depression, neurogenesis, and serotonin Signaling. CNS Neurol Disord Drug Targets. 2015;14:1307–11.

8. Sharma AN, da Costa e Silva BF, Soares JC, Carvalho AF, Quevedo J. Role of trophic factors GDNF, IGF-1 and VEGF in major depressive disorder: a comprehensive review of human studies. J Affect Disord. 2016;197:9–20.
9. Murrell CJ, Cotter JD, Thomas KN, Lucas SJ, Williams MJ, Ainslie PN. Cerebral blood flow and cerebrovascular reactivity at rest and during sub-maximal exercise: effect of age and 12-week exercise training. Age. 2013;35:905–20.
10. Ainslie PN, Cotter JD, George KP, Lucas S, Murrell C, Shave R, Thomas KN, Williams MJ, Atkinson G. Elevation in cerebral blood flow velocity with aerobic fitness throughout healthy human ageing. J Physiol. 2008;586:4005–10.
11. Fong SS, Ng GY. Does taekwondo training improve physical fitness? Phys Ther Sport. 2011;12:100–6.
12. Lee B, Kim K. Effect of taekwondo training on physical fitness and growth index according to IGF-1 gene polymorphism in children. Korean J Physiol Pharmacol. 2015;19:341–7.
13. Lakes KD, Bryars T, Sirisinahal S, Salim N, Arastoo S, Emmerson N, Kang D, Shim L, Wong D, Kang CJ. The healthy for life taekwondo pilot study: a preliminary evaluation of effects on executive function and BMI, feasibility, and acceptability. Ment Health Phys Act. 2013;6:181–8.
14. Kim YJ, Cha EJ, Kim SM, Kang KD, Han DH. The effects of taekwondo training on brain connectivity and body intelligence. Psychiatry Investig. 2015;12:335–40.
15. Bartkowska K, Turlejski K, Djavadian RL. Neurotrophins and their receptors in early development of the mammalian nervous system. Acta Neurobiol Exp. 2010;70:454–67.
16. Kure CE, Rosenfeldt FL, Scholey AB, Pipingas A, Kaye DM, Bergin PJ, Croft KD, Wesnes KA, Myers SP, Stough C. Relationships among cognitive function and cerebral blood flow, oxidative stress, and inflammation in older heart failure patients. J Card Fail. 2016;22:548–59.
17. Specker B, Thiex NW, Sudhagoni RG. Does exercise influence pediatric bone? A systematic review. Clin Orthop Relat Res. 2015;473:3658–72.
18. Pacifico L, Anania C, Martino F, Poggiogalle E, Chiarelli F, Arca M, Chiesa C. Management of metabolic syndrome in children and adolescents. Nutr Metab Cardiovasc Dis. 2011;21:455–66.
19. Gómez-Palacio-Schjetnan A, Escobar ML. Neurotrophins and synaptic plasticity. Curr Top Behav Neurosci. 2013;15:117–36.
20. Fabel K, Fabel K, Tam B, Kaufer D, Baiker A, Simmons N, Kuo CJ, Palmer TD. VEGF is necessary for exercise-induced adult hippocampal neurogenesis. Eur J Neurosci. 2003;18:2803–12.
21. Palmer TD, Willhoite AR, Gage FH. Vascular niche for adult hippocampal neurogenesis. J Comp Neurol. 2000;425:479–94.
22. Whiteman AS, Young DE, He X, Chen TC, Wagenaar RC, Stern CE, Schon K. Interaction between serum BDNF and aerobic fitness predicts recognition memory in healthy young adults. Behav Brain Res. 2014;259:302–12.
23. Dietz P. The influence of sports on cognitive task performance—a critical overview. Cogn Enhanc. 2013:67–72.
24. Lautenschlager NT, Almeida OP. Physical activity and cognition in old age. Curr Opin Psychiatry. 2006;19:190–3. https://doi.org/10.1097/01.yco.0000214347.38787.37.
25. Kramer AF, Erickson KI. Capitalizing on cortical plasticity: influence of physical activity on cognition and brain function. Trends Cogn Sci. 2007;11:342–8. https://doi.org/10.1016/j.tics.2007.06.009.
26. Bademli K, Lok N, Canbaz M, Lok S. Effects of physical activity program on cognitive function and sleep quality in elderly with mild cognitive impairment: a randomized controlled trial. Perspect Psychiatr Care. 2019;55:401–8. https://doi.org/10.1111/ppc.12324.
27. Erickson KI, Hillman CH, Kramer AF. Physical activity, brain, and cognition. Curr Opin Behav Sci. 2015;4:27–32. https://doi.org/10.1016/j.cobeha.2015.01.005.
28. Zhaang Y, Shi W, Wang H, Liu M, Tang D. The impact of acute exercise on implicit cognitive reappraisal in association with left dorsolateral prefronta activation: a fNIRS study. Behav Brain Res. 2021;406:113233. https://doi.org/10.1016/j.bbr.2021.113233.

29. Chaddock-Heyman L, Erickson KI, Voss M, Knecht A, Pontifex MB, Castelli D, Hillman C, Kramer A. The effects of physical activity on functional MRI activation associated with cognitive control in children: a randomized controlled intervention. Front Hum Neurosci. 2013;7:72. https://doi.org/10.3389/fnhum.2013.00072.
30. Haeger A, Costa AS, Schulz JB, Reetz K. Cerebral changes improved by physical activity during cognitive decline: a systematic review on MRI studies. NeuroImage Clin. 2019;23:101933. https://doi.org/10.1016/j.nicl.2019.101933.
31. Furlano JA, Nagamatsu LS. Feasibility of a 6-month pilot randomised controlled trial of resistance training on cognition and brain health in Canadian older adults at-risk for diabetes: study protocol. BMJ Open. 2019;9:e032047. https://doi.org/10.1136/bmjopen-2019-032047.
32. Douris P, Douris C, Balder N, LaCasse M, Rand A, Tarapore F, Zhuchkan A, Handrakis J. Martial art training and cognitive performance in middle-aged adults. J Hum Kinet. 2015;47:277. https://doi.org/10.1515/hukin-2015-0083.
33. Gothe NP, Hayes JM, Temali C, Damoiseaux JS. Differences in brain structure and function among yoga practitioners and controls. Front Integr Neurosci. 2018;12:26. https://doi.org/10.3389/fnint.2018.00026.

Chapter 14
Functional Neurological Disorders

Osman Sinanović, Sanela Zukić, Silva Banović, Emina Sinanović,
and Mirsad Muftić

Introduction

Functional neurological symptoms (FNSs) refer to neurological disorders that are not explained by disease. They may also be called psychogenic, hysterical, non-organic, somatoform, dissociative or conversion symptoms or medically unexplained [1].

Functional neurological disorder (FND) is defined by motor and sensory symptoms (e.g., tremor, dystonia, limb weakness, numbness, and seizures) that demonstrate clinical features incompatible with other neurological/medical diagnoses and that can be associated with significant distress and functional impairment [2–4].

According to the American Psychiatric Association Diagnostic and Statistical Manual of Mental Disorders (DSM-5), Functional neurologycal symptom disorder

O. Sinanović (✉)
Medical Faculty and Faculty of Education and Rehabilitation, University of Tuzla,
Tuzla, Bosnia and Herzegovina

Sarajevo Medical School of Medicine, University of Sarajevo School of Science and
Technology, Sarajevo, Bosnia and Herzegovina

S. Zukić · S. Banović
Medical Faculty and Faculty of Education and Rehabilitation, University of Tuzla,
Tuzla, Bosnia and Herzegovina

E. Sinanović
Psychotherapy Training Institute for Integrative Psychotherapy,
Sarajevo, Bosnia and Herzegovina

M. Muftić
Sarajevo Medical School of Medicine, University of Sarajevo School of Science and
Technology, Sarajevo, Bosnia and Herzegovina

Faculty for Health Sciences, University of Sarajevo, Sarajevo, Bosnia and Herzegovina

© The Author(s), under exclusive license to Springer Nature
Switzerland AG 2023
V. Demarin et al. (eds.), *Mind, Brain and Education*,
https://doi.org/10.1007/978-3-031-33013-1_14

(FNSD) is described as neurological symptoms that are not consistently explained by neurological or medical conditions [4]. Specific examples of FNSD include psychogenic non-epileptic seizures (PNES), psychogenic coma, conversion paralysis, functional movement disorder, blindness, and non-dermatomal sensory deficits [5, 6].

Functional neurological disorders (FNDs) are common in neurology wards with the levels of disability similar to epilepsy or multiple sclerosis [7].

The most common FNSs are non-epileptic attacks and functional weakness. These are common in neurology and general medical practice, especially in emergencies, where they can be mistaken for epilepsy, stroke, or another neurological disease. Around 20% of patients brought into the hospital in apparent status epilepticus and about one in seven patients attending a 'first fit' clinic have a diagnosis of dissociative (non-epileptic) attacks. Patients with functional weakness are at least as common as patients with multiple sclerosis and represent the leading misdiagnosis in patients wrongly given thrombolysis for presumed stroke. Physicians are often uncertain about how to approach patients with these problems.

Etiology

Despite the long-standing interest in FND and the growing body of neuroimaging research in the last decade, the etiology of FND remains elusive. However, volumetric MRI studies show evidence of differences in both cortical and subcortical brain anatomy in patients with functional neurological disorders. A group studying patients with motor functional neurological disorders with unilateral limb weakness found decreased volumes in the lentiform, thalamic, and caudate nuclei [8]. A variety of cortical changes have been seen. For instance, patients with motor functional neurological disorders have been found to exhibit an increased thickness of the premotor cortex bilaterally, whereas those with psychogenic nonepileptic seizures showed cortical atrophy of the motor and premotor regions in the right hemisphere and bilateral cerebellar atrophy [9, 10]. Whether these changes are primary or secondary to changes in neural circuitry is unknown [11].

Neural network dysfunction may underlie the symptomatic manifestation of functional neurological disorders. This is evidenced by multiple studies across different functional neurological symptoms. Functional MRI studies have shown that patients with functional neurological disorders activate their brains differently than healthy subjects simulating a similar condition [11]. For example, patients with motor functional neurological disorders show hypoactivation of cortical and subcortical motor pathways, and those with visual functional neurological disorders show hypoactivation in either the visual association or the primary visual cortex [12]. Additional studies point to dysfunction in sensorimotor integration, with hypoactivation of the right temporal-parietal region possibly resulting in loss of agency, validating the patient's experience that functional neurological disorders are involuntary [13, 14]. Functional neurological disorders are associated with high rates of trauma, particularly in childhood [15]. Abnormal emotion regulation and cognitive control patterns (possibly facilitated by past trauma history) may secondarily lead to brain activation changes and interfere with motor planning and self-regulation in individuals who are vulnerable to experiencing functional motor manifestations [11, 16, 17]. In one recent meta-analysis which included 34 case-control studies, with 1405 patients it was shown that stressful life events and maltreatment are substantially more common in people with functional neurological disorders than in healthy controls and patient controls. Emotional neglect had a higher risk than traditionally emphasized sexual and physical abuse, but many cases report no stressors [18].

A 'functional' model of the symptoms is useful both in thinking about the problem and when explaining the symptoms to the patient. In the past conversion disorder became the prominent term, reflecting Sigmund Freud's theory that psychological traumas are converted into physical symptoms [19].

Functional neurological disorder (FND) became an official term in DSM-5 and is becoming dominant as a result of patient preference, with widespread use in online support groups. This nomenclature is causally neutral, and this is reflected in the DSM-5 classification, which no longer requires a psychological precipitant as an essential diagnostic criterion for an important development [4]. Patients with functional neurological disorder present with neurological symptoms inconsistent or incongruent with typical pathophysiological disease; they are presumed to be psychological in origin.

The onset of FNDs has been reported throughout the life course. The onset of non-epileptic attacks peaks in the third decade, and motor symptoms have their peak onset in the fourth decade. The symptoms can be transient or persistent. The prognosis may be better in younger children than in adolescents and adults [4].

Frequency

Doctors in nearly all medical specialties see patients with physical symptoms that are genuine but cannot be explained based on a recognized 'organic' disease. Around 30–50% of outpatient visits in primary and secondary care are for this reason [20, 21].

FND is a common cause of disability and distress, especially in neurological practice. Functional disorders represent the second commonest reason to see a neurologist after headache [20]. More tightly defined FND still accounts for at least 5–10% of new neurological consultations. Estimates of incidence are conservatively 12 per 100,000 per year [22]. Based on this, around 8000 new diagnoses of FND are made per year in the UK and around 50,000–100,000 people have it in the community. FND disproportionately affects women (around 3:1) although, as age of onset increases, the proportion of men affected increases. Incident cases demonstrate that FND can occur across all ages, from young children (although it is rare before 10 years old) up to patients in their 80 s [23].

Carson et al. did interesting the prospective cohort study in the regional neurology service in Lothian, Scotland with 300 newly-referred outpatients. Neurologists rated the degree to which patients' symptoms were explained by organic disease or organicity. Out off 300 new patients 11% (95% confidence interval (95% CI) 7%–14%) had symptoms that were rated as "not at all explained" by organic disease, 19% (15% to 23%) "somewhat explained", 27% (22% to 32%) "largely explained", and 43% (37% to 49%) "completely explained" by organic disease. "Depressive and anxiety disorders were more common in patients with symptoms of lower organicity (70% of patients in the "not at all" group had an anxiety or depressive disorder compared with 32% in the "completely explained" group ($p < 0.0005$). Based on this research they concluded that one-third of new referrals to general neurology clinics have symptoms that are poorly explained by identifiable organic disease, and that these patients were disabled and distressed [24].

This finding that one-third of referred patients had symptoms that were not well explained by organic disease agrees with previous reports. For example, Perkin [25] in his review of 7836 consecutive new neurological patients whom he had seen personally at Charing Cross Hospital, reported that 26% of patients received no medical diagnosis and a further 4% had symptoms of conversion hysteria. Similar findings have been shown in female in-patient neurological populations by Creed et al. [26].

According to DSM-5 manual transient conversion (functional) symptoms are common, but the precise prevalence of the disorder is unknown. This is partly because the diagnosis usually requires assessment in secondary care, where it is found in approximately 5% of referrals to neurology clinics. The incidence of individual persistent conversion (functional) symptoms is estimated to be 2–5/100,000 per year. According to the UK National Organization for Rare Disorders, functional neurological symptom disorder (FND) is thought to occur in 14–22 cases per 100,000 people.

The most common FNSs are non-epileptic attacks and functional weakness. These are common in neurology and general medical practice, especially in emergencies, where they can be mistaken for epilepsy, stroke, or another neurological disease. Around 20% of patients brought into the hospital in apparent status epilepticus and about one in seven patients attending a 'first fit' clinic have a diagnosis of dissociative (non-epileptic) attacks. Patients with functional weakness are at least as common as patients with multiple sclerosis and represent the leading misdiagnosis in patients wrongly given thrombolysis for presumed stroke [5].

The overall prevalence of psychogenic non-epileptic seizures (PNES) is between 1/3000 and 1/50,000 [27]. One study of 3781 neurology patients in Scotland found that about 5% had a primary diagnosis of functional neurological symptoms, such as non-epileptic attacks, functional weakness, and movement disorder [5, 10, 27].

Stefánsson et al. [28] using data from a large psychiatric registry as well as a psychiatric consultation service, found the annual incidence of functional neurological disorders to be 22 per 100,000 in patients from Monroe, New York. Studies show that a majority (78–93%) of patients with functional neurological disorders are women [29, 30]. Up to two-thirds of these patients have a significant psychiatric disease, with a history of depression and a history of trauma being the most common psychiatric comorbid conditions [27]. Dissociative symptoms and personality disorders are also prevalent in this patient population [31].

Diagnosis

As we already stressed Functional neurological disorder (FND) became an official term in DSM-5 [4] and is becoming dominant as a result of patient preference, with widespread use in online support groups (Table 14.1).

The diagnosis of a functional neurological disorder is based on findings in the neurological examination (or other tests) demonstrating that the symptom is incompatible with a structural neurological illness [32]. In the case of psychogenic nonepileptic seizures, a subtype of functional neurological disorder, an electrophysiological study that demonstrates normal electrophysiological function at the time of a seizure-like episode (via video EEG monitoring), coupled with signs suggestive of psychogenic nonepileptic seizures, would satisfy the diagnostic criteria [33].

The diagnostic criteria for functional neurological disorders in the DSM-5 emphasize the importance of positive physical criteria in making the diagnosis (see below) [4]. Patients no longer must have had recent psychological stressors (even though some will have). These new criteria bring the diagnosis of functional disorders back into a form that neurologists should be comfortable with using. In

Table 14.1 Diagnostic criteria conversion disorder (functional neurological symptom disorder)

Diagnostic Criteria
A. One or more symptoms of altered voluntary motor or sensory function.
B. Clinical findings provide evidence of incompatibility between the symptom and recognized neurological or medical conditions.
C. The symptom or deficit is not better explained by another medical or mental disorder.
D. The symptom or deficit causes clinically significant distress or impairment in social, occupational, or other important areas of functioning or warrants medical evaluation.
Specify symptom type:
 (F44.4) With weakness or paralysis
 (F44.4) With abnormal movement (e.g., tremor, dystonic movement, myoclonus, gait disorder)
 (F44.4) With swallowing symptoms
 (F44.4) with speech symptoms (e.g., dysphonia, slurred speech)
 (F44.5) With attacks or seizures
 (F44.6) With anesthesia or sensory loss
 (F44.6) With special sensory symptoms (e.g., visual, olfactory, or hearing disturbance)
 (F44.7) With mixed symptoms Specify if:
Acute episode; Symptoms present for less than 6 months.
Persistent: Symptoms occurring for 6 months or more.
Specify if:
With psychological stressor (specify stressor)
Without psychological stressor

addition, the International Classification of Diseases (ICD), the tenth revision is also being revised for its 11th edition in 2017. For the first time, functional neurological disorders should appear in the neurology section as well as in the psychiatry section) [9, 34–36]. One of the hopes of these revised international criteria is that they will encourage greater confidence in the diagnosis from neurologists and better interdisciplinary working between neurology and psychiatry [37].

In DSM-5 manual [4] diagnostic features and associated features supporting diagnosis are described as follows:

Diagnostic Features

Many clinicians use the alternative names of "functional" (referring to abnormal central nervous system functioning) or "psychogenic" (referring to an assumed etiology) to describe the symptoms of conversion disorder (functional neurological symptom disorder). In conversion disorder, there may be one or more symptoms of various types. Motor symptoms include weakness or paralysis; abnormal movements, such as tremors or dystonic movements; gait abnormalities; and abnormal limb posturing. Sensory symptoms include altered, reduced, or absent skin sensation, vision, or hearing. Episodes of abnormal generalized limb shaking with apparent impairment or loss of consciousness may resemble epileptic seizures (also called psychogenic or non-epileptic seizures). There may be episodes of unresponsiveness resembling syncope or coma. Other symptoms include reduced or absent speech volume (dysphonia/aphonia), altered articulation (dysarthria), a sensation of a lump in the throat (globus), and diplopia. Although the diagnosis requires that the symptom is not explained by neurological disease, it should not be made simply because results from investigations are normal or because the symptom is "bizarre." There must be clinical findings that show clear evidence of incompatibility with neurological disease. Internal inconsistency at examination is one way to demonstrate incompatibility (i.e., demonstrating that physical signs elicited through one examination method are no longer positive when tested a different way). Examples of such examination findings include:

1. Hoover's sign, in which weakness of hip extension returns to normal strength with contralateral hip flexion against resistance.
2. Marked weakness of ankle plantar-flexion when tested on the bed in an individual who is able to walk on tiptoes;
3. Positive findings on the tremor entrainment test. On this test, a unilateral tremor may be identified as functional if the tremor changes when the individual is distracted away from it. This may be observed if the individual is asked to copy the examiner in making a rhythmical movement with their unaffected hand and this causes the functional tremor to change such that it copies or "entrains" to the rhythm of the unaffected hand or the functional tremor is suppressed, or no longer makes a simple rhythmical movement.

4. In attacks resembling epilepsy or syncope ("psychogenic" non-epileptic attacks), the occurrence of closed eyes with resistance to opening or a normal simultaneous electroencephalogram (although this alone does not exclude all forms of epilepsy or syncope).
5. For visual symptoms, a tubular visual field (i.e., tunnel vision). It is important to note that the diagnosis of conversion disorder should be based on the overall clinical picture and not on a single clinical finding.

Associated Features Supporting Diagnosis

Several associated features can support the diagnosis of conversion disorder. There may be a history of multiple similar somatic symptoms. Onset may be associated with stress or trauma, either psychological or physical. The potential etiological relevance of this stress or trauma may be suggested by a close temporal relationship. However, while assessment for stress and trauma is important, the diagnosis should not be withheld if none is found.

Conversion disorder is often associated with dissociative symptoms, such as depersonalization, derealization, and dissociative amnesia, particularly at symptom onset or during attacks. The phenomenon of labelle indifférence (i.e., lack of concern about the nature or implications of the symptom) has been associated with conversion disorder but it is not specific for conversion disorder and should not be used to make the diagnosis. Similarly, the concept of secondary gain (i.e., when individuals derive external benefits such as money or release from responsibilities) is also not specific to conversion disorder and particularly in the context of definite evidence for feigning, the diagnoses that should be considered instead would include factitious disorder or simulation [4].

A literature review showed that the diagnosis of a functional neurological disorder can be made reliably, with a misdiagnosis rate of about 4% [1, 38].

Treatment

The treatment of the FND is very complex. As we already pointed out, despite the long-standing interest in FND and the growing body of neuroimaging research in the last decade, the etiology of FND remains elusive. A 'functional' model of the symptoms is useful both in thinking about the problem and when explaining the symptoms to the patient. Patients who are given a diagnosis of FND frequently have difficulty understanding or accepting the diagnosis, and may verbalize this as such.

Giving a clear diagnosis can itself be therapeutic for some patients, and it diminishes the concern they may have that they are suffering from some obscure neurological illness [39].

The explanation of the diagnosis needs to make sense to the patient. Showing the patient the inconsistencies on examination and explaining how the preserved physiological function is ascertained is extremely helpful [40]. It is not helpful simply to inform patients which conditions they do not suffer from; rather, it is preferable to focus the discussion on the uncovered diagnosis. Clinicians must carefully explain the term "psychogenic" so the patient does not misinterpret it to imply that he or she is feigning or malingering, which is not correct [11].

Patients with a functional neurological disorder have an increased number of psychiatric conditions, including depressive disorders, anxiety disorders, PTSD, and hypochondriasis [41]. The evidence for antidepressant treatment in functional neurological disorders comes from a few uncontrolled studies. No randomized placebo-controlled trials have demonstrated the efficacy of any particular antidepressant for this disorder. Therefore, choosing which antidepressant or anxiolytic medication is most appropriate if any should be based on the treatment of the identified psychiatric comorbidities. Clinical judgment should be used as to when it is appropriate to initiate medications to treat comorbid psychiatric conditions [11, 42].

There is evidence that physical therapy (PT) helps treat the motor and gait manifestations of functional neurological disorders [43]. In a randomized controlled crossover study of 60 patients with a psychogenic gait disorder who were randomly assigned to receive immediate treatment with inpatient PT or treatment after 4 weeks (control group), the benefit of the 3-week inpatient course of PT was substantial, and it was maintained even a year after PT was completed [44]. Other studies have validated the efficacy of PT in motor manifestations of functional neurological disorders [45]. These studies have led to the development of an expert consensus recommendation supporting the use of PT in the treatment of motor functional neurological disorders [11, 46].

Cognitive-behavioral therapy (CBT) has also been shown to be beneficial in treating functional neurological disorders [47]. CBT includes education about functional neurological disorders and the stress response cycle, trains patients in stress management techniques and new behavioral responses, and helps patients identify and change unhelpful thought patterns that reinforce their symptoms. CBT is the preferred treatment modality for patients with episodic symptoms, such as psychogenic nonepileptic seizures.

Functional neurological disorders are truly at the intersection of neurology and psychiatry: patients present with neurological symptoms that are a manifestation of a neuropsychiatric disorder. Dualistic thinking is not helpful for these patients, as neurological symptoms and emotional functioning need to be viewed as influencing each other. Integration of care is needed for this patient group [11].

Speech and language professionals have the skills to evaluate and treat communication problems and swallowing disorders caused by functional neurological disorders. The goal of speech and language therapy for people with FND is primarily the treatment of dysphonia/aphonia, dysfluency that is similar to stuttering, dysarthria (articulation deficits), prosodic disturbances, and language symptoms [48–50]. One of the goals is also to explain diseases and related speech and language symptoms and swallowing disorders in a simple, understandable, but serious way [51].

When communicating the diagnosis to the patient, it is extremely important to "say what it is" - You have FND. After that, it is necessary to share with the patient the positives of the diagnosis. Explain what is known about the condition and what can be done, and also emphasize if certain symptoms are reversible and if treatment procedures can reduce or minimize the consequences [52]. The general principles in therapy that need to be adhered to are: individualized communication of the diagnosis, which means adjusting the terminology if a clear and concise explanation is needed; breaking down prejudices about the diagnosis; encouraging adequate behavior; conducting a conversation to raise readiness to participate in therapeutic procedures; involving family members in treatment procedures and encouraging adequate independence in the process of self-care [48]. There is a need to expand treatment procedures for the rehabilitation of speech, language, and swallowing difficulties for persons with FND. It is also necessary to initiate additional studies to generate adequate diagnostic criteria for difficulties in these areas, as well as longitudinal studies to assess the therapeutic effects on speech, language, and swallowing of people with FND [49].

References

1. Sojka P, Bareš M, Kašpárek T, Světlák M. Processing of emotion in functional neurological disorder. Front Psych. 2018;9:479.
2. Pick S, Anderson DG, Asadi-Pooya AA, Aybek S, Baslet G, Bloem BR, et al. Outcome measurement in functional neurological disorder: a systematic review and recommendations. J Neurol Neurosurg Psychiatry. 2020;91(6):638–49.
3. Espay AJ, Aybek S, Edwards MJ CA, Goldstein LH, Hallett M, et al. Current concepts in diagnosis and treatment of functional neurological disorders. JAMA Neurol. 2018;75:1132–41.
4. American Psychiatric Association. Diagnostic and statistical manual of mental disorders. 5th ed. Washington (DC): American Psychiatric Association; 2013.
5. Fobian AD, Elliott L. A review of functional neurological symptom disorder etiology and the integrated etiological summary model. J Psychiatry Neurosci. 2019;44(1):8–18.
6. Stone J. Functional neurological symptoms. JR Coll Physicians Edinb. 2011;41:38–42.
7. Stone J. Functional neurological symptoms. Clin Med (Lond). 2013;13(1):80–3.
8. Atmaca M, Aydin A, Tezcan E, Poyraz AK, Kara B. Volumetric investigation of brain regions in patients with conversion disorder. Prog Neuro-Psychopharmacol Biol Psychiatry. 2006;30:708–13.
9. Aybek S, Nicholson TR, Draganski B, Daly E, Murphy DG, David AS, Kanaan RA. Grey matter changes in motor conversion disorder. J Neurol Neurosurg Psychiatry. 2014;85:236–8.
10. Labate A, Cerasa A, Mula M, Mumoli L, Gioia MC, Aguglia U, et al. Neuroanatomic correlates of psychogenic nonepileptic seizures: a cortical thickness and VBM study. Epilepsia. 2012;53:377–85.
11. O'Neal MA, Baslet G. Treatment for patents with functional neurological disorders (conversion disorder): an integrated approach. Am J Psychiatry. 2018;175(4):307–14.
12. Aybek S, Vuilleumier P. Imaging studies of functional neurologic disorders. Handb Clin Neurol. 2016;139:73–84.
13. Voon V, Cavanna AE, Coburn K, Sampson S, LaFrance RA, WC. Functional neuroanatomy and neurophysiology of functional neurological disorders (conversion disorder). J Neuropsychiatry Clin Neurosci. 2016;28:168–90.

14. Roelofs K, Pasman J. Stress, childhood trauma, and cognitive functions in functional neurologic disorders. Handb Clin Neurol. 2016;139:139–55.
15. Nimnuan C, Hotopf M, Wessely S. Medically unexplained symptoms: an epidemiological study in seven specialities. J Psychosom Res. 2001;51:361–7.
16. Aybek S, Nicholson TR, O'Daly O, Zelaya E, Kanaan RA, David AS. Emotion-motion interactions in conversion disorder: an fMRI study. PLoS One. 2015;10:e0123273.
17. Aybek S, Nicholson TR, Zelaya F, O'Daly OG, Craig TJ, David AS, Kanaan RA. Neural correlates of recall of life events in conversion disorder. JAMA Psychiatry. 2014;71:52–60.
18. Ludwig L, Pasman JA, Nicholson T, Aybek S, David AS, Tuck S, et al. Stressful life events and maltreatment in conversion (functional neurological) disorder: systematic review and meta-analysis of case-control studies. Lancet Psychiatry. 2018;5(4):307–20.
19. Stone J. Functional neurological disorders: the neurological assessment as treatment. Pract Neurol. 2016;16:7–17.
20. Stone J, Carson A, Duncan R, Roberts R, Warlow C, Hibberd C, et al. Who is referred to neurology clinics?—the diagnoses made in 3781 new patients. Clin Neurol Neurosurg. 2010;112:747–51.
21. Benbadis SR, Hauser WA. An estimate of the prevalence of psychogenic non-epileptic seizures. Seizure. 2000;9(4):280–1.
22. Carson A, Lehn A. Epidemiology. In: Hallett M, Stone J, Carson A, editors. Handbook of clinical neurology: vol 139: functional neurologic disorders. Elsevier; 2016. p. 47–60.
23. Bennett K, Diamond C, Hoeritzauer I, Gardiner P, McWhirter L, Carson A, Stone J. A practical review of functional neurological disorder (FND) for the general physician. Clin Med. 2021;21(1):28–36.
24. Carson AJ, Ringbauer B, Stone J, McKenzie L, Warlow C, Sharpe M. Do medically unexplained symptoms matter? A prospective cohort study of 300 new referrals to neurology outpatients. J Neurol Neurosurg Psychiatry. 2000;68:207–10.
25. Perkin GD. An analysis of 7836 successive new outpatient referrals. J Neurol Neurosurg Psychiatry. 1989;52:447–8.
26. Creed F, Firth D, Timol M, Metcalfe R, Pollock S. Somatization and illness behaviour in a neurology ward. J Psychosom Res. 1990;34:427–37.
27. Stone J, Carson AJ, Duncan R, Coleman R, Roberts R, Warlow C, et al. Symptoms 'unexplained by organic disease' in 1144 new neurology out-patients: how often does the diagnosis change at follow-up? Brain. 2009;132:2878–88.
28. Edwards MJ, Bhatia KP. Functional (psychogenic) movement disorders: merging mind and brain. Lancet Neurol. 2012;11:250–60.
29. Stefánsson JG, Messina JA, Meyerowitz S. Hysterical neurosis, conversion type: clinical and epidemiological considerations. Acta Psychiatr Scand. 1976;53:119–38.
30. Deka K, Chaudhury PK, Bora K, Kalita P. A study of clinical correlates and socio-demographic profile in conversion disorder. Indian J Psychiatry. 2007;49:205–7.
31. Carson AJ, Ringbauer B, MacKenzie L, Warlow C, Sharpe M. Neurological disease, emotional disorder, and disability: they are related: a study of 300 consecutive new referrals to a neurology outpatient department. J Neurol Neurosurg Psychiatry. 2000;68:202–6.
32. Sar V, Akyüz G, Kundakçi T, Kiziltan E, Dogan O. Childhood trauma, dissociation, and psychiatric comorbidity in patients with conversion disorder. Am J Psychiatry. 2004;161:2271–6.
33. Stone J, Carson A, Sharpe M. Functional symptoms and signs in neurology: assessment and diagnosis. J Neurol Neurosurg Psychiatry. 2005;76:i2–i12.
34. LaFrance WC Jr, Baker GA, Duncan R, Goldstein LH, Reuber M. Minimum requirements for the diagnosis of psychogenic nonepileptic seizures: a staged approach: a report from the international league against epilepsy nonepileptic seizures task force. Epilepsia. 2013;54:2005–18.
35. Stone J, Hallett M, Carson A, Bergen D, Shakir R. Functional disorders in the neurology section of ICD-11: a landmark opportunity. Neurology. 2014;83:2299–301.
36. World Health Organisation. ICD11 Beta Draft. http://apps.who.int/classifications/icd11/browse/f/en

37. Sinanović O. Funkcionalni neurološki poremećaji: klasifikacija i dijagnostički kriteriji. Knjiga sažetaka 2. Simpozija Funkcionalni poremećaji u neurološkoj kliničkoj praksi – spona između neurologije i psihijatrije, Zagreb, Hrvatska, 2017: 7–8.
38. Stone J, Smyth R, Carson A, Lewis S, Prescott R, Warlow C, et al. Systematic review of misdiagnosis of conversion symptoms and "hysteria". BMJ. 2005;331(7523):989. https://doi.org/10.1136/bmj.38628.466898.55.
39. Stone J. Functional neurological disorders: the neurological assessment as treatment. Neurophysiol Clin. 2014;44:363–73.
40. Stone J, Edwards M. Trick or treat? Showing patients with functional (psychogenic) motor symptoms their physical signs. Neurology. 2012;79:282–4.
41. Hoge CW, Terhakopian A, Castro CA, Messer SC, Engel CC. Association of posttraumatic stress disorder with somatic symptoms, health care visits, and absenteeism among Iraq war veterans. Am J Psychiatry. 2007;164:150–3.
42. Kleinstäuber M, Witthöft M, Steffanowski A, van Marwijik H, Hiller W, Lambert MJ. Pharmacological interventions for somatoform disorders. Cochrane Database Syst Rev. 2014;11:CD010628. https://doi.org/10.1002/14651858.CD010628.pub2.
43. Kaur J, Garnawat D, Ghimiray D, et al. Conversion disorder and physical therapy. Delphi Psychiatry J. 2012;15:394–7.
44. Jordbru AA, Smedstad LM, Klungsøyr O, Martinsen EW. Psychogenic gait disorder: a randomized controlled trial of physical rehabilitation with one-year follow-up. J Rehabil Med. 2014;46:181–7.
45. Nielsen G, Ricciardi L, Demartini Bhunter R, Loyce E, Edwards MJ. Outcomes of a 5-day physiotherapy programme for functional (psychogenic) motor disorders. J Neurol. 2015;262:674–81.
46. Nielsen G, Stone J, Matthews A, Brown M, Sparkes C, Farmer R, et al. Physiotherapy for functional motor disorders: a consensus recommendation. J Neurol Neurosurg Psychiatry. 2015;86:1113–9.
47. Sharpe M, Walker J, Williams C, Stone J, Cavanagh J, Murray G, et al. Guided self-help for functional (psychogenic) symptoms: a randomized controlled efficacy trial. Neurology. 2011;77:564–72.
48. Espay AJ, Aybek S, Carson A, Edwards MJ, Goldstein LH, Hallett M, et al. Current concepts in diagnosis and treatment of functional neurological disorders. JAMA Neurol. 2020;75:1132–41.
49. Barnett C, Armes J, Smith S. Speech, language and swallowing impairments in functional neurological disorder: a scoping review. IJLCD. 2018;54:309–20.
50. Duffy JR. Functional speech disorders: clinical manifestations, diagnosis, and management. Handb Clin Neurol. 2016;139:379–88.
51. Baker J, Barnett C, Cavalli L, Dietrich M, Dixon L, Duffy JR, et al. Management of functional communication, swallowing, cough and related disorders: consensus recommendations for speech and language therapy. J Neurol Neurosurg Psychiatry. 2021;92:1112–25.
52. Bennett K, Diamond C, Hoeritzauer I, Gardiner P, McWhirter L, Carson A, Stone J. A practical review of functional neurological disorder (FND) for the general physician. Clin Med (Lond). 2021;21:28–36.

Chapter 15
Psychoneuroimmunology in Pula: A Hot Topic with a Long Past

Norbert Müller

As part of the former Yugoslavia, Croatia played an important role in progress in psychoneuroimmunology. As early as 1986, an important international congress, the second International Workshop on Neuroimmunomodulation, was held in Dubrovnik from June 1 to 6, 1986, by the International Working Group of Immunomodulation and included more than 140 papers or posters and 182 scientists from 26 countries. It followed the first International Workshop on Neuroimmunomodulation, held in Bethesda, Maryland, USA, and was the first congress in the field in Europe. The organizers in Dubrovnik, B. D. Jankovic, B. M. Marcovic (Yugoslavia), and N. H. Spector (USA), were all pioneers in this field. At this time, neuroimmunology was a marginal topic, but the enthusiasm at this meeting was an important basis for further success in the field, and the meeting was an outstanding starting point for my own career in the field of psychoneuroimmunology. At the Dubrovnik congress, the International Society of NeuroImmunoModulation (ISNIM) was founded, an important and well-reputed society that has regularly held scientific congresses every other year.

Over the last six decades, research on the neurobiology of psychiatric disorders such as depression, bipolar disorder, and schizophrenia has focused overwhelmingly on disturbances of serotonergic and dopaminergic neurotransmission. This focus was well justified because the pathogenesis of these disorders undoubtedly involves dysfunction of these neurotransmitters. However, despite all the research, the pathogenetic mechanisms remain unclear. Many findings point beyond the neurotransmitter hypothesis, including the fact that the therapeutic effects of antipsychotic, antidepressant, and mood stabilizing drugs are still unsatisfactory. What are

N. Müller (✉)
Department of Psychiatry and Psychotherapy, Ludwig Maximilian University, Munich, Germany
e-mail: norbert.mueller@med.uni-muenchen.de

V. Demarin et al. (eds.), *Mind, Brain and Education*,
https://doi.org/10.1007/978-3-031-33013-1_15

the mechanisms that lead to the neurotransmitter disturbance? Psychoneuroimmunological research indicates that the disturbed neurotransmission may have an immune or inflammatory background.

Interactions between the immune system, hormones, and neurotransmitters became a growing area of interest in biological psychiatry in the 1980s, especially among basic researchers, for example because of Robert Ader's work on the conditioning of immunity and his overview of the field [1]. However, at that time groups of researchers interested in the field worked separately, the methodological resources of immunological research were limited, the interfering factors confusing, and the interpretation of the findings often speculative because of the lack of detailed knowledge about the interactions within the immune system. Interest and knowledge in the field improved when, in the early 1990s, the German Volkswagen Foundation founded a large international research program in psychoneuroimmunology. As a result of this endeavor, the first German university chair in psychoneuroimmunology was established at the University Marburg, and Hugo Besedovsky, a pioneer in research on the interaction between the immune system, the glucocorticoid hormone system, and neurotransmission, held this chair for many years [2]. Moreover, this program contributed to both a lively young scientists scene in the field of psychoneuroimmunology in Germany that continues to this day and to the founding of the German Endocrine Brain Immune Network (GEBIN), a scientific association that continues to hold regular biannual congresses.

In the late 1980s and early 1990s, psychoneuroimmunological research in psychiatric disorders was concentrated in Europe. The group of Michael Maes in Antwerp/Maastricht, which focused on major depression, and the group in Munich, which focused on schizophrenia, contributed actively to the field and showed that pro-inflammatory cytokines and an activated immune state play a role in major depression, schizophrenia, and bipolar disorder [3–7].

The evaluation of immunological mechanisms in psychiatric disorders concentrated on the one hand on developmental aspects and the role of immune activation, in particular in animal models [8], and on the other hand on the role of the blood-brain barrier [9, 10], the blood-brain communication between pro- and anti-inflammatory cytokines [2, 11], and the immune effects of psychopharmacological treatment.

Because of the results of these funding programs, the field also found its way to Pula. Sporadic lectures on the above-mentioned aspects of psychoneuroimmunology were given in Pula in the early 2000s, some of which placed a special emphasis on the basic mechanisms of the interaction between the immune system and mental processes, including clinical aspects of this research. In those years, progress in psychoneuroimmunology was based on findings from" pure" neuroimmunology because research in neurological disorders such as multiple sclerosis contributed many insights into basic mechanisms of the immune system in the central nervous system (CNS). Compared with psychoneuroimmunology, neuroimmunology was methodologically advanced. Those mechanisms, e.g., in multiple sclerosis, were regularly discussed at the Pula meetings. Moreover, Karl Bechter consistently made contributions on inflammatory aspects of diagnosis and treatment of psychiatric disorders, e.g., on the role of cerebrospinal fluid (CSF) analysis in the diagnosis of psychosis and depression and on new therapeutic techniques based on immunological findings. He developed the interesting CSF filtration technique as a therapeutic tool in certain cases of severe psychosis and treatment-resistant, severe depression and reported on his findings and therapeutic advances.

Mood Disorders and Immunity

Psychoneuroimmunology was a key focus in Pula for the first time at the 48th Neuropsychiatric Congress Pula 2008. The meeting included two key symposia with speakers such as Brian Leonard (Ireland), Francesco Benedetti (Milan), Hemmo Drexhage (Rotterdam), Karl Bechter (Ulm), and Markus Schwarz, Aye Mu Myint, and Norbert Müller (Munich). The symposia discussed the scientific prerequisites and theoretical background of the role of the immune system, including immune genetics and the role of inflammation in mood disorders. They presented the large multicenter Europe-wide research program MOODINFLAME, which included a study on anti-inflammatory treatment in mood disorders.

At that time, research was establishing a more advanced view of the role of the immune system in the pathogenesis of mood disorders such as major depression or bipolar disorder, and several findings supported the research. A major starting point was sickness behavior as an animal model for major depression: The model included the behavioral, vegetative, cognitive, and emotional reaction of an organism to infection and inflammation [12]. Increasing evidence from animal models and clinical studies indicated that sickness behavior represents a highly differentiated adjustment reaction with the aim to specifically fight an infection and promote survival [13, 14]. The pro-inflammatory cytokines interleukin 1 beta (IL-1β), IL-6, and tumor necrosis factor alpha (TNF-α), which are released by macrophages during the early innate immune response, play a key role in sickness behavior and are involved in the communication between the peripheral immune system and the CNS [12, 14]. A peripheral immune response, e.g., induced by macrophages, results in increased production and release of cytokines in the CNS. Cytokines can convey signals to the CNS via different routes, including activation of afferent vagal fibers that project to the core of the solitary tract and higher viscerosensory centers via cytokine-specific transport molecules that are expressed in CNS endothelial tissue. Moreover, the circumventricular organs lack a blood-brain barrier [12]. As soon as they reach the CNS, signals from cytokines can be amplified through the network of central cytokines, which has important effects on neurotransmitter metabolism, neuroendocrine functions, synaptic plasticity, and behavior [15].

In humans, the involvement of cytokines in the regulation of sickness behavior has been studied by administering the bacterial endotoxin lipopolysaccharide to healthy volunteers [13, 14]. Studies showed that levels of anxiety, depression, and cognitive impairment are related to the levels of circulating cytokines [16, 17].

A high blood level of C-reactive protein (CRP) is a valid and common marker for an inflammatory process. Higher than normal CRP levels have been repeatedly observed in depression, for example in severely depressed inpatients [18], and high

CRP levels have been found to be associated with the severity of depression [19]. Elevated CRP levels were also observed in remitted patients after a depressive state, in both men [20, 21] and women [22, 23]. In a sample of older healthy persons, CRP levels (and IL-6 levels) were predictive of cognitive symptoms of depression 12 years later [24].

Characteristics of immune activation in major depression include increased numbers of circulating lymphocytes and phagocytic cells; upregulated serum levels of markers of immune activation (neopterin, soluble IL-2 receptors); higher serum concentrations of positive acute phase proteins (APPs), coupled with reduced levels of negative APPs; and increased release of proinflammatory cytokines, such as IL-1β, IL-2, IL-6, and TNF-α through activated macrophages and interferon-γ through activated T cells [25–28].

In major depression, increased numbers of peripheral mononuclear cells have been described by various research groups [29, 30]. In agreement with the findings of increased monocytes and macrophages, an increased level of neopterin has also been described [31–35]. The role of cellular immunity, cytokines, and the innate and adaptive immune systems in depression has been reviewed elsewhere [35].

Recent interesting studies from the MOODINFLAME project could show, that premature ageing of certain immune cells might be characteristic for major depression [36, 37].

Tryptophan/Kynurenine Metabolism and Psychoneuroimmunology

A second symposium at the Pula meeting in 2008 focused on research into the role of tryptophan/kynurenine metabolism. This metabolism is driven by pro- and anti-inflammatory cytokines and has a direct influence on melatonergic, serotonergic, and glutamatergic neurotransmission through different neuroactive metabolites, such as kynurenic acid (KYNA) and quinolinic acid (QUIN) [38, 39]. The production of KYNA is regulated by indoleamine 2,3-dioxygenase (IDO) and tryptophan 2,3-dioxygenase. The relationship between components of kynurenine metabolism and psychiatric disorders were discussed in Pula and elsewhere by using the example of schizophrenia: Tryptophan metabolism influences serotonergic and glutamatergic neurotransmission via the activation or inhibition of the enzyme IDO. Cytokines of the activated immune system activate IDO in tryptophan/kynurenine metabolism, whereas other cytokines have an inhibitory effect. The inhibition of IDO by the catalyzing step from serotonin to formyl-5-hydroxykynuramine increases the availability of serotonin, whereas the neuroactive metabolites of tryptophan/kynurenine metabolism KYNA and QUIN act as an N-methyl-D-aspartate (NMDA) receptor antagonist and agonist, respectively. KYNA is one of the neuroactive intermediate products of the kynurenine pathway and the only known naturally occurring NMDA receptor antagonist in the human CNS. Kynurenine is the primary major degradation product of tryptophan. It is excitatory, and its metabolites 3-hydroxykynurenine and QUIN are synthesized from kynurenine during metabolism to nicotinamide adenine dinucleotide, but KYNA is formed in a dead-end side arm of the pathway [40]. KYNA acts both as a blocker of the glycine co-agonistic site of the NMDA receptor [41] and as a noncompetitive inhibitor of the $\alpha 7$ nicotinic acetylcholine receptor [42]. Both enzymes catalyze the first step in the pathway, the degradation from tryptophan to kynurenine. Stimulating cytokines such as interferon-γ or TNF-α increase the activity of IDO [Grohmann et al., 2001]. However, QUIN exhibits neurotoxic effects. In accordance with the glutamatergic depletion hypothesis of schizophrenia, studies point to an overweight of KYNA over QUIN in the CSF [43] and microglial cells of the hippocampus [44]. The function of the KYN metabolites and the findings in schizophrenia and major depression make these metabolites interesting candidates for therapeutic interventions in major psychiatric disorders. However, so far only studies in animal models have been reported. One topic of interest is whether a direct effect on inflammation may have a better therapeutic impact than the indirect effects of the KYN metabolites.

Neuroimaging and Psychoneuroimmunology

The described loss of CNS volume and the activation of microglia, both of which have been clearly demonstrated in neuroimaging studies in schizophrenia and major depression, match the assumption of a (low level) inflammatory neurotoxic process. The group of Francesco Benedetti from Milan presented many cutting-edge data on the relationship between immune components such as cytokines and neuroimaging results in psychiatric disorders, with a special emphasis on bipolar disorder [45].

In the subsequent years, lectures, seminars, and workshops on the topic of psychoneuroimmunology were part of the program in Pula. A pro-inflammatory immune state was discussed not only in schizophrenia, but also in major depression and bipolar disorder. On the one hand, clinical influences such as body mass index, smoking, sleep, acuity and chronicity of the psychiatric disorder, aspects of pharmacological treatment, disease course and stage, and age, were topics of research and provoked a more differentiated view on immunological findings. On the other hand, a better insight into the highly complex interactions of the cellular and humoral immune systems, cytokines, the role of the blood-brain barrier, and the influence of pro- and anti-inflammatory components of the immune system, including immune genetics, advanced the knowledge in the field and were reflected in the Pula program.

The results were broadly discussed, e.g., at the 2016 Mind and Brain Congress in Pula, where neuroinflammation and neurorepair were core themes. The role of neuroinflammation in the autonomous nervous system and the mild encephalitis hypothesis were discussed by Jürgen Bär from Jena and Karl Bechter from Ulm [46]. Francesco Benedetti presented important new results on the role of inflammation in bipolar disorder [45]. His work showed a clear, strong relationship between findings from new neuroimaging methods and inflammatory markers in the blood in some psychiatric disorders.

Anti-Inflammatory Treatment in Psychiatric Disorders

The Munich group, including myself [47, 48], showed data from clinical trials on anti-inflammatory effects of therapeutic compounds in psychiatric disorders. First clinical treatment trials were performed in the early 2000s. The research focused on therapeutic studies in major depression, schizophrenia, and bipolar disorder and focused on immunomodulatory, anti-infective, or anti-inflammatory compounds. The therapeutic trials in these psychiatric disorders were mainly funded by the Theodore and Vada Stanley Research Foundation program (USA), while other funding programs focused on the elucidation of the pathophysiological mechanisms of the psycho-immune interaction.

Successful therapeutic studies on modulation of the immune system in psychiatric disorders validated the psychoneuroimmunological approach from a different perspective when they proved a benefit of anti-inflammatory drugs in patients with schizophrenia. For example, the cyclooxygenase-2 (COX-2) inhibitor celecoxib showed a therapeutic benefit in schizophrenia. Interestingly, COX-2 inhibition also affected cognition in schizophrenia [48]. The efficacy of treatment with a COX-2 inhibitor seems to be most pronounced in the first years of the schizophrenic disease process [47]. Further studies demonstrated a beneficial effect of acetylsalicylic acid and other anti-inflammatory compounds in schizophrenia and schizophrenic spectrum disorders [49]. A meta-analysis of the clinical effects of nonsteroidal anti-inflammatory drugs in schizophrenia showed significant effects on schizophrenic total, positive, and negative symptoms [50], although this effect appeared to be dependent on the duration of the disease and was most pronounced in first-episode schizophrenia [51], findings that are in accordance with an inflammatory origin of the disease.

In parallel, many studies on anti-inflammatory compounds were performed in major depression. As in schizophrenia, a benefit of anti-inflammatory treatment was also found for celecoxib [52, 53]. In a subgroup of patients with major depression who showed signs of inflammation, i.e., increased CRP, the TNF-alpha antagonist infliximab showed advantages over placebo [54].

Outlook

To develop more targeted, personalized therapies for psychiatric disorders, further research needs to not only evaluate the exact immunological mechanisms of the pathophysiology of a disease but also optimize the anti-inflammatory treatment approach.

In the last few years, an important role of immunological and inflammatory mechanisms has also been discussed for dementia, in particular Alzheimer's disease. An example of the interdisciplinary character and benefit of the Pula congresses were the main themes of the 57th Mind and Brain congress in 2017:

depression, multiple sclerosis, and dementia. Immune therapy and prophylaxis with interferons and other immunomodulators play a central role in multiple sclerosis, and in Alzheimer's disease, treatment with monoclonal antibodies against amyloid plaques is being developed and an immune pathogenesis is being discussed. In depression, anti-inflammatory treatment trials have shown encouraging results. Moreover, other disorders, such as Tourette's syndrome and obsessive-compulsive disorder—both syndromes that are partly triggered by infections— were topics in Pula. Pediatric acute-onset neuropsychiatric syndrome is a disorder triggered by *Streptococcus* that presents with tics, obsessive-compulsive symptoms, or both in children; however, this mechanism may also play a role in adults. The syndrome is treated by antibiotics or immunomodulating therapies, such as corticosteroids, plasmapheresis, or intravenous immunoglobulins.

In conclusion, neuroimmunology has been a topic in Pula for decades. In the beginning, it was included mainly as a neurological topic, particularly with respect to multiple sclerosis research. Subsequently, psychoneuroimmunology increasingly becoming a focus of interest as a result of increasing knowledge about immunology and the interactions between the immune and nervous systems, methodological progress, and increasing scientific funding. I am sure that the progress in the field will continue to be enthusiastically presented by leading researchers and experts and critically discussed at future meetings in Pula, and I look forward to following the future developments and contributing actively at Pula congresses to the dissemination of recent scientific results from the field.

References

1. Ader R, Felten DL. Psychoneuroimmunology. San Diego, New York, Toronto: Academic; 1991.
2. Besedovsky H, del Rey A, Sorkin E, Da Prada M, Burri R, Honegger C. The immune response evokes changes in brain noradrenergic neurons. Science. 1983;221(4610):564–6.
3. Maes M, Bosmans E, Suy E, Vandervorst C, De Jonckheere C, Raus J. Immune disturbances during major depression: upregulated expression of interleukin-2 receptors. Neuropsychobiology. 1990;24(3):115–20.
4. Maes M, Lambrechts J, Bosmans E, Jacobs J, Suy E, Vandervorst C, et al. Evidence for a systemic immune activation during depression: results of leukocyte enumeration by flow cytometry in conjunction with monoclonal antibody staining. Psychol Med. 1992;22(1):45–53.
5. Muller N, Ackenheil M, Eckstein R, Hofschuster E, Mempel W. Reduced suppressor cell function in psychiatric patients. Ann N Y Acad Sci. 1987;496:686–90.
6. Muller N, Ackenheil M, Hofschuster E, Mempel W, Eckstein R. Cellular immunity in schizophrenic patients before and during neuroleptic treatment. Psychiatry Res. 1991;37(2):147–60.
7. Muller N, Hofschuster E, Ackenheil M, Mempel W, Eckstein R. Investigations of the cellular immunity during depression and the free interval: evidence for an immune activation in affective psychosis. Prog Neuro-Psychopharmacol Biol Psychiatry. 1993;17(5):713–30.
8. Meyer U, Schwendener S, Feldon J, Yee BK. Prenatal and postnatal maternal contributions in the infection model of schizophrenia. Exp Brain Res. 2006;173(2):243–57. https://doi.org/10.1007/s00221-006-0419-5.

9. Muller N, Ackenheil M. Immunoglobulin and albumin content of cerebrospinal fluid in schizophrenic patients: relationship to negative symptomatology. Schizophr Res. 1995;14(3):223–8. https://doi.org/10.1016/0920-9964(94)00045-a.

10. Muller N, Riedel M, Hadjamu M, Schwarz MJ, Ackenheil M, Gruber R. Increase in expression of adhesion molecule receptors on T helper cells during antipsychotic treatment and relationship to blood-brain barrier permeability in schizophrenia. Am J Psychiatry. 1999;156(4):634–6.

11. Dantzer R, Kelley KW. Stress and immunity: an integrated view of relationships between the brain and the immune system. Life Sci. 1989;44(26):1995–2008.

12. Dantzer R, O'Connor JC, Freund GG, Johnson RW, Kelley KW. From inflammation to sickness and depression: when the immune system subjugates the brain. Nat Rev Neurosci. 2008;9(1):46–56. https://doi.org/10.1038/nrn2297.

13. Reichenberg A, Kraus T, Haack M, Schuld A, Pollmacher T, Yirmiya R. Endotoxin-induced changes in food consumption in healthy volunteers are associated with TNF-alpha and IL-6 secretion. Psychoneuroendocrinology. 2002;27(8):945–56.

14. Reichenberg A, Yirmiya R, Schuld A, Kraus T, Haack M, Morag A, et al. Cytokine-associated emotional and cognitive disturbances in humans. Arch Gen Psychiatry. 2001;58(5):445–52.

15. Raison CL, Capuron L, Miller AH. Cytokines sing the blues: inflammation and the pathogenesis of depression. Trends Immunol. 2006;27(1):24–31. https://doi.org/10.1016/j.it.2005.11.006.

16. Danner M, Kasl SV, Abramson JL, Vaccarino V. Association between depression and elevated C-reactive protein. Psychosom Med. 2003;65(3):347–56.

17. Hafner S, Baghai TC, Eser D, Schule C, Rupprecht R, Bondy B, et al. C-reactive protein is associated with polymorphisms of the angiotensin-converting enzyme gene in major depressed patients. J Psychiatr Res. 2008;42(2):163–5. https://doi.org/10.1016/j.jpsychires.2007.02.002.

18. Ford DE, Erlinger TP. Depression and C-reactive protein in US adults: data from the third National Health and Nutrition Examination Survey. Arch Intern Med. 2004;164(9):1010–4. https://doi.org/10.1001/archinte.164.9.1010.

19. Kling MA, Alesci S, Csako G, Costello R, Luckenbaugh DA, Bonne O, et al. Sustained low-grade pro-inflammatory state in unmedicated, remitted women with major depressive disorder as evidenced by elevated serum levels of the acute phase proteins C-reactive protein and serum amyloid A. Biol Psychiatry. 2007;62(4):309–13. https://doi.org/10.1016/j.biopsych.2006.09.033.

20. Cizza G, Eskandari F, Coyle M, Krishnamurthy P, Wright EC, Mistry S, et al. Plasma CRP levels in premenopausal women with major depression: a 12-month controlled study. Horm Metab Res. 2009;41(8):641–8. https://doi.org/10.1055/s-0029-1220717.

21. Gimeno D, Marmot MG, Singh-Manoux A. Inflammatory markers and cognitive function in middle-aged adults: the Whitehall II study. Psychoneuroendocrinology. 2008;33(10):1322–34. https://doi.org/10.1016/j.psyneuen.2008.07.006.

22. Maes M, Meltzer HY, Bosmans E, Bergmans R, Vandoolaeghe E, Ranjan R, et al. Increased plasma concentrations of interleukin-6, soluble interleukin-6, soluble interleukin-2 and transferrin receptor in major depression. J Affect Disord. 1995;34(4):301–9.

23. Maes M, Meltzer HY, Buckley P, Bosmans E. Plasma-soluble interleukin-2 and transferrin receptor in schizophrenia and major depression. Eur Arch Psychiatry Clin Neurosci. 1995;244(6):325–9.

24. Irwin M. Immune correlates of depression. Adv Exp Med Biol. 1999;461:1–24. https://doi.org/10.1007/978-0-585-37970-8_1.

25. Nunes SO, Reiche EM, Morimoto HK, Matsuo T, Itano EN, Xavier EC, et al. Immune and hormonal activity in adults suffering from depression. Braz J Med Biol Res. 2002;35(5):581–7.

26. Müller N, Schwarz MJ. Immunology in anxiety and depression. In: Kasper S, den Boer JA, Sitsen JMA, editors. Handbook of depression and anxiety. New York: Marcel Dekker; 2002. p. 267–88.

27. Mikova O, Yakimova R, Bosmans E, Kenis G, Maes M. Increased serum tumor necrosis factor alpha concentrations in major depression and multiple sclerosis. Eur Neuropsychopharmacol. 2001;11(3):203–8.

28. Herbert TB, Cohen S. Depression and immunity: a meta-analytic review. Psychol Bull. 1993;113(3):472–86.
29. Seidel A, Arolt V, Hunstiger M, Rink L, Behnisch A, Kirchner H. Major depressive disorder is associated with elevated monocyte counts. Acta Psychiatr Scand. 1996;94(3):198–204.
30. Rothermundt M, Arolt V, Fenker J, Gutbrodt H, Peters M, Kirchner H. Different immune patterns in melancholic and non-melancholic major depression. Eur Arch Psychiatry Clin Neurosci. 2001;251(2):90–7.
31. Duch DS, Woolf JH, Nichol CA, Davidson JR, Garbutt JC. Urinary excretion of biopterin and neopterin in psychiatric disorders. Psychiatry Res. 1984;11(2):83–9.
32. Dunbar PR, Hill J, Neale TJ, Mellsop GW. Neopterin measurement provides evidence of altered cell-mediated immunity in patients with depression, but not with schizophrenia. Psychol Med. 1992;22(4):1051–7.
33. Maes M, Scharpe S, Meltzer HY, Okayli G, Bosmans E, D'Hondt P, et al. Increased neopterin and interferon-gamma secretion and lower availability of L-tryptophan in major depression: further evidence for an immune response. Psychiatry Res. 1994;54(2):143–60.
34. Bonaccorso S, Lin AH, Verkerk R, Van Hunsel F, Libbrecht I, Scharpe S, et al. Immune markers in fibromyalgia: comparison with major depressed patients and normal volunteers. J Affect Disord. 1998;48(1):75–82.
35. Muller N, Myint AM, Schwarz MJ. Inflammatory biomarkers and depression. Neurotox Res. 2011;19(2):308–18. https://doi.org/10.1007/s12640-010-9210-2.
36. Schiweck C, Valles-Colomer M, Arolt V, Muller N, Raes J, Wijkhuijs A, et al. Depression and suicidality: a link to premature T helper cell aging and increased Th17 cells. Brain Behav Immun. 2020;87:603–9. https://doi.org/10.1016/j.bbi.2020.02.005.
37. Simon MS, Schiweck C, Arteaga-Henriquez G, Poletti S, Haarman BCM, Dik WA, et al. Monocyte mitochondrial dysfunction, inflammaging, and inflammatory pyroptosis in major depression. Prog Neuro-Psychopharmacol Biol Psychiatry. 2021;111:110391. https://doi.org/10.1016/j.pnpbp.2021.110391.
38. Kim YK, Myint AM, Verkerk R, Scharpe S, Steinbusch H, Leonard B. Cytokine changes and tryptophan metabolites in medication-naive and medication-free schizophrenic patients. Neuropsychobiology. 2009;59(2):123–9. https://doi.org/10.1159/000213565.
39. Linderholm KR, Skogh E, Olsson SK, Dahl ML, Holtze M, Engberg G, et al. Increased levels of kynurenine and kynurenic acid in the CSF of patients with schizophrenia. Schizophr Bull. 2012;38(3):426–32. https://doi.org/10.1093/schbul/sbq086.
40. Schwarcz R, Pellicciari R. Manipulation of brain kynurenines: glial targets, neuronal effects, and clinical opportunities. J Pharmacol Exp Ther. 2002;303(1):1–10. https://doi.org/10.1124/jpet.102.034439.
41. Kessler M, Terramani T, Lynch G, Baudry M. A glycine site associated with N-methyl-D-aspartic acid receptors: characterization and identification of a new class of antagonists. J Neurochem. 1989;52(4):1319–28.
42. Hilmas C, Pereira EF, Alkondon M, Rassoulpour A, Schwarcz R, Albuquerque EX. The brain metabolite kynurenic acid inhibits alpha7 nicotinic receptor activity and increases non-alpha7 nicotinic receptor expression: physiopathological implications. J Neurosci. 2001;21(19):7463–73.
43. Kegel ME, Bhat M, Skogh E, Samuelsson M, Lundberg K, Dahl ML, et al. Imbalanced kynurenine pathway in schizophrenia. Int J Tryptophan Res. 2014;7:15–22. https://doi.org/10.4137/IJTR.S16800.
44. Gos T, Myint AM, Schiltz K, Meyer-Lotz G, Dobrowolny H, Busse S, et al. Reduced microglial immunoreactivity for endogenous NMDA receptor agonist quinolinic acid in the hippocampus of schizophrenia patients. Brain Behav Immun. 2014;41:59–64. https://doi.org/10.1016/j.bbi.2014.05.012.
45. Benedetti F, Aggio V, Pratesi ML, Greco G, Furlan R. Neuroinflammation in Bipolar Depression. Front Psychiatry. 2020;11:71. https://doi.org/10.3389/fpsyt.2020.00071.

46. Bechter K. Schizophrenia--a mild encephalitis? Fortschr Neurol Psychiatr. 2013;81(5):250–9. https://doi.org/10.1055/s-0033-1335253.
47. Muller N, Krause D, Dehning S, Musil R, Schennach-Wolff R, Obermeier M, et al. Celecoxib treatment in an early stage of schizophrenia: results of a randomized, double-blind, placebo-controlled trial of celecoxib augmentation of amisulpride treatment. Schizophr Res. 2010;121(1–3):118–24. https://doi.org/10.1016/j.schres.2010.04.015.
48. Muller N, Riedel M, Schwarz MJ, Engel RR. Clinical effects of COX-2 inhibitors on cognition in schizophrenia. Eur Arch Psychiatry Clin Neurosci. 2005;255(2):149–51. https://doi.org/10.1007/s00406-004-0548-4.
49. Laan W, Grobbee DE, Selten JP, Heijnen CJ, Kahn RS, Burger H. Adjuvant aspirin therapy reduces symptoms of schizophrenia spectrum disorders: results from a randomized, double-blind, placebo-controlled trial. J Clin Psychiatry. 2010;71(5):520–7. https://doi.org/10.4088/JCP.09m05117yel.
50. Zheng W, Cai DB, Yang XH, Ungvari GS, Ng CH, Muller N, et al. Adjunctive celecoxib for schizophrenia: a meta-analysis of randomized, double-blind, placebo-controlled trials. J Psychiatr Res. 2017;92:139–46. https://doi.org/10.1016/j.jpsychires.2017.04.004.
51. Nitta M, Kishimoto T, Muller N, Weiser M, Davidson M, Kane JM, et al. Adjunctive use of nonsteroidal anti-inflammatory drugs for schizophrenia: a meta-analytic investigation of randomized controlled trials. Schizophr Bull. 2013;39(6):1230–41. https://doi.org/10.1093/schbul/sbt070.
52. Muller N, Schwarz MJ, Dehning S, Douhe A, Cerovecki A, Goldstein-Muller B, et al. The cyclooxygenase-2 inhibitor celecoxib has therapeutic effects in major depression: results of a double-blind, randomized, placebo controlled, add-on pilot study to reboxetine. Mol Psychiatry. 2006;11(7):680–4. https://doi.org/10.1038/sj.mp.4001805.
53. Kohler O, Benros ME, Nordentoft M, Farkouh ME, Iyengar RL, Mors O, et al. Effect of anti-inflammatory treatment on depression, depressive symptoms, and adverse effects: a systematic review and meta-analysis of randomized clinical trials. JAMA Psychiatry. 2014;71(12):1381–91. https://doi.org/10.1001/jamapsychiatry.2014.1611.
54. Raison CL, Rutherford RE, Woolwine BJ, Shuo C, Schettler P, Drake DF, et al. A randomized controlled trial of the tumor necrosis factor antagonist infliximab for treatment-resistant depression: the role of baseline inflammatory biomarkers. JAMA Psychiatry. 2013;70(1):31–41. https://doi.org/10.1001/2013.jamapsychiatry.4.

Chapter 16
Therapeutic Interventions in Psycho-Neuro-Endocrino-Immunology (PNEI)

Sanja Toljan

Introduction

Established 40 years ago by Robert Ader and Nicholas Cohen, a psychoneuroimmunology is defined as the study of brain–immune system interactions. That is, psychoneuroimmunology addresses the integrated nature of the relationships among behavioral, neural, endocrine, and immune responses that enable an organism to adapt to the environment in which it lives. This is what medical schools used to call systemic physiology [1–4]. Science and technology moved their thinking towards interconnectivity, which at that time was revolutionary that they had to establish the journal to publish scientific results of their experiments. Brain, Behavior and Immunity was meant to freely publish academic work which used the new paradigm of interconnectivity and networks, what is today very well-known and established, but unfortunately still therapeutically neglected [1].

If not translated into clinical practice, what is the purpose of basic scientific research? However, many clinicians are concerned that most translational discoveries may remain hypothetical if they are not translated into clinical trials [5]. To overcome this common problem, the term Translational Medicine (TM) was coined in 1990s: TM science should be defined as any activity that generates new discoveries or observations, which help to form our knowledge of the human body and its interactions with the environment. It should also carry a clear hope of attaining novel achievements for the benefit of human health. TM science must aim for principles, such as objectivity and reproducibility [6].

S. Toljan (✉)
Clinics for Otorhinolaryngology, Surgery and Anesthesiology "Orlando", Zagreb, Croatia
e-mail: info@poliklinika-orlando.hr

V. Demarin et al. (eds.), *Mind, Brain and Education*,
https://doi.org/10.1007/978-3-031-33013-1_16

Psychoneuroimmunology is a classic example of scientific field still waiting to become clinical discipline. Despite enormous scientific facts explaining the multi-directional link between neural, endocrine, and immune systems and personal ecology of individuum, there is still lack of bedside application. Clinical psycho-neuro-endocrino-immunology (PNEI) would be an example how to propagate the science to clinics and improve human health, which should be the ultimate goal of medical research.

Clinical psycho-neuro-endocrino-immunology takes human being as a whole, with the environmental impact, but also with memory, emotions and feelings. This scientific paradigm can deeply change medical and psychological sciences and clinical practice, integrating psychology and medicine, being artificially divided. Therefore, every patient, when seeing a doctor, a surgeon, a psychologist, or another therapist, should receive an integrated approach, an analysis considering and combining examinations with biological and psychological evaluations.

At the same time, the therapist should know the effectiveness, limits, and risks of each of the possible therapeutic proposals and, therefore, know that not only drugs, but, social support, psychotherapy, meditation, physical exercise, and body manipulations can also affect care. Indeed, their combination, adapted to the individual patient, may have surprising synergistic effects [7, 8].

Diagnosis in PNEI

The ultimate diagnostic criterion in PNEI is to evaluate the „wear and tear "of individual organism, known as allostatic load. Using the word "stress" does not really recognize all of the underlying biology, while the word "allostasis" focuses on the active process of adaptation to many challenges, whether or not we call them stressful. "Allostatic load" is a term that refers to the cumulative changes in the body and brain that are produced by dysregulation and overuse of the "mediators" of allostasis [8–10].

More recently, clinical criteria for the determination of allostatic load, that provide information on the underlying individual experiential causes, have been developed and used in a number of investigations. These clinimetric tools can increase the number of people screened, while putting the use of biomarkers in a psychosocial context. The criteria allow the personalization of interventions to prevent or decrease the negative impact of toxic stress on health, with particular reference to lifestyle modifications and cognitive behavioral therapy [11].

The biological model of allostatic load focuses on glucocorticoid dysregulation as part of a network of mediators involving autonomic, endocrine, metabolic, and inflammatory parameters, such as: resting systolic and diastolic blood pressure, body mass index, waist-hip ratio, high-density lipoprotein and low-density lipoprotein cholesterol, triglycerides, glycosylated hemoglobin, fasting glucose, plasma C-reactive protein, fibrinogen, serum measures of interleukin-6, the soluble adhesion molecules E-selectin, intracellular adhesion molecule-1, levels of urinary epinephrine, norepinephrine, cortisol diurnal curve and a serum measure of the hormone dehydroepiandrosterone sulfate (DHEA-S) [11].

Besides, determination of allostatic load should deal with specification of the stressor, uncontrolled or undesirable events in patient's life and their impact on patient's living conditions, social and family circle and work. This index of allostatic load was found to be better predictor of mortality and decline in physical functioning than individual biomarkers alone [12].

Chronic Low Grade Inflammation Modulation

PNEI deals with general and widespread inflammation, which is recognized by general and specific symptoms, known as sickness behavior and lack of optimal function. The mechanism behind chronic low grade inflammation gives us insight into complicated neuro-humoral interaction and biochemistry of maintaining balance between pro-inflammatory and anti-inflammatory states, both being equally important.

Inflammation is dynamic process which is highly modifiable, and thus represents the core of therapeutic interventions in PNEI. To modify inflammatory response, one needs to know the variables. According to Furman et al., the most prominent factors in modulating inflammation across the life span are physical activity, diet, nighttime blue light exposure, tobacco smoking, environmental and industrial toxicants exposure and psychological stress [13].

Modifiable therapeutic targets in PNEI
1. Circadian rhythm
2. Nutrition
3. Physical activity
4. Stress resilience

Strategies to obtain goals
1. Circadian rhythm maintenance
   Regulation of wake/sleep, eating/fasting, rest/activity cycle
2. Adequate nutrition
   Nutritional requirements to sustain healthy microbiome and gut - brain axis.
3. Upgrade resilience
   Coping with daily stressors, psychosocial therapies
4. Low dose medicine

Circadian Rhythm Maintenance

Human tissues and cells have an autonomous circadian oscillator with a period of roughly 24 h, dependent by external stimulations or "zeitgebers". The most important control of circadian rhythm is obtained by light, food and physical activity [14]. External timing cues, also called "synchronizers," "zeitgebers," or "entraining agents" can reset the body's circadian clock and place all cells at the same phase of circadian oscillation, a process called circadian rhythm synchronization. Moreover, internal signals such as circulating hormones, cytokines and metabolites are significant timing cues that regulate peripheral clocks [14, 15].

In mammals, many physiological functions under the regulation of the circadian clock are affected by external cues such as sleep and wakefulness, alertness and motor ability, body temperature fluctuations, the urinary system, hormone secretion, immune regulation, cytokine release, and cell cycle progression [15, 16]. When circadian systems are disrupted by various environmental or genetic defects, the result is dysfunction of various physiological processes which modulate the allostatic processes in the body [17, 18].

The incidence of pathologies such as depression, obesity, diabetes, cancer are connected to shift work and the reason why shift work was classified as probable human carcinogen [19, 20].

Control of proper function of circadian rhythm is obtained by regulating "zeitgebers". Meal timing, activity timing and sleep timing are key points in therapeutic aspect in psycho-neuro-endocrino-immunology.

Complex organizational structure of biochemical functioning in each and every human cell is cyclic in nature. The processes are being conducted according to set of peripheral proteins which form a hierarchy of oscillators that function at the cellular, tissue and systems level and are composed of at least three feedback loops [21].

The interplay between these three regulatory loops is at the core of circadian rhythmicity and clock-related gene expression [22].

When we think about epigenetics, as practitioners, we think about one's potential to promote change, to accept change and finally to do changing process *per se*.

One essential aspect of allostasis and allostatic load/overload is how the brain responds. We now know that genes are turned on and off epigenetically by experiences over the life course, and that there is an adaptive structural plasticity of synapses, some of which are eliminated while others are formed during the daily circadian day-night cycle, as well as following acute and chronic stressors [23].

Plasticity of human body and capacity to adapt is fundamental to therapeutic approach in PNEI. Acknowledging the mechanism of epigenetics and human organs plasticity, we set therapeutic procedures to enable organism to balance allostasis.

Setting the Wake Sleep Cycle

Adequate wake-sleep cycle is fundamental to a person's emotional and physical health. Inadequate sleep is a known risk factor for obesity, diabetes, heart disease and depression, inflammatory diseases with ever growing prevalence.

In recent years the role of chronic low-grade inflammation has been recognized as a biological driver of ill health associated with stress. Sleep deprivation and sleep loss have been reported to be associated with changes in multiple immune parameters that favor a pro-inflammatory state, which may underlie the negative health effects of sleep disruption. Many scientific evidences of sleep-induced immune perturbations along with evidence that pathogen-induced immune responses or damage-induced immune responses can negatively impact sleep architecture, highlight a bidirectional relationship between sleep and immunity [24].

Sleep disorders are a broad category of disorders that encompass all types of dysfunctions involving sleep, including difficulty falling asleep at night, poor sleep quality, early waking, circadian rhythm disorders, parasomnias, sleep-related movement disorders and sleep-related breathing disorders [25].

Conventional pharmacological treatments for various types of sleep disorders have substantial side effects, including excessive daytime sleepiness, poor tolerance to the medication, cognitive impairment, dependency and withdrawal, with no results in acquiring control over wake-sleep cycle. Conversely, drugs that are capable to influence the circadian rhythm positively are referred as chronobiotics. The prototype is melatonin, whose secretion is an 'arm' of the biologic clock, but also a chemical code of night: the longer the night, the longer the duration of its secretion [26].

Melatonin, natural chronobiotic, is widely used as a dietary supplement to alleviate the symptoms of disrupted circadian rhythms, such as jet lag. It is also available in a prolonged-release formulation for the treatment of primary insomnia. Melatonin is extremely safe in terms of side effects, but with poor pharmacokinetic properties, such as low oral bioavailability, due to high first pass metabolism and a short plasma half-life (~ 30 min). For this reason, a number of melatonin analogues have been synthesized, allowing at the same time to acquire information on structure-activity relationships and to obtain compounds selective for its binding sites. Currently, three MT1/MT2 nonselective melatonin receptor agonists were given marketing authorization: Ramelteon is approved for insomnia therapy, tasimelteon for the treatment of Non-24-Hour Sleep-Wake Disorder in blind people and agomelatine as antidepressant.

The systematic review and meta-analysis by Moon et al. (2022) was to summarize current knowledge about the effects of melatonin supplements and melatonin agonists on the sleep-wake cycle as well as on the circadian rhythm of melatonin in healthy participants and in patients with psychiatric disorders.

Trials on healthy participants demonstrated that specific melatonergic supplements and agonists advanced the phase of sleep-wake and circadian rhythms that were originally within the normal range. The meta-analysis of studies with sleep-wake cycle parameters showed that the exogenous melatonin and melatonergic agonists significantly advanced the phase of circadian melatonin rhythm. Given an advancing effect of melatonin on circadian parameters, shortening sleep latency might be caused by a chronobiotic effect such as a phase advance. Alternatively, these effects of exogenous melatonin and melatonergic agents on sleep parameters could be mediated by a hypnotic effect and/or sleep consolidation.

Meanwhile, studies with psychiatric patients reported stabilizing effects on circadian and sleep-wake rhythms, similar to the results in healthy participants. Among the three trials in patients with schizophrenia, one RCT with exogenous melatonin for 3 weeks improved sleep-wake cycle, and another RCT with add-on ramelteon medication enhanced circadian melatonin rhythm. A trial with ADHD also reported that ramelteon advanced the timing of sleep-wake cycle. However, the authors found that antipsychotics actually might be more harmful and rhythm disrupting and because of the negative impact of selective serotonergic agents which has been recognized earlier, and the contraindications of the use of benzodiazepines, they made conclusion that melatonergic agents have shown potential efficacy in and are the most promising agents for correcting disrupted sleep-wake and circadian rhythms [27].

Sleep Training and Sleep Hygiene

Sleep hygiene (SH) is a set of recommendations for the promotion of healthy sleep that includes behavioral and environmental components. Although SH is not an optimal choice for sleep health improvement, it is commonly used and has the potential to be an effective strategy. There is not a uniform consensus on what behaviors constitute SH, but recommendations mostly include limit daytime naps, avoid alcohol, caffeine and nicotine use, regularly participate in physical activity, reduce technology use before bed, and establish a relaxing sleep environment. Sleep hygiene education (SHE) programs tend to focus on the dissemination of knowledge related to the importance of sleep (e.g., implications for health) and how sleep hygiene practices can improve sleep quality.

Astonishing results are shown in the study by Mindell et al. (2017) indicating that sleep disturbances in infants and toddlers can be quickly ameliorated within just a few nights after implementation of a consistent bedtime routine, including a bath, massage, and quiet activities [28]. Household chaos is very often the reason of pediatric sleep disorders [29]. There are still significant gaps among pediatricians both in basic knowledge about pediatric sleep disorders, and in the translation of that knowledge into clinical practice. Despite their acknowledgment of the importance of sleep problems, many pediatricians fail to screen adequately for them, especially in older children and adolescents [30].

Among the common problems related to aging is sleep quality; over half of older adults suffer from symptoms of insomnia. Age-related changes in circadian sleep/wake regulation constitute a major underlying factor. Constant and organized lifestyle may moderate the effects of circadian rhythm changes on sleep. Maintenance of daily routines is associated with a reduced rate of insomnia in the elderly [31].

Cognitive Behavioral Therapy for Insomnia

Cognitive behavioral therapy for insomnia (CBT-I) is an effective nonpharmacological treatment that improves sleep outcomes with minimal adverse effects and is preferred by patients to drug therapy. The approach to CBT-I has been refined in recent years, and it is now most commonly studied as a combined cognitive and behavioral treatment incorporating some or all of five components. (cognitive therapy, stimulus control, sleep restriction, sleep hygiene, relaxation). Meta-analysis showed that it significantly improved all sleep parameters (sleep onset latency, wake after sleep onset and sleep efficiency) and should be now commonly recommended as first-line treatment for chronic insomnia [32].

Circadian rhythm disruption is a condition which needs extensive treatment due to its enormous contribution to aggravating general inflammation. Pharmacological and nonpharmacological measures should be equally applied in trying to set the rhythm and improve wake-sleep control.

Setting the Fasting-Eating Cycle

The natural rhythm of most living organisms is to spend one phase of a 24-hour day in an active state and feeding and the other in a resting and fasting state. Humans should naturally spend the light phase in the active and feeding state and the dark phase in the resting and fasting state. Overall global electrification and artificial light at night syndrome (ALAN), deviated people from the original pattern of eating only during the light phase. Furthermore, night shifts workers experience an almost complete reversal of food intake, with intake occurring primarily during the night and rest and fasting occurring during daylight hours. Dietary intake that is misaligned with the natural rhythms of the circadian clock has been shown to negatively impact human metabolic health [33].

A misaligned circadian rhythm is one in which the normal schedule of feeding and fasting is disrupted [34]. Timed meals play a role in synchronizing peripheral circadian rhythms in humans and may have particular relevance for patients with circadian rhythm disorders (shift work, jet leg, social jet leg) [35].

The circadian rhythm of the body is established by the central clock located in the suprachiasmatic nucleus of the hypothalamus but also by clocks of peripheral organs. Although the master clock is strongly entrained by light, clocks of peripheral organs are additionally responsive to food supply, and temporal restriction of food can reset clock gene rhythms. Therefore, time of eating and nutrient delivery may have cardiometabolic health implications via alterations in peripheral clocks, most notably that of the liver.

All major diseases, including cardiovascular disease, diabetes, neurodegenerative disorders, arthritis, and cancers involve chronic inflammation in the affected tissues and, in many cases, systemically. Local tissue inflammation involves production of proinflammatory cytokines and reactive oxygen species. Overweight and obesity promote inflammation, and intermitent energy restriction (IER) suppresses inflammation in human subjects and animal models of diseases. Obese women who changed their diet from multiple daily meals to alternate-day energy restriction exhibited significant reductions in levels of circulating TNF and IL-6. In asthma patients, 2 month of alternate-day energy restriction reduced circulating TNF and markers of oxidative stress, and improved asthma symptoms and airway resistance. Multiple studies have shown that fasting can lessen symptoms in patients with rheumatoid arthritis, and data from animal studies suggest that the pathogenesis of other autoimmune disorders may also be counteracted by IER, including multiple sclerosis, lupus erythematosus, and type I diabetes. In a mouse model of stroke, IER suppressed elevations of TNF and IL-1β in the ischemic cerebral cortex and striatum, which was associated with improved functional outcome. Inflammation is increasingly recognized as a contributing factor for cancer cell growth and, because excessive energy intake promotes inflammation, it is likely that suppression of inflammation plays a role in the inhibition of tumor growth by IER [36].

The data reviewed by Scientific Statement From the American Heart Association [37] suggest that irregular patterns of total energy intake appear less favorable for the maintenance of body weight and optimal cardiometabolic health. Clinicians may be able to use this information to suggest to patients that a more intentional approach to eating that focuses on the timing and frequency of meals and snacks could be the basis of a healthier lifestyle and improved risk factor management. An intentional approach to eating requires eating at planned intervals to distribute total energy intake throughout the day:

1. Develop an intentional approach to eating that focuses on the timing and frequency of meals and snacks as the basis of a healthier lifestyle and improved risk factor management
2. Understand the patient's frame of reference in how he or she may define meals and snacks
3. Recommend distributing calories over a defined portion of the day
4. Recommend eating a greater share of the total calorie intake earlier in the day to have positive effects on risk factors for heart disease and diabetes mellitus
5. Promote consistent overnight fast periods
6. Link eating episodes to influence subsequent energy intake (eg, place snacks strategically before meals that might be associated with overeating)
7. Include intermittent fasting approaches as an option to help lower calorie intake and to reduce body weight
8. Use added eating episodes to introduce a wider variety of healthful food options and to displace less healthful foods
9. Use planned meals and snacks timed throughout the day to help manage hunger and to achieve portion control

Diet composition, but also food timing influences gut microbiota composition and activity. There is a bidirectional relationship between microbiota and food timing. Intestinal epithelia cells' internal circadian clock influences activity of the pituitary-adrenal axis and cortisol secretion, and this rhythm is influenced by microbiota status; and at the same time, an alteration of microbiota could lead to a disrupted corticosteroid circadian rhythm influencing food uptake. However, microbiota composition has its variability during the day that could be disrupted by a variety of conditions, for example, jet-lag or high-fat diets [38].

Recently it has been demonstrated that a chronic circadian misalignment in mice and a time shift jet–lag in humans induces a dysbiosis. On the other hand, maintaining a correct eating phase and increasing the fasting period could positively affect the gut microbiome, reducing gut permeability and improving systemic inflammation.

Much research in recent years suggests a positive health effect of a wide temporal fasting window during the day, i.e., limiting daily food intake to a ~ 6–8 h time window seems to induce, in humans, many health benefits compared to the normal daily meal distribution (i.e., three to five meals, spread from breakfast to late dinner), even in isocaloric conditions [38, 39]. It is clear that fasting, in general, exerts many positive effects on health, with some features in common with the caloric restriction approach (protects against diabetes, cancers, heart disease, and neurodegeneration; reduces obesity, hypertension, asthma, and rheumatoid arthritis) [40].

Based on the medical literature, several health-promoting recommendations can be delivered:

1. consuming a greater proportion of calories earlier in the day, which often involves breakfast consumption, as compared to consuming a large number of calories later at night.
2. There may also be benefits to extending the daily fasting period beyond a standard overnight fast or implementing occasional fasting periods.

In order to reconcile these two strategies, an individual could eat from breakfast until mid- to late-afternoon each day, thus aligning peripheral clocks with central clock.

Setting the Rest- Activity Cycle

Besides the influence of normal circadian variations of CBT(core body temperature) on exercise performance, the peripheral muscle clock might have a crucial impact on diurnal energy metabolism. Initial results on metabolomic and transcriptomic investigations demonstrated that exercise in the early active phase influences other metabolic processes than exercise in the early rest phase [41].

Anti-inflammatory effects of physical exercise were first noticed in chronic inflammatory diseases such as autoimmune arthritis or chronic obstructive lung disease. It was found that physical exercise decreases IL-6 and CRP levels in these patients.

Physical exercise was found to have slight to moderate effects on improving some biomarkers in breast and colon cancer patients including insulin, leptin, estrogens, inflammation, immune function and apoptosis regulation. There is a 25% average risk reduction among physically active women compared to the least active ones. Physical activity reduces lung cancer by 20–30% in women and 20–50% in men, possibly via improved pulmonary function, reduced concentrations of carcinogenic agents in the lungs, enhanced immune function, reduced inflammation, enhanced DNA repair capacity, changes in growth factor levels and possible gene-PA interactions. Similar mechanisms may also be involved in the preventive effect of physical exercise against colon cancer [42, 43].

Concerning exercise as "zeitgeber", it is an external time cue directly acting on the molecular clock as have been shown that skeletal muscle contractions have an impact on skeletal muscle circadian rhythmicity [44]. Exercise controls the molecular clock in skeletal muscle, and, in turn, the clock modulates physiological activity of this tissue. In fact, the time of day is important regarding skeletal muscle performances. In humans, studies have shown that skeletal muscle torque, strength and power is higher in the late afternoon, between 16:00 and 18:00 hours, compared to the morning. Determining temporal-specific training responses has major potential to improve sport performances, based on sleep-onset, sleep duration and wake up time. By classifying individuals into three groups, early circadian phenotype (ECT), intermediate circadian phenotype (ICT) and late circadian phenotype (LCT), the highest performances are obtained at different times of the day depending on the chronotype, i.e. around 12 h for ECT, 16 h for ICT and 19 h for LCT. This concept points out the importance of chronobiology in skeletal muscle activity. Running at the 'right' time is essential for optimal improvement of skeletal muscle performance and to simultaneously limit damage [45].

Skeletal muscle clocks may contribute to the circadian rhythms of human resting energy expenditure, with lowest levels seen in the late biological night and highest in the afternoon and evening [46].

In addition, prolonged restriction of physical activity can induce abnormal changes in biological rhythm. Therefore, appropriate exercise plays a vital role in maintaining regular circadian rhythm in the body. Although resistance exercise-induced skeletal muscle hypertrophy but its training time should be individually set. Exercise at the "right" time is essential for improving the functions of skeletal muscle. Abnormal timing for exercise can induce the shift change of circadian phase in skeletal muscle and even increase the risk of skeletal muscle atrophy and injury. Normal circadian rhythm can positively affect changes in the mass and function of skeletal muscle, and exercise combined with regulation of circadian rhythm can better improve the quality and function of skeletal muscle, and highly positively impact the overall health.

Adequate Nutrition

Anti-Inflammatory Diet

Nutrition is very potent modulator of inflammatory processes, exerted by nutrients and non-nutrient food components [47]. Inflammation is an innate defense response which activates immune cells and releases various soluble mediators such as chemokines, cytokines, eicosanoids, free radicals, and vasoactive amines. There is substantial amount of studies connecting diet with inflammation etiology, especially SAD standard american diet) or WD (western diet), characterized by high intakes of refined sugars, animal fats, processed meats, refined grains, conventionally-raised animal products, salt, eggs, potatoes, corn, and low consumption of complex carbohydrates from fruits, vegetables, beans, legumes, and whole grains, which are also sources of essential vitamins, minerals and essential phytochemicals [48]. The high intake of saturated fats, including excessive amounts of omega-6 polyunsaturated fatty acids (PUFA), small amounts of omega-3 PUFA, and an unhealthy omega-6/omega-3 ratio of 20:1 is especially harmful in terms of the metabolic consequences and immunomodulation.

Furthermore, food additives like emulsifiers or sweeteners, that are widely used in fat-based foods, are proven to impair gut barrier function [49], promote alterations in gut microflora, resulting in dysbiosis, gut barrier dysfunction, increased intestinal permeability, and leakage of toxic bacterial metabolites into the circulation, all of which contribute to the development of low-grade systemic inflammation. This phenomenon is also considered to be associated with inflammation-related ageing of organisms called "inflammaging" [50].

High fat diet (HFD), not only disturbs the peripheral circadian clock, but also significantly increases lypopolysacharide (LPS)-containing bacteria abundance in the gut [51], resulting in higher plasma endotoxin concentration, known as "metabolic endotoxemia". LPS on its own has multiple adverse effects on gut function, as it promotes intestinal inflammation, disrupts tight junction (TJ) organization via specific signaling pathways, directly causes intestinal epithelial cells shedding without compensatory TJ-resealing, and may induce oxidative stress in intestinal epithelial cells, mitophagy, and mitochondrial failure [52].

Gut microbiota obtains energy mainly by fermentation of non-digested carbohydrates like fiber, generating short chain fatty acids, SCFA. Acetate, propionate, and butyrate comprise 95% of SCFA in the colon and feces in humans. SCFA have multiple physiological functions in signaling brain-gut axis. They are an essential energy source for the colonic epithelium and liver gluconeogenesis and play a role in regulating energy metabolism and immune system modulation [53].

Moreover, dysbiosis and high dietary fat intake lead to bile acid (BA) alterations. BAs are mostly reabsorbed from the gut lumen, undergoing enterohepatic recirculation, with only 5 to 10% of BAs not reabsorbed and undergo biotransformation to secondary BAs. For this reason, the composition of BAs in the small intestine is similar to the biliary pool. In contrast, the BA profile in the colon differs significantly, as it comprises secondary BAs [54].

It has been shown that HFD increases total BAs and total secondary BAs along with elevated BA concentration in the caecum, while chronically high gut and fecal BA concentrations can reduce gut barrier integrity. Thus, the HFD-associated gut hyperpermeability can be related to increased BA secretion [55].

Humans rely on microbes to metabolize the indigestible components of food like dietary fibers [56]. Through interactions with the GI microbiota, the presence of fermentable dietary fibers in the diet may strengthen gut barrier function, thus preventing paracellular translocation of LPSs [57].

A reduction in endotoxemia and inflammation may be one mechanism by which increased consumption of dietary fiber improves metabolic health. An analysis of data from the 1999–2000 NHANES reported an inverse relation between dietary fiber intake and the risk of having elevated CRP, defined by the American Heart Association as a serum concentration > 3.0 mg/L, and a modest direct association of saturated dietary fat intake and the risk of elevated serum CRP, a potent pro-inflammatory cytokine [58].

A large prospective single-center cohort study examined the relation between consumption of a Mediterranean diet pattern and endotoxemia—the authors reported that plasma LPS concentrations were negatively associated with adherence to the Mediterranean diet pattern (measured by validated questionnaire) and with fruit and legume intake in adults with atrial fibrillation [59].

In addition to dietary fiber, high consumption of whole plant foods as part of the Mediterranean diet pattern provides an abundance of phytochemicals that also affect the GI microbiome, inflammation, and the development of metabolic disease. Phytochemicals, which include polyphenols, phenolic acids, flavonoids, carotenoids, and lignans [60], have been shown to decrease the risk of coronary heart disease, diabetes, and nonalcoholic fatty liver disease [61–63]. This reduction in risk may be due to interactions with the GI microbiome and immune system, but also in circadian rhythm allignment [64, 65].

Orally administered phytochemicals isolated from natural sources are strong external signals (zeitgebers) for the cell autonomous molecular system that are more convenient and stable than other zeitgebers for entraining circadian rhythms. Based on the close relationship between the circadian clock and metabolic diseases, as supported by various findings, the circadian clock has become a new target for the treatment of metabolic diseases by phytochemicals. Hence, the occurrence and development of metabolic diseases can be modulated through complex interactions between the circadian clock and dietary phytochemicals.

Food source chronobiotics may be readily absorbed in the small intestine. However, many of them remain almost unchanged in the colon depending on the structural complexity and polymerization, such as resveratrol, curcumin, capsaicin and catechins. Thus, dietary chronobiotics with poor bioavailability are probably give play to its health benefits via gut microbiota regulation [65].

Psycho-neuro-endocrino-immunological intervention by manipulating nutrition are possible, largely thanks to human microbiota function, thus feeding process should be regulated and medically controlled. Intestinal microbiom through its metabolomics strongly affects immunity modulation, which then affects the organism *in toto*.

Antiinflamatory diet is *conditio sine qua non* in PNEI therapeuticall approach, and along with circadian rhythm alignment, the main staple of health.

Psychosocial Interventions Addressing Stress Resilience

The stress was best defined by S. Cohen: "*The experience of negative events or the perceptions of distress and negative affect that are associated with the inability to cope with them*" [66]. Individual reaction to stress is not only the perception but also the individual response to the stressors, the way how individual copes the stressors shows the lack of uniformity in reaction to the same event. According to this, McEwen and his team have recognized that protective and damaging effects of the biologic response to stressors should be named allostasis and allostatic overload, respectively [67]. In particular, allostasis, distinguished from homeostasis, is an adaptive process that tries to maintain homeostasis by promoting the release of glucocorticoids, catecholamines, and cytokines. On the other hand, allostatic overload refers to the response to prolonged stress, mediated by many neuroendocrine mediators [68]. Finally, neural mechanisms influence how an individual copes with this situation determining either vulnerability or resilience [69, 70].

Stress has two different effects on immune responses: immune suppression to cause mortality by infection and cancer, and excessive immune activation to induce chronic inflammation and autoimmune disease. Consistently, stress-induced glucocorticoids strongly suppress cell-mediated immunity and cause viral infection and tumor development. They may also enhance the development of pathogenic helper T cells and cause tissue damage through neural and intestinal inflammation. Past studies have reported the positive and negative effects of glucocorticoids on the immune system. These opposing properties of glucocorticoids may regulate the immune balance between the responsiveness to antigens and excessive inflammation in steady-state and stress conditions [71].

The first systematic review and meta-analysis of randomized clinical trials that have examined the effects of a psychosocial intervention on immune system outcomes focused on 8 psychosocial intervention types: behavior therapy, cognitive therapy, CBT, CBT plus additive treatment or mode of delivery that augmented the CBT (eg, CBT plus benzodiazepines or phone/video sessions), bereavement or supportive therapy, multiple or combined interventions, other psychotherapy, and psychoeducation. In addition, study examined 7 immune outcomes that could be influenced by these interventions: proinflammatory cytokines (eg, interleukin-6) and markers (eg, C-reactive protein), anti-inflammatory cytokines (eg, interleukin-10), antibodies (eg, IgA), immune cell counts (eg, CD4), natural killer cell activity (eg, cytotoxicity), viral load (eg, HIV RNA), and other immune outcomes (eg, blastogenesis, number of postoperative infectious diseases). Finally, they investigated 9 factors that could potentially moderate associations between psychosocial interventions and immune system function: type of psychosocial intervention, intervention format (no group vs group sessions), intervention length, type of immune marker, whether the immune marker represented basal or stimulated levels, time from treatment cessation to immune marker measurement, participants' disease state or reason for receiving treatment, age, and sex. This meta-analysis thus addresses the critical question of which types of psychosocial interventions are most consistently associated with enhanced immune system function, under what conditions, and for whom, which may in turn inform research efforts and public policy aimed at using psychosocial interventions to improve immune-related health.

This comprehensive review of 56 RCTs revealed that psychosocial interventions were significantly associated with enhanced immune system function, as indexed most consistently by intervention-related decreases in levels of proinflammatory cytokines or markers (eg, interleukin-6, C-reactive protein) and, secondarily, by increases in immune cell counts (eg, CD56, CD4) over time. These associations were most consistent for CBT and for interventions incorporating multiple psychotherapies. Moreover, they did not differ by participants' age, sex, or intervention duration. Finally, we found that these associations persisted for at least 6 months following treatment cessation. Considered together, these results suggest that psychosocial interventions in general—and especially CBT and multiple or combined psychotherapeutic interventions—enhance immune system function and may thus represent a viable strategy for improving immune-related health outcomes [72].

Converted to percentages, these data reveal that, relative to the control group, psychosocial interventions were associated with an 18.0% (95% CI, 7.2%–28.8%) reduction in harmful immune system function as indexed, for example, by proinflammatory cytokine activity. In comparison, an RCT [53] found that, relative to a control group, treatment with a 40-mg dose of darapladib for reducing cardiovascular disease risk decreased interleukin-6 levels by 7.8% and C-reactive protein levels by 6.0%, whereas a 160-mg dose of darapladib decreased interleukin-6 levels by 12.3% and C-reactive protein levels by 13.0%. Psychosocial interventions thus appear to reduce systemic inflammatory activity in a manner that is similar to using darapladib for treating atherosclerosis.

The investigators also found that psychosocial interventions were associated with improvements in immune system function over time—in particular, with decreased proinflammatory cytokines or markers and increased immune cell counts—and that these associations were most consistent for interventions that incorporate CBT or multiple interventions. Given the effectiveness and relative affordability of psychosocial interventions for treating chronic disease, we suggest that psychosocial interventions may represent a viable strategy for reducing disease burden and improving human health. Looking forward, additional research is needed to elucidate the mechanisms through which psychosocial interventions exert relatively long-lasting, beneficial effects on the immune system and health [72].

Low Dose Medicine (LDM)

Low dose medicine (LDM) originated from the combination of PNEI research and low dose pharmacology. Last 40 years, the development of the concepts expressed by PNEI resulted in a change of perspective in the interpretation of the biological functions of the human body and its diseases, shifting from reductionism (each disease affects a single organ or tissue) to the cellular network, until recognizing the importance of continuous dialogue – cross talk – between cells, organs and systems both in physiological and pathological conditions.

Starting from these assumptions, pharmacological research has been focusing on the role played by special biological molecules, thus paving the way for what could have been a new therapeutic solution, the use of the same organic molecules as medicaments to restore the original physiological conditions - neuropeptides, hormones, and cytokines, but in physiological (low) concentration.

The pharmaceutical technology called SKA (Sequential Kinetic Activation) has made it possible to "duplicate" this specific concentration and incorporate the molecules into medicaments. This novel and scientifically substantiated perception paved the way for the therapeutic exploration of the Treg-IL-2 axis in the setting of immune-mediated and inflammatory diseases with the aim to upregulate the Treg population directly in the patient thereby modulating pathogenic responses and re-establishing immune function. The principle of using low doses of IL-2 for the treatment of immunological diseases, instead of high doses as long time approved for cancer therapy, was introduced because of the assumption, and meanwhile convincingly proven fact, that Treg are more sensitive to IL-2 and require by far much lower doses of IL-2 for their stimulation compared to anti-tumor T cells and NK cells.

Low-dose IL-2 therapy aims to strengthen the Treg population in order to be more effective in counter-regulating inflammation while avoiding global immunosuppression.

Data from several pilot studies and clinical trials, including first randomized trials, broadly and reproducibly prove that low-dose IL-2 therapy is very safe and capable to selectively expand a functionally competent Treg population independent of the underlying disease. In addition, these trials provided preliminary evidence for the clinical efficacy of low-dose IL-2-therapy in a large variety of inflammatory and autoimmune diseases [73].

Those agents would indeed allow to re-establish the body's homeostasis, both at a systemic and local level. The micro-immunotherapy medicine, a sort of low- dose medicine has been developed to attenuate the symptoms of rheumatic diseases including RA, within a holistic immunomodulatory approach. Micro-immunotherapy (MI) is a pioneering modulative therapeutic approach that can be employed alone or in association with other therapies, providing clinical benefits while reducing side effects. This strategy combines the use of several immune system mediators such as cytokines, growth factors, hormones, nucleic acids and specific nucleic acids, in order to orchestrate both the innate and the adaptive/acquired immune responses [74].

Altogether, these active ingredients are independently employed in MI medicines at either low doses (LD), to stimulate their own production by the organism, or on the opposite side of the spectrum, at ultra-low doses (ULD), in order to modulate/inhibit their own biological effects.

These evidences trace a new path for the management of problems, especially inflammation-related one, often very complex, for which the classic pharmacological intervention can show limits of safety and compliance. Moreover, the particular propensity of low dose drugs to control inflammatory conditions and chronic immunological dysregulations allows us to include some of these drugs among the potential tools for the management of low grade chronic inflammation, such as the long-COVID syndrome, which is currently difficult to manage due to its particular symptomatologic complexity [75].

Conclusion

Therapeutic interventions in PNEI are intended to balance allostatic load of individual organism and fight chronic low grade systemic inflammation. The factors which emerged to be relevant in achieving this high goals are basically non-pharmacological: circadian rhythm maintenance, nutrition and psychosocial conditioning, but also low dose medicine as novel pharmaceutical approach. PNEI is easily applicable as adjuvant to contemporary practicing medicine and its strong evidence –based background makes it necessary in *armarium* of modern health care provider.

References

1. Ader R, Kelley KW. A global view of twenty years of brain, behavior, and immunity. Brain Behav Immun. 2007;21(1):20–2. https://doi.org/10.1016/j.bbi.2006.07.003. Epub 2006 Sep 25. PMID: 16996715; PMCID: PMC1899234
2. Ader R, Cohen N. Behaviorally conditioned immunosuppression. Psychosom Med. 1975;37(4):333–40. https://doi.org/10.1097/00006842-197507000-00007.
3. Ader R, Cohen N, Felten D. Psychoneuroimmunology: interactions between the nervous system and the immune system. Lancet. 1995;345(8942):99–103. https://doi.org/10.1016/s0140-6736(95)90066-7.
4. Ader R, Cohen N. Psychoneuroimmunology: conditioning and stress. Annu Rev Psychol. 1993;44:53–85. https://doi.org/10.1146/annurev.ps.44.020193.000413.
5. Chabner BA, Boral AL, Multani P. Translational research: walking the bridge between idea and cure--seventeenth Bruce F. Cain Memorial Award Lecture Cancer Res. 1998;58(19):4211–6.
6. Hegyi P, Petersen OH, Holgate S, Erőss B, Garami A, Szakács Z, Dobszai D, Balaskó M, Kemény L, Peng S, Monteiro J, Varró A, Lamont T, Laurence J, Gray Z, Pickles A, FitzGerald GA, Griffiths CEM, Jassem J, Rusakov DA, Verkhratsky A, Szentesi A. Academia europaea position paper on translational medicine: the cycle model for translating scientific results into community benefits. J Clin Med. 2020;9(5):1532. https://doi.org/10.3390/jcm9051532. PMID: 32438747; PMCID: PMC7290380
7. Bottaccioli AG, Bologna M, Bottaccioli F. Psychic life-biological molecule bidirectional relationship: pathways, mechanisms, and consequences for medical and psychological sciences-a narrative review. Int J Mol Sci. 2022;23(7):3932. https://doi.org/10.3390/ijms23073932. PMID: 35409300; PMCID: PMC8999976
8. McEwen CA. Connecting the biology of stress, allostatic load and epigenetics to social structures and processes. Neurobiol Stress. 2022;17:100426. https://doi.org/10.1016/j.ynstr.2022.100426.
9. Juster RP, McEwen BS, Lupien SJ. Allostatic load biomarkers of chronic stress and impact on health and cognition. Neurosci Biobehav Rev. 2010;35(1):2–16. https://doi.org/10.1016/j.neubiorev.2009.10.002. Epub 2009 Oct 12
10. McEwen BS. Stress, adaptation, and disease. Allostasis and allostatic load. Ann N Y Acad Sci. 1998;840:33–44. https://doi.org/10.1111/j.1749-6632.1998.tb09546.x.
11. Fava GA, McEwen BS, Guidi J, Gostoli S, Offidani E, Sonino N. Clinical characterization of allostatic overload. Psychoneuroendocrinology. 2019;108:94–101. https://doi.org/10.1016/j.psyneuen.2019.05.028. Epub 2019 May 31
12. Buckwalter JG, Castellani B, McEwen B, Karlamangla AS, Rizzo AA, John B, O'Donnell K, Seeman T. Allostatic load as a complex clinical construct: a case-based computational modeling approach. Complexity. 2016;21(Suppl 1):291–306. https://doi.org/10.1002/cplx.21743. Epub 2015 Dec 23. PMID: 28190951; PMCID: PMC5300684
13. Furman D, Campisi J, Verdin E, Carrera-Bastos P, Targ S, Franceschi C, Ferrucci L, Gilroy DW, Fasano A, Miller GW, Miller AH, Mantovani A, Weyand CM, Barzilai N, Goronzy JJ, Rando TA, Effros RB, Lucia A, Kleinstreuer N, Slavich GM. Chronic inflammation in the etiology of disease across the life span. Nat Med. 2019;25(12):1822–32. https://doi.org/10.1038/s41591-019-0675-0. Epub 2019 Dec 5. PMID: 31806905; PMCID: PMC7147972

14. Xie Y, Tang Q, Chen G, Xie M, Yu S, Zhao J, Chen L. New insights into the circadian rhythm and its related diseases. Front Physiol. 2019;10:682. https://doi.org/10.3389/fphys.2019.00682. PMID: 31293431; PMCID: PMC6603140

15. Panda S, Hogenesch JB, Kay SA. Circadian rhythms from flies to human. Nature. 2002;417(6886):329–35. https://doi.org/10.1038/417329a.

16. Bass J, Takahashi JS. Circadian integration of metabolism and energetics. Science. 2010;330(6009):1349–54. https://doi.org/10.1126/science.1195027. PMID: 21127246; PMCID: PMC3756146

17. Fishbein AB, Knutson KL, Zee PC. Circadian disruption and human health. J Clin Invest. 2021;131(19):e148286. https://doi.org/10.1172/JCI148286. PMID: 34596053; PMCID: PMC8483747

18. McEwen BS. The untapped power of allostasis promoted by healthy lifestyles. World Psychiatry. 2020;19(1):57–8. https://doi.org/10.1002/wps.20720. PMID: 31922670; PMCID: PMC6953580

19. Zhou L, Zhang Z, Nice E, Huang C, Zhang W, Tang Y. Circadian rhythms and cancers: the intrinsic links and therapeutic potentials. J Hematol Oncol. 2022;15(1):21. https://doi.org/10.1186/s13045-022-01238-y. PMID: 35246220; PMCID: PMC8896306

20. Erren TC, Morfeld P, Groß JV, Wild U, Lewis P. IARC 2019: "night shift work" is probably carcinogenic: what about disturbed chronobiology in all walks of life? J Occup Med Toxicol. 2019;14:29. https://doi.org/10.1186/s12995-019-0249-6. PMID: 31798667; PMCID: PMC6882045

21. Piggins HD. Human clock genes. Ann Med. 2002;34(5):394–400. https://doi.org/10.1080/078538902320772142.

22. Mohawk JA, Green CB, Takahashi JS. Central and peripheral circadian clocks in mammals. Annu Rev Neurosci. 2012;35:445–62. https://doi.org/10.1146/annurev-neuro-060909-153128. Epub 2012 Apr 5. PMID: 22483041; PMCID: PMC3710582

23. McEwen BS, Nasca C, Gray JD. Stress effects on neuronal structure: hippocampus, amygdala, and prefrontal cortex. Neuropsychopharmacology. 2016;41(1):3–23. https://doi.org/10.1038/npp.2015.171. Epub 2015 Jun 16. PMID: 26076834; PMCID: PMC4677120

24. West NP, Hughes L, Ramsey R, Zhang P, Martoni CJ, Leyer GJ, Cripps AW, Cox AJ. Probiotics, anticipation stress, and the acute immune response to night shift. Front Immunol. 2021;11:599547. https://doi.org/10.3389/fimmu.2020.599547. Erratum in: Front Immunol. 2021 Jun 21;12:713237. PMID: 33584665; PMCID: PMC7877220

25. Amihaesei IC, Mungiu OC. Main neuroendocrine features and therapy in primary sleep troubles. Rev Med Chir Soc Med Nat Iasi. 2012;116:862–6.

26. Cardinali DP, Furio AM, Reyes MP, Brusco LI. The use of chronobiotics in the resynchronization of the sleep-wake cycle. Cancer Causes Control. 2006;17(4):601–9. https://doi.org/10.1007/s10552-005-9009-2.

27. Moon E, Partonen T, Beaulieu S, Linnaranta O. Melatonergic agents influence the sleep-wake and circadian rhythms in healthy and psychiatric participants: a systematic review and meta-analysis of randomized controlled trials. Neuropsychopharmacology. 2022;47(8):1523–36. https://doi.org/10.1038/s41386-022-01278-5. Epub 2022 Feb 4. PMID: 35115662; PMCID: PMC9206011

28. Mindell JA, Leichman ES, Lee C, Williamson AA, Walters RM. Implementation of a nightly bedtime routine: how quickly do things improve? Infant Behav Dev. 2017;49:220–7. https://doi.org/10.1016/j.infbeh.2017.09.013. Epub 2017 Oct 3. PMID: 28985580; PMCID: PMC6587179

29. Whitesell CJ, Crosby B, Anders TF, Teti DM. Household chaos and family sleep during infants' first year. J Fam Psychol. 2018;32(5):622–31. https://doi.org/10.1037/fam0000422. Epub 2018 May 21. PMID: 29781634; PMCID: PMC6072580

30. Owens JA. The practice of pediatric sleep medicine: results of a community survey. Pediatrics. 2001;108(3):E51. https://doi.org/10.1542/peds.108.3.e51.

31. Zisberg A, Gur-Yaish N, Shochat T. Contribution of routine to sleep quality in community elderly. Sleep. 2010;33(4):509–14. https://doi.org/10.1093/sleep/33.4.509. PMID: 20394320; PMCID: PMC2849790

32. Trauer JM, Qian MY, Doyle JS, Rajaratnam SM, Cunnington D. Cognitive Behavioral therapy for chronic insomnia: a systematic review and meta-analysis. Ann Intern Med. 2015;163(3):191–204. https://doi.org/10.7326/M14-2841.

33. Taetzsch A, Roberts SB, Bukhari A, Lichtenstein AH, Gilhooly CH, Martin E, Krauss AJ, Hatch-McChesney A, Das SK. Eating timing: associations with dietary intake and metabolic health. J Acad Nutr Diet. 2021;121(4):738–48. https://doi.org/10.1016/j.jand.2020.10.001. Epub 2020 Nov 10

34. Kaczmarek JL, Thompson SV, Holscher HD. Complex interactions of circadian rhythms, eating behaviors, and the gastrointestinal microbiota and their potential impact on health. Nutr Rev. 2017;75(9):673–82. https://doi.org/10.1093/nutrit/nux036.

35. Wehrens SMT, Christou S, Isherwood C, Middleton B, Gibbs MA, Archer SN, Skene DJ, Johnston JD. Meal timing regulates the human circadian system. Curr Biol. 2017;27(12):1768-1775.e3. https://doi.org/10.1016/j.cub.2017.04.059. Epub 2017 Jun 1. PMID: 28578930; PMCID: PMC5483233

36. Mattson MP, Allison DB, Fontana L, Harvie M, Longo VD, Malaisse WJ, Mosley M, Notterpek L, Ravussin E, Scheer FA, Seyfried TN, Varady KA, Panda S. Meal frequency and timing in health and disease. Proc Natl Acad Sci U S A. 2014;111(47):16647–53. https://doi.org/10.1073/pnas.1413965111. Epub 2014 Nov 17. PMID: 25404320; PMCID: PMC4250148

37. St-Onge MP, Ard J, Baskin ML, Chiuve SE, Johnson HM, Kris-Etherton P, Varady K, American Heart Association Obesity Committee of the Council on Lifestyle and Cardiometabolic Health, Council on Cardiovascular Disease in the Young; Council on Clinical Cardiology, Stroke Council. Meal timing and frequency: implications for cardiovascular disease prevention: a scientific statement from the American Heart Association. Circulation. 2017;135(9):e96–e121. https://doi.org/10.1161/CIR.0000000000000476. Epub 2017 Jan 30. PMID: 28137935; PMCID: PMC8532518

38. Paoli A, Tinsley G, Bianco A, Moro T. The influence of meal frequency and timing on health in humans: the role of fasting. Nutrients. 2019;11(4):719. https://doi.org/10.3390/nu11040719. PMID: 30925707; PMCID: PMC6520689

39. Rothschild J, Hoddy KK, Jambazian P, Varady KA. Time-restricted feeding and risk of metabolic disease: a review of human and animal studies. Nutr Rev. 2014;72:308–18. https://doi.org/10.1111/nure.12104.

40. Longo VD, Mattson MP. Fasting: molecular mechanisms and clinical applications. Cell Metab. 2014;19:181–92. https://doi.org/10.1016/j.cmet.2013.12.008.

41. Nader GA, Lundberg E. Exercise as an anti-inflammatory intervention to combat inflammatory diseases of muscle. Curr Opin Rheumatol. 2009;21:599–603.

42. Winzer BM, Whiteman DC, Reeves MM, Paratz JD. Physical activity and cancer prevention: a systematic review of the clinical trials. Cancer Causes Control. 2011;22:811–26.

43. Ertek S, Cicero A. Impact of physical activity on inflammation: effects on cardiovascular disease risk and other inflammatory conditions. Arch Med Sci. 2012;8(5):794–804. https://doi.org/10.5114/aoms.2012.31614. Epub 2012 Nov 7. PMID: 23185187; PMCID: PMC3506236

44. Zambon AC, McDearmon EL, Salomonis N, et al. Time- and exercise-dependent gene regulation in human skeletal muscle. Genome Biol. 2003;4:R61.

45. Mayeuf-Louchart A, Staels B, Duez H. Skeletal muscle functions around the clock. Diabetes Obes Metab. 2015;17(Suppl 1):39–46. https://doi.org/10.1111/dom.12517.

46. Gutierrez-Monreal MA, Harmsen JF, Schrauwen P, Esser KA. Ticking for metabolic health: the skeletal-muscle clocks. Obesity (Silver Spring). 2020;28(Suppl 1):S46–54. https://doi.org/10.1002/oby.22826. Epub 2020 May 28. PMID: 32468732; PMCID: PMC7381376

47. Minihane AM, Vinoy S, Russell WR, Baka A, Roche HM, Tuohy KM, Teeling JL, Blaak EE, Fenech M, Vauzour D, et al. Low-grade inflammation, diet composition and health: current research evidence and its translation. Br J Nutr. 2015;114:999–1012.

48. Mozaffarian D. Dietary and policy priorities for cardiovascular disease, diabetes, and obesity: a comprehensive review. Circulation. 2016;133:187–225. https://doi.org/10.1161/CIRCULATIONAHA.115.01858551.
49. Cani PD. Metabolism: dietary emulsifiers--sweepers of the gut lining? Nat Rev Endocrinol. 2015;11:319–20. https://doi.org/10.1038/nrendo.2015.59.
50. Calder PC, Bosco N, Bourdet-Sicard R, Capuron L, Delzenne N, Doré J, Franceschi C, Lehtinen MJ, Recker T, Salvioli S, et al. Health relevance of the modification of low grade inflammation in ageing (Inflammageing) and the role of nutrition. Ageing Res Rev. 2017;40:95–119. https://doi.org/10.1016/j.arr.2017.09.001.
51. Velasquez MT. Altered gut microbiota: a link between diet and the metabolic syndrome. Metab Syndr Relat Disord. 2018;16:321–8. https://doi.org/10.1089/met.2017.0163.
52. Guo S, Al-Sadi R, Said HM, Ma TY. Lipopolysaccharide causes an increase in intestinal tight junction permeability in vitro and in vivo by inducing enterocyte membrane expression and localization of TLR-4 and CD14. Am J Pathol. 2013;182:375–87. https://doi.org/10.1016/j.ajpath.2012.10.014.
53. Ratajczak W, Ryl A, Mizerski A, Walczakiewicz K, Sipak O, Laszczyńska M. Immunomodulatory potential of gut microbiome-derived short-chain fatty acids (SCFAs). Acta Biochim Pol. 2019;66:1–12. https://doi.org/10.18388/abp.2018_2648.
54. Ridlon JM, Kang D-J, Hylemon PB. Bile salt biotransformations by human intestinal bacteria. J Lipid Res. 2006;47:241–59. https://doi.org/10.1194/jlr.R500013-JLR200.
55. De La Serre CB, Ellis CL, Lee J, Hartman AL, Rutledge JC, Raybould HE. Propensity to high-fat diet-induced obesity in rats is associated with changes in the gut microbiota and gut inflammation. Am J Physiol Gastrointest Liver Physiol. 2010;299:G440–8. https://doi.org/10.1152/ajpgi.00098.2010.
56. Ley RE, Peterson DA, Gordon JI. Ecological and evolutionary forces shaping microbial diversity in the human intestine. Cell. 2006;124(4):837–48.
57. Wang HB, Wang PY, Wang X, Wan YL, Liu YC. Butyrate enhances intestinal epithelial barrier function via up-regulation of tight junction protein Claudin-1 transcription. Dig Dis Sci. 2012;57(12):3126–35.
58. King DE, Egan BM, Geesey ME. Relation of dietary fat and fiber to elevation of C-reactive protein. Am J Cardiol. 2003;92(11):1335–9.
59. Pastori D, Carnevale R, Nocella C, Novo M, Santulli M, Cammisotto V, Menichelli D, Pignatelli P, Violi F. Gut-derived serum lipopolysaccharide is associated with enhanced risk of major adverse cardiovascular events in atrial fibrillation: effect of adherence to Mediterranean diet. J Am Heart Assoc. 2017;6(6):e005784.
60. Saura-Calixto F, Goni I. Antioxidant capacity of the Spanish Mediterranean diet. Food Chem. 2006;94(3):442–7.
61. Arai Y, Watanabe S, Kimira M, Shimoi K, Mochizuki R, Kinae N. Dietary intakes of flavonols, flavones and isoflavones by Japanese women and the inverse correlation between quercetin intake and plasma LDL cholesterol concentration. J Nutr. 2000;130(9):2243–50.
62. Xiao JB, Hogger P. Dietary polyphenols and type 2 diabetes: current insights and future perspectives. Curr Med Chem. 2015;22(1):23–38.
63. Rodriguez-Ramiro I, Vauzour D, Minihane AM. Polyphenols and non-alcoholic fatty liver disease: impact and mechanisms. Proc Nutr Soc. 2016;75(1):47–60.
64. Cires MJ, Wong X, Carrasco-Pozo C, Gotteland M. The gastrointestinal tract as a key target organ for the health-promoting effects of dietary proanthocyanidins. Front Nutr. 2017;3:57.
65. Huang JQ, Lu M, Ho CT. Health benefits of dietary chronobiotics: beyond resynchronizing internal clocks. Food Funct. 2021;12(14):6136–56. https://doi.org/10.1039/d1fo00661d. Epub 2021 May 31
66. Cohen S, Miller GE, Rabin BS. Psychological stress and antibody response to immunization: a critical review of the human literature. Psychosom Med. 2001;63(1):7–18. https://doi.org/10.1097/00006842-200101000-00002.

67. McEwen BS, Seeman T. Protective and damaging effects of mediators of stress. Elaborating and testing the concepts of allostasis and allostatic load. Ann N Y Acad Sci. 1999;896:30–47. https://doi.org/10.1111/j.1749-6632.1999.tb08103.x.
68. Charney DS. Psychobiological mechanisms of resilience and vulnerability: implications for successful adaptation to extreme stress. Am J Psychiatr. 2004;161(2):195–216. https://doi.org/10.1176/appi.ajp.161.2.195.
69. Hodes GE, Pfau ML, Leboeuf M, Golden SA, Christoffel DJ, Bregman D, Rebusi N, Heshmati M, Aleyasin H, Warren BL, Lebonte B, Horn S, Lapidus KA, Stelzhammer V, Wong EH, Bahn S, Krishnan V, Bolanos-Guzman CA, Murrough JW, et al. Individual differences in the peripheral immune system promote resilience versus susceptibility to social stress. Proc Natl Acad Sci. 2014;111(45):16136–41. https://doi.org/10.1073/pnas.141519111.
70. Shimba A, Ikuta K. Control of immunity by glucocorticoids in health and disease. Semin Immunopathol. 2020;42(6):669–80. https://doi.org/10.1007/s00281-020-00827-8. Epub 2020 Nov 20
71. Shields GS, Spahr CM, Slavich GM. Psychosocial interventions and immune system function: a systematic review and meta-analysis of randomized clinical trials. JAMA Psychiatry. 2020;77(10):1031–43. https://doi.org/10.1001/jamapsychiatry.2020.0431. PMID: 32492090; PMCID: PMC7272116
72. Mohler ER III, Ballantyne CM, Davidson MH, Darapladib Investigators, et al. The effect of darapladib on plasma lipoprotein-associated phospholipase A2 activity and cardiovascular biomarkers in patients with stable coronary heart disease or coronary heart disease risk equivalent: the results of a multicenter, randomized, double-blind, placebo-controlled study. J Am Coll Cardiol. 2008;51(17):1632–41. https://doi.org/10.1016/j.jacc.2007.11.079.
73. Graßhoff H, Comdühr S, Monne LR, Müller A, Lamprecht P, Riemekasten G, Humrich JY. Low-dose IL-2 therapy in autoimmune and rheumatic diseases. Front Immunol. 2021;12:648408. https://doi.org/10.3389/fimmu.2021.648408. PMID: 33868284; PMCID: PMC8047324
74. Jacques C, Floris I, Lejeune B. Ultra-low dose cytokines in rheumatoid arthritis, three birds with one stone as the rationale of the 2LARTH® micro-immunotherapy treatment. Int J Mol Sci. 2021;22(13):6717. https://doi.org/10.3390/ijms22136717. PMID: 34201546; PMCID: PMC8268272
75. Fioranelli M, Roccia MG, Flavin D, Cota L. Regulation of inflammatory reaction in health and disease. Int J Mol Sci. 2021;22(10):5277. https://doi.org/10.3390/ijms22105277. PMID: 34067872; PMCID: PMC8157220

Chapter 17
Secondary Stroke Prevention in Patients with Atrial Fibrillation: From no Treatment to Direct Oral Anticoagulants

Hrvoje Budinčević, Latica Friedrich, Petra Črnac Žuna, Dorotea Vidaković, and Ivanka Maduna

Introduction

Stroke represents the leading cause of disability and it is among the leading causes of mortality in the world [1]. It is estimated that 30% of ischemic strokes are caused by cardioembolism [2]. Atrial fibrillation is the leading cause of cardioembolic stroke and it represents the cause of about 15% of all strokes [2]. The most common cardiac arrhythmia is atrial fibrillation, its prevalence and incidence increase with age [3]. Atrial fibrillation is an independent risk factor for stroke, and stroke is the most common consequence of atrial fibrillation [4]. Strokes caused by atrial fibrillation are more often associated with greater disability, mortality, and a higher risk of recurrence [5, 6].

The stroke risk is five times higher in patients with atrial fibrillation, and 17 times higher in patients with valvular atrial fibrillation [4, 6]. Valvular atrial fibrillation refers to the presence of prosthetic heart valves and rheumatic valvular (heart) disease (usually moderate to severe mitral stenosis) [3]. CHADS$_2$ rating scale is most commonly used to assess the risk of ischemic stroke in patients with atrial fibrillation, and more recently its modified version of the so-called CHA$_2$DS 2-VASc

H. Budinčević (✉)
Department of Neurology, Sveti Duh University Hospital, Zagreb, Croatia

Department of Neurology and Neurosurgery, Faculty of Medicine, J.J. Strossmayer University of Osijek, Osijek, Croatia

L. Friedrich · P. Črnac Žuna
Department of Neurology, Sveti Duh University Hospital, Zagreb, Croatia

D. Vidaković · I. Maduna
Health Center of Osijek-Baranja County, Osijek, Croatia

V. Demarin et al. (eds.), *Mind, Brain and Education*,
https://doi.org/10.1007/978-3-031-33013-1_17

[3]. The most often used scale to assess the risk of bleeding is HAS-BLED [3]. The goals of the treatment of atrial fibrillation include: (1) prevention of thromboembolic incidents, and (2) heart rate and rhythm control [3].

In the prevention of thromboembolism, including stroke in patients with atrial fibrillation, anticoagulant therapy still represents the leading treatment of choice [7]. Conditions and diseases in which oral anticoagulant therapy is considered for stroke prevention are: (1) atrial fibrillation, (2) myocardial infarction with intramural thrombus in the left heart ventricle, (3) dilated cardiomyopathy, (4) rheumatic mitral valve disease, (5) prosthetic heart valves, (6) arterial dissection, (7) patent foramen ovale, and (8) cerebral venous sinus thrombosis [8].

"Old" Anticoagulant Therapy—The Story of Warfarin

Apart from stroke prevention, oral anticoagulant therapy is indicated for several other conditions such as: deep venous thrombosis, pulmonary embolism, ischemic heart disease, dilated cardiomyopathy, left ventricular aneurysm, inherited thrombophilias, and antiphospholipid syndrome.

Several years ago, vitamin K antagonists were the gold standard for the prevention of cardioembolic stroke [9]. The most commonly prescribed vitamin K antagonist is warfarin [9, 10]. Warfarin blocks the regeneration of the reduced form of vitamin K, which is an essential coenzyme in the carboxylation of glutamate residues into procoagulant forms of factor II, VII, IX, and X in the coagulation cascade, which ultimately leads to reduced production of thrombin and fibrin and inhibition of coagulation [11].

The history of warfarin starts in the prairies of Canada and the Northern Plains of America in the 1920s with cattle haemorrhagic disease [12]. This disease was named "Sweet clover disease" since cattle ingested mouldy silage made from sweet clover [12]. Schofield and Roderick demonstrated that sweet clover disease can be potentially reversible by removing offending mouldy hay, or by transfusion of fresh blood into the bleeding animals [12]. Without this treatment, cattle would manifest the disease within 15 days of ingestion and die within 50 days [13]. Karl Link and his team set out to isolate haemorrhagic agents from spoiled hay in 1933 [13, 14]. They found that natural coumarin became oxidised in mouldy hay and formed the substance that would become better known as dicoumarol in 1940 [14]. The patent rights for dicoumarol were given in 1941 to the Wisconsin Alumni Research Foundation, which funded the project [12]. A dicoumarol derivate, warfarin was firstly used as rodenticide in 1948 [12]. In 1951, a US Army inductee attempted suicide with multiple doses of warfarin in rodenticide, but fully recovered after being treated with vitamin K in hospital [15]. Eventually, warfarin has been transitioned into clinical use under the trade name Coumadin, and was approved for use in humans in 1954 [15]. Thereupon prothrombin time (PT) test was established for warfarin dosage control [12]. Due to high variability of PT testing, in 1982 the World Health Organisation (WHO) adopted a model to convert the PT obtained with any reagent to an International Normalised Ratio (INR), that is the PT that would have been obtained if an International Reference Preparation (IRP) had been used [16]. The first clinical trial was obtained in 1960 in pulmonary embolism patients, and in the 1990s started studies in stroke prevention [9, 12]. The mechanism of action of warfarin was discovered and reported in 1978 when Suttie and colleagues showed that warfarin disrupts vitamin K metabolism by inhibiting the epoxide reductase enzyme [17].

Following the 1990s warfarin clinical trials showed superiority over placebo and acetylsalicylic acid in terms of stroke prevention in patients with atrial fibrillation [9]. Thus, for example, in primary stroke prevention, the use of warfarin with the INR target between 2 and 3 contributes to the reduction of the relative risk for stroke by up to 62%, and in secondary prevention by up to 67%, as opposed to acetylsalicylic acid, which reduces the relative risk for stroke in primary prevention by 22% and in secondary prevention by 21% [18] . Despite such good effectiveness, oral anticoagulant therapy for vitamin K antagonists was underapplied even in the most developed countries in the world [19]. The reasons for the insufficient use of vitamin K antagonists were associated with numerous limitations, mostly due to the interactions with food and medications and a narrow therapeutic window, which required dose titration with repeated measurement of level of anticoagulation [11].

Novel Oral Anticoagulants in Stroke Prevention

In order to eliminate the numerous limitations of vitamin K antagonists, clinical trials were conducted on novel (direct) oral anticoagulants [20–25]. Direct thrombin inhibitors (ximelagatran and dabigatran) and direct inhibitors of factor Xa (rivaroxaban, apixaban and edoxaban) were investigated [20–25].

In *The Stroke Prevention using an Oral Thrombin Inhibitor in Atrial Fibrillation III and V* (SPORTIF-III and SPORTIF V) studies, ximelagatran has been shown, in relation to warfarin, to be equally effective with an approximate number of major haemorrhages and a lower incidence of total bleeding, but due to hepatotoxicity it has not been accepted by regulatory authorities in the United States and Europe [20]. In *The Randomised Evaluation of Long-Term Anticoagulation Therapy* (RE-LY), dabigatran was compared with warfarin at a dose of 2 × 150 mg and showed a lower incidence of stroke and systemic embolism, but a similar incidence of major haemorrhages. Dabigatran at a dose of 2 × 110 mg showed a similar incidence of strokes and systemic embolisms as warfarin, but a lower incidence of major bleeding. Among the side effects of dabigatran, gastrointestinal problems with dyspepsia are most often reported [22].

The *Apixaban Versus Acetylsalicylic Acid [ASA] to Prevent Stroke in Atrial Fibrillation Patients Who Have Failed or Are Unsuitable for Vitamin K Antagonist Treatment* (AVERROES) compared the effect of apixaban in relation to acetylsalicylic acid in patients who were not suitable for anticoagulant therapy in stroke prevention. The study was previously discontinued due to the superiority of apixaban.

The study showed that the risk for ischemic stroke on apixaban therapy was reduced by 50%, and without a significant increase in major bleeding [21]. The *Apixaban for Reduction in Stroke and Other Thromboembolic Events in Atrial Fibrillation* (ARISTOTLE) study showed the superiority of apixaban in the prevention of stroke and systemic embolism over warfarin. In addition, patients on apixaban had a statistically significant lower number of haemorrhages (large and intracranial) which all led to a decrease in mortality [23].

The *Rivaroxaban Once Daily Oral Direct Factor Xa Inhibition Compared with Vitamin K Antagonism for Prevention of Stroke and Embolism Trial in Atrial Fibrillation* (ROCKET AF) compared the effect of rivaroxaban in relation to warfarin in stroke prevention. The study showed that rivaroxaban is not inferior in stroke prevention or systemic embolism. It was also shown that there is no statistically significant difference in large haemorrhages, however, intracranial and fatal haemorrhages occurred somewhat less frequently in patients on rivaroxaban [24].

The study The Effective Anticoagulation with Factor Xa Next Generation in Atrial Fibrillation–Thrombolysis in Myocardial Infarction 48 (ENGAGE AF-TIMI 48) compared the effect of edoxaban in relation to warfarin in stroke prevention. Edoxaban has been shown to be non-inferior in the prevention of stroke and systemic embolism with a higher incidence of bleeding and mortality than other cardiovascular causes [25].

Guidelines in the Prevention of Stroke

Figure 17.1 presents an algorithm for the prevention of stroke in patients with atrial fibrillation [26, 27].

The selection of antithrombotic therapy should be individualised on the basis of risk factors, the safety and submission of the drug, patient selection, potential drug interactions and other clinical characteristics including renal function and times in therapeutic values (TTR- Time in Therapeutic Range) for patients taking vitamin K antagonists [28–30].

The results of the mentioned studies allowed novel (direct) oral anticoagulants to be included in international guidelines for the treatment of patients with non-valvular atrial fibrillation. Warfarin is still a possible treatment option and it is indicated to be administered in patients with valvular atrial fibrillation (rheumatic mitral stenosis, artificial (prosthetic) heart valves and after mitral valve surgery, and does not include valvular aortic and tricuspid valve disease and mitral regurgitation where direct oral anticoagulants can be applied) [31–33]. The latest guidelines from the American Heart Association, American Stroke Association, European Society of Cardiology and European Stroke Organisation recommend the use of direct oral anticoagulants over vitamin K antagonists for patients with non-valvular atrial fibrillation [7, 30, 34–37]. Previous guidelines recommended acetylsalicylic acid in patients with contraindications to oral anticoagulants, with the possible addition of clopidogrel [34, 35], similar to the American Academy of Neurology and the

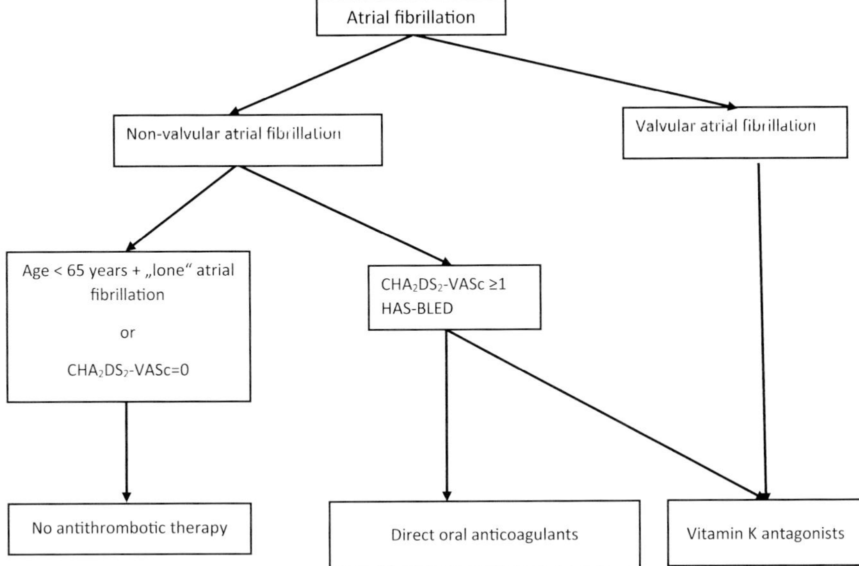

Fig. 17.1 Algorithm for usage of oral anticoagulant therapy in stroke prevention. Modified by: Camm AJ et al. Eur Heart J. 2012 Nov; 33(21):2719-47

European Society of Cardiology [8, 36]. Novel oral anticoagulants should not be used in case of severe renal insufficiency with creatinine clearance <15 mL/min [27], with the exception of apixaban [38].

The current literature lacks clear statements or recommendations regarding the optimal time to start oral anticoagulation therapy after an ischemic stroke or transient ischemic attack [30, 39]. Although, without clear scientific evidence, the prevailing opinion is that therapy can be started in the first 2 weeks after ischemic stroke [8]. An individual assessment is required based on the size of the ischemic lesion, but also other clinical factors such as poorly controlled arterial hypertension, liver disease, etc. [26]. In accordance with the common guidelines of the American Heart Association and the American Stroke Association (AHA/ASA), it is possible to use low-molecular heparin as a "bridging" therapy until the introduction of warfarin in patients at high risk for stroke [8], but European guidelines do not recommend it [30]. In accordance with the latest ESC guidelines, the initiation of anticoagulant therapy can be roughly started in transient ischemic attack after the first day, in a mild stroke after 3 days, in a moderate stroke after 6 days, and in a severe stroke after 12 days [39]. In any case, other parameters such as: stroke severity, heart attack size, age, need for surgical care and control of arterial hypertension should be taken into account [7]. According to the latest European guidelines the use of antiplatelet therapy with acetylsalicylic acid is recommended prior to the introduction of an oral anticoagulant agent and bridging with low molecular weight heparin is not required for stroke prevention [30]. Low molecular weight heparins should be used for prevention of deep venous thrombosis [38].

The decision of which oral anticoagulant therapy should be introduced in secondary stroke prevention is mostly based on experts' opinions as it is shown in Table 17.1 [40–44].

Individual assessment is also necessary in clinical situations with anticoagulant-related intracerebral haemorrhage [45, 46]. In cases of intracerebral, subarachnoid or subdural haemorrhage, anticoagulant and antiplatelet therapy should be discontinued for a period of 1 to 2 weeks after haemorrhage, and anticoagulant effects should be immediately reversed with appropriate medications (e.g. idarucizumab for dabigatran reversal, and andexanet alfa for rivaroxaban and apixaban reversal, prothrombin complex concentrate for edoxaban (and rivaroxaban or apixaban if

Table 17.1 Selection of oral anticoagulant therapy depending on the clinical scenario [40–44]

Scenario	Therapeutic choice
Valvular Afib (mechanical prosthetic valve, rheumatic mitral stenosis, moderate to severe mitral stenosis)	Warfarin
Ischaemic stroke on oral anticoagulant agent	Dabigatran 150
High thromboembolic and bleeding risk	Apixaban
High thromboembolic risk and low bleeding risk	Dabigatran 150
Low thromboembolic risk and high bleeding risk	Edoxaban 30 Apixaban
Moderate thromboembolic and bleeding risk	Apixaban Edoxaban 30 Dabigatran 110
Recurrent stroke	Rivaroxaban
High overall bleeding risk	Dabigatran 110 Apixaban
Coronary disease	Rivaroxaban
High gastrointestinal bleeding risk / prior GI bleeding	Apixaban Dabigatran
Dyspepsia	Apixaban Rivaroxaban
Elderly people	Apixaban Rivaroxaban Edoxaban
Renal impairment (CrCl>15 ml/min)	Apixaban Rivaroxaban
End-stage renal disease	Warfarin Apixaban (in selected cases)
Liver dysfunction	Warfarin Child-Pugh category A – all NOACs Child-Pugh category B – no rivaroxaban Child-Pugh category C – warfarin
Patient preference	Rivaroxan Edoxaban

andexanet alfa is not available) or vitamin K antagonists (plus vitamin K)). If PCC is not available, fresh frozen plasma may be used for vitamin K antagonist associated ICH [6, 47]. It is then necessary to assess the risks of a new ischemic stroke and the risk of recurrent intracerebral bleeding [45, 46]. In particular, several points should be addressed in the assessment: indication and selection of anticoagulant therapy, age, magnitude of ischemic or hemorrhagic lesion and prior use of reperfusion procedures such as intravenous or intraarterial thrombolysis and mechanical thrombectomy [45, 46]. If the risk for ischemic stroke is assessed as extremely high, in accordance with the 2007 AHA/ASA recommendations, warfarin resumption may be considered 7–10 days after intracerebral haemorrhage [45]. It is recommended that oral anticoagulant therapy can be resumed 3 to 4 weeks after ICH with rigorous INR monitoring and with lower INR targets [6]. According to the guidelines of the European Society of Cardiology, oral anticoagulant therapy should be resumed 4 to 8 weeks after intracerebral haemorrhage, and in the case of contraindication to oral anticoagulant therapy, left atrial appendage closure would be a therapeutic choice [39].

Conclusion

In conclusion, oral anticoagulant therapy still represents a gold standard for secondary stroke prevention in patients with atrial fibrillation. Studies on direct oral anticoagulants - direct thrombin inhibitor (dabigatran) and factor Xa inhibitors (rivaroxaban, apixaban and edoxaban) showed at least non-inferiority in treatment efficacy in comparison to warfarin, and a better safety profile, therefore these agents are included in international guidelines and became a preferred treatment option.

References

1. Demarin V. The burden of stroke: a growing health care and economy problem. Acta Clin Croat. 2004;43:9–13.
2. Kolominsky-Rabas PL, Weber M, Gefeller O, Neundoerfer B, Heuschmann PU. Epidemiology of ischemic stroke subtypes according to TOAST criteria: incidence, recurrence, and long-term survival in ischemic stroke subtypes: a population-based study. Stroke. 2001;32(12):2735–40. Epub 2001/12/12
3. Camm AJ, Kirchhof P, Lip GY, Schotten U, Savelieva I, Ernst S, et al. Guidelines for the management of atrial fibrillation: the Task Force for the Management of Atrial Fibrillation of the European Society of Cardiology (ESC). Eur Heart J. 2010;31(19):2369–429. Epub 2010/08/31
4. Wolf PA, Abbott RD, Kannel WB. Atrial fibrillation as an independent risk factor for stroke: the Framingham study. Stroke. 1991;22(8):983–8. Epub 1991/08/01
5. Dulli DA, Stanko H, Levine RL. Atrial fibrillation is associated with severe acute ischemic stroke. Neuroepidemiology. 2003;22(2):118–23. Epub 2003/03/12
6. Fuster V, Ryden LE, Cannom DS, Crijns HJ, Curtis AB, Ellenbogen KA, et al. ACC/AHA/ESC 2006 guidelines for the management of patients with atrial fibrillation-executive summary:

a report of the American College of Cardiology/American Heart Association Task Force on practice guidelines and the European Society of Cardiology Committee for practice guidelines (writing committee to revise the 2001 guidelines for the management of patients with atrial fibrillation). Eur Heart J. 2006;27(16):1979–2030. Epub 2006/08/04

7. Kirchhof P, Benussi S, Kotecha D, Ahlsson A, Atar D, Casadei B, et al. 2016 ESC guidelines for the management of atrial fibrillation developed in collaboration with EACTS. Eur Heart J. 2016;37(38):2893–962. Epub 2016/08/28

8. Sacco RL, Adams R, Albers G, Alberts MJ, Benavente O, Furie K, et al. Guidelines for prevention of stroke in patients with ischemic stroke or transient ischemic attack: a statement for healthcare professionals from the American Heart Association/American Stroke Association Council on Stroke: co-sponsored by the Council on Cardiovascular Radiology and intervention: the American Academy of Neurology affirms the value of this guideline. Stroke. 2006;37(2):577–617. Epub 2006/01/25

9. Hart RG, Benavente O, McBride R, Pearce LA. Antithrombotic therapy to prevent stroke in patients with atrial fibrillation: a meta-analysis. Ann Intern Med. 1999;131(7):492–501. Epub 1999/10/03

10. Hammwohner M, Goette A. Will warfarin soon be passe? New approaches to stroke prevention in atrial fibrillation. J Cardiovasc Pharmacol. 2008;52(1):18–27. Epub 2008/07/03

11. Moyer TP, O'Kane DJ, Baudhuin LM, Wiley CL, Fortini A, Fisher PK, et al. Warfarin sensitivity genotyping: a review of the literature and summary of patient experience. Mayo Clin Proc. 2009;84(12):1079–94. Epub 2009/12/04

12. Wardrop D, Keeling D. The story of the discovery of heparin and warfarin. Br J Haematol. 2008;141(6):757–63. Epub 2008/03/22

13. Duxbury BM, Poller L. The oral anticoagulant saga: past, present, and future. Clin Appl Thromb Hemost. 2001;7(4):269–75. Epub 2001/11/08

14. Last JA. The missing link: the story of Karl Paul link. Toxicol Sci. 2002;66(1):4–6. Epub 2002/02/28

15. Lim GB. Milestone 2: warfarin: from rat poison to clinical use. Nat Rev Cardiol. 2017; Epub 2017/12/15

16. Kirkwood TB. Calibration of reference thromboplastins and standardisation of the prothrombin time ratio. Thromb Haemost. 1983;49(3):238–44. Epub 1983/06/28

17. Whitlon DS, Sadowski JA, Suttie JW. Mechanism of coumarin action: significance of vitamin K epoxide reductase inhibition. Biochemistry. 1978;17(8):1371–7. Epub 1978/04/18

18. Olsen TS, Langhorne P, Diener HC, Hennerici M, Ferro J, Sivenius J, et al. European stroke initiative recommendations for stroke management-update 2003. Cerebrovasc Dis. 2003;16(4):311–37. Epub 2003/10/31

19. Deplanque D, Leys D, Parnetti L, Schmidt R, Ferro J, de Reuck J, et al. Secondary prevention of stroke in patients with atrial fibrillation: factors influencing the prescription of oral anticoagulation at discharge. Cerebrovasc Dis. 2006;21(5-6):372–9. Epub 2006/02/24

20. Halperin JL. Ximelagatran compared with warfarin for prevention of thromboembolism in patients with nonvalvular atrial fibrillation: rationale, objectives, and design of a pair of clinical studies and baseline patient characteristics (SPORTIF III and V). Am Heart J. 2003;146(3):431–8. Epub 2003/08/30

21. Connolly SJ, Eikelboom J, Joyner C, Diener HC, Hart R, Golitsyn S, et al. Apixaban in patients with atrial fibrillation. N Engl J Med. 2011;364(9):806–17. Epub 2011/02/12

22. Connolly SJ, Ezekowitz MD, Yusuf S, Eikelboom J, Oldgren J, Parekh A, et al. Dabigatran versus warfarin in patients with atrial fibrillation. N Engl J Med. 2009;361(12):1139–51. Epub 2009/09/01

23. Granger CB, Alexander JH, McMurray JJ, Lopes RD, Hylek EM, Hanna M, et al. Apixaban versus warfarin in patients with atrial fibrillation. N Engl J Med. 2011;365(11):981–92. Epub 2011/08/30

24. Patel MR, Mahaffey KW, Garg J, Pan G, Singer DE, Hacke W, et al. Rivaroxaban versus warfarin in nonvalvular atrial fibrillation. N Engl J Med. 2011;365(10):883–91. Epub 2011/08/13

25. Giugliano RP, Ruff CT, Braunwald E, Murphy SA, Wiviott SD, Halperin JL, et al. Edoxaban versus warfarin in patients with atrial fibrillation. N Engl J Med. 2013;369(22):2093–104. Epub 2013/11/21

26. Budinčević H. Utjecaj antikoagulantne terapije na ishod ishemijskoga moždanoga udara u bolesnika s fibrilacijom atrija [Disertacija]. Zagreb: Medicinski fakultet, Sveučilište u Zagrebu; 2014.

27. Camm AJ, Lip GY, De Caterina R, Savelieva I, Atar D, Hohnloser SH, et al. 2012 focused update of the ESC guidelines for the management of atrial fibrillation: an update of the 2010 ESC guidelines for the management of atrial fibrillation--developed with the special contribution of the European heart rhythm association. Europace: European pacing, arrhythmias, and cardiac electrophysiology : journal of the working groups on cardiac pacing, arrhythmias, and cardiac cellular electrophysiology of the European Society of Cardiology. Epub 2012/08/28. 2012;14(10):1385–413.

28. Kernan WN, Ovbiagele B, Black HR, Bravata DM, Chimowitz MI, Ezekowitz MD, et al. Guidelines for the prevention of stroke in patients with stroke and transient ischemic attack: a guideline for healthcare professionals from the American Heart Association/American Stroke Association. Stroke. 2014;45(7):2160–236. Epub 2014/05/03

29. Demarin V, Rundek T, Budincevic H. Kaj je novega v smernicah obravnave ishemične možganske kapi/ What is new in the guidelines for ischemic stroke management. In: Žvan BZM, editor. Akutna možganska kap. Ljubljana, Slovenia: Društvo za preprečevanje možganskih in žilnih bolezni; 2015. p. 167–83.

30. Klijn CJ, Paciaroni M, Berge E, Korompoki E, Korv J, Lal A, et al. Antithrombotic treatment for secondary prevention of stroke and other thromboembolic events in patients with stroke or transient ischemic attack and non-valvular atrial fibrillation: a European Stroke Organisation guideline. Eur Stroke J. 2019;4(3):198–223. Epub 2020/01/28

31. Siontis KC, Yao X, Gersh BJ, Noseworthy PA. Direct oral anticoagulants in patients with atrial fibrillation and valvular heart disease other than significant mitral stenosis and mechanical valves: a meta-analysis. Circulation. 2017;135(7):714–6. Epub 2017/02/15

32. Di Biase L. Use of direct oral anticoagulants in patients with atrial fibrillation and valvular heart lesions. J Am Heart Assoc. 2016;5:2. Epub 2016/02/20

33. Baumgartner H, Falk V, Bax JJ, De Bonis M, Hamm C, Holm PJ, et al. 2017 ESC/EACTS guidelines for the management of valvular heart disease. Eur Heart J. 2017;38(36):2739–91. Epub 2017/09/10

34. Furie KL, Kasner SE, Adams RJ, Albers GW, Bush RL, Fagan SC, et al. Guidelines for the prevention of stroke in patients with stroke or transient ischemic attack: a guideline for healthcare professionals from the american heart association/american stroke association. Stroke. 2011;42(1):227–76. Epub 2010/10/23

35. Fuster V, Ryden LE, Cannom DS, Crijns HJ, Curtis AB, Ellenbogen KA, et al. 2011 ACCF/AHA/HRS focused updates incorporated into the ACC/AHA/ESC 2006 guidelines for the management of patients with atrial fibrillation: a report of the American College of Cardiology Foundation/American Heart Association Task Force on Practice Guidelines developed in partnership with the European Society of Cardiology and in collaboration with the European Heart Rhythm Association and the Heart Rhythm Society. J Am Coll Cardiol. 2011;57(11):e101–98. Epub 2011/03/12

36. Hindricks G, Potpara T, Dagres N, Arbelo E, Bax JJ, Blomstrom-Lundqvist C, et al. 2020 ESC guidelines for the diagnosis and management of atrial fibrillation developed in collaboration with the European Association for Cardio-Thoracic Surgery (EACTS): the task force for the diagnosis and management of atrial fibrillation of the European Society of Cardiology (ESC) developed with the special contribution of the European Heart Rhythm Association (EHRA) of the ESC. Eur Heart J. 2021;42(5):373–498. Epub 2020/08/30

37. Katsanos AH, Kamel H, Healey JS, Hart RG. Stroke prevention in atrial fibrillation: looking forward. Circulation. 2020;142(24):2371–88. Epub 2020/12/15

38. Kleindorfer DO, Towfighi A, Chaturvedi S, Cockroft KM, Gutierrez J, Lombardi-Hill D, et al. 2021 guideline for the prevention of stroke in patients with stroke and transient ischemic attack: a guideline from the American Heart Association/American Stroke Association. Stroke. 2021;52(7):e364–467. Epub 2021/05/25

39. Kirchhof P, Benussi S, Kotecha D, Ahlsson A, Atar D, Casadei B, et al. 2016 ESC guidelines for the management of atrial fibrillation developed in collaboration with EACTS. Europace : European pacing, arrhythmias, and cardiac electrophysiology : journal of the working groups on cardiac pacing, arrhythmias, and cardiac cellular electrophysiology of the European Society of Cardiology. Epub 2016/11/04. 2016;18(11):1609–78.

40. Weitz JI, Gross PL. New oral anticoagulants: which one should my patient use? Hematology Am Soc Hematol Educ Program. 2012;2012:536–40. Epub 2012/12/13

41. Schaefer JK, McBane RD, Wysokinski WE. How to choose appropriate direct oral anticoagulant for patient with nonvalvular atrial fibrillation. Ann Hematol. 2016;95(3):437–49. Epub 2015/12/15

42. Savelieva I, Camm AJ. 'Preferred' management of atrial fibrillation in Europe. Europace: European pacing, arrhythmias, and cardiac electrophysiology : journal of the working groups on cardiac pacing, arrhythmias, and cardiac cellular electrophysiology of the European Society of Cardiology. Epub 2013/10/01. 2014;16(1):1–3.

43. Hammersley D, Signy M. Navigating the choice of oral anticoagulation therapy for atrial fibrillation in the NOAC era. Ther Adv Chronic Dis. 2017;8(12):165–76. Epub 2017/12/05

44. Gish RG, Flamm SL. Anticoagulation in patients with chronic liver disease. Gastroenterol Hepatol. 2021;17(1 Suppl 1):10–5. Epub 2021/06/18

45. Broderick J, Connolly S, Feldmann E, Hanley D, Kase C, Krieger D, et al. Guidelines for the management of spontaneous intracerebral hemorrhage in adults: 2007 update: a guideline from the American Heart Association/American Stroke Association Stroke Council, High Blood Pressure Research Council, and the Quality of Care and Outcomes in Research Interdisciplinary Working Group. Stroke. 2007;38(6):2001–23. Epub 2007/05/05

46. Jauch EC, Saver JL, Adams HP Jr, Bruno A, Connors JJ, Demaerschalk BM, et al. Guidelines for the early management of patients with acute ischemic stroke: a guideline for healthcare professionals from the American Heart Association/American Stroke Association. Stroke. 2013;44(3):870–947. Epub 2013/02/02

47. Christensen H, Cordonnier C, Korv J, Lal A, Ovesen C, Purrucker JC, et al. European Stroke Organisation Guideline on Reversal of oral anticoagulants in acute intracerebral haemorrhage. Eur Stroke J. 2019;4(4):294–306. Epub 2020/01/07

Chapter 18
Telestroke: Illusion or Reality

Bojana Žvan, Matija Zupan, and Marjan Zaletel

Introduction

Telemedicine refers to the use of telecommunication technology to provide health care from the distance. In 1999, Levine and Gorman introduced the term 'telestroke' as the use of telemedicine to provide neurological consultation for stroke patients in hospitals lacking the appropriate level of expertise [1].

It is well known that intravenous thrombolysis (IVT) for acute ischaemic stroke (AIS), when applied early after symptom initiation, improves patients` outcomes in a time-dependent manner [2].

Since late 1990s, multiple telestroke projects have been developed worldwide for the management of stroke patients. Several studies have demonstrated increasing IVT rates after telestroke implementation without significant differences in safety or efficacy (class of recommendation IIb, level of evidence B) [3–5]. Additionally, telestroke, an application of telemedicine, allows a rapid and reliable recognition and treatment of AIS without on-site neurologic expertise. Eligible patients receive IVT in the emergency department and are admitted either to the stroke unit (SU) of the admitting hospital (drip and keep) or to a comprehensive stroke centre (CSC) (drip and ship) to complete the infusion [6].

Within telestroke networks, a comprehensive organisation of acute stroke care could lead to many other benefits which so far are heavily underused in neurology. These benefits include: shortening hospital stay of patients through advanced care, avoiding a large number of unnecessary patient transfers, identifying specific stroke patients who require urgent interventions or surgery (subarachnoid haemorrhage

B. Žvan (✉) · M. Zupan · M. Zaletel
Department of Vascular Neurology and Intensive Therapy, Neurology Clinic, University Medical Centre, Zaloska 2, Ljubljana, Slovenia
e-mail: bojana.zvan@kclj.si

V. Demarin et al. (eds.), *Mind, Brain and Education*,
https://doi.org/10.1007/978-3-031-33013-1_18

(SAH), intracerebral and intraventricular haemorrhage (ICH, IVH), candidates for craniectomy), leading to establishment of stroke units and stroke teams in spoke hospitals and overall improvement of stroke care in spoke hospitals, early diagnosis and proper treatment of stroke and non-stroke patients. Further benefits may be: to facilitate staff recruitment to spoke hospitals, to deliver expertise to developing countries, participation of spoke hospitals to acute stroke treatment trials and stroke prevention trials, and environmental effects [7].

Telestroke is valuable also for the management of intracranial haemorrhages in rural hospitals and hospitals lacking neurosurgical departments, given that surgical/interventional therapy is only recommended for a subgroup of patients. It also helps select patients who need advanced neurological and/or neurosurgical care [8].

Telestroke (TeleKap) in Slovenia

In Slovenia, the national IVT rate was estimated to be <3% before 2015. Available stroke services before 2015 included general hospitals (GHs) with around-the-clock neurologic staff (24 h, 7 days per week—24/7), GHs with limited availability of neurologists only during regular working hours (8 h, 5 days per week—8/5), and GHs without neurologists (0/0). The only comprehensive stroke centre (CSC) was located in the capital city of Ljubljana. Only three GHs had a stroke unit (SU). The other GHs were equipped with standard intensive care units for critically ill medical patients. The system as it existed at that time did not allow appropriate coordination. To overcome these inconsistencies, we launched the national telestroke network, named TeleKap. TeleKap has been established as an extension of the existing system merely offering optional support to attending neurologists and internists in treating AIS patients. The main aim of our network has been to enhance IVT use in Slovenia by using telemedicine. Thus, we established a hub-and-spoke system covering the entire nation. Shortly after the TeleKap system was established, physicians from spoke hospitals were found to be using the network for many patients with other neurological disorders, not just for AIS.

Slovenia is a mostly rural country with a population of two million inhabitants and covers a surface area of 20,273 km^2. Population density is very variable, ranging from over 1000/km^2 in Ljubljana, the capital town, to only 5/km^2 in some Alpine regions. Given the relatively small size of Slovenia, we set up a nationwide stroke care telemedicine system covering the entire country. The network started operating on September 15, 2014, and is approved by the appropriate authorities. The service is fully reimbursed by the Health Insurance Institute of Slovenia. A map of the national TeleKap network is depicted in Fig. 18.1.

Spokes are classified according to the availability of a neurologist. There are two around-the-clock neurologist spokes (24/7). Around-the-clock neurologic expertise is also available at the University medical centres (UMCs) in both Ljubljana and Maribor. There are six spokes with 8/5 neurologic expertise. Outside regular

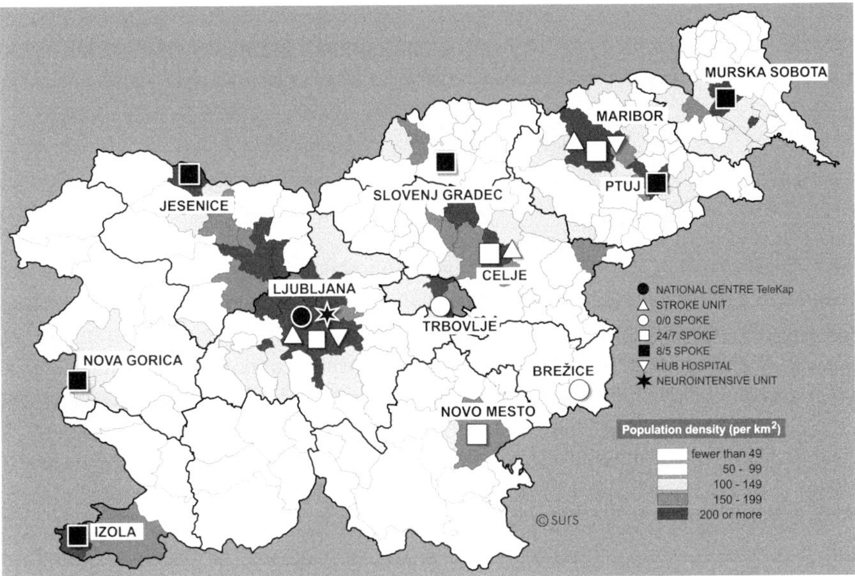

Fig. 18.1 The Slovenian national TeleKap network (adapted with permission from www.stat.si)

working hours, these spokes have access to neurologic expertise via telemedicine only. There are two 0/0 spokes without a neurologist, which rely entirely on neurologic expertise via telemedicine. UMC Ljubljana (UMCL) is the foremost hub hospital covering the entire nation in regard to 24/7 comprehensive stroke care, including endovascular treatment (EVT). Additionally, UMCL offers the only neurologic critical care unit, 24/7 neurosurgical expertise, and an SU. The emergency neurologic department (END) at the UMCL is also one of the beneficiaries of teleconsultations in the network, especially if a junior neurologist or a neurology resident is the attending physician. UMC Maribor (UMCM) has not been asking for teleconsultations in TeleKap regarding IVT in their own patients but has been participating as the hub for neurosurgical patients from its spokes. At the same time, UMCM is not yet the hub for EVT in AIS patients from its spokes, who have been transferred to UMCL. The technical equipment in the TeleKap centre consists of a dedicated stationary workstation with an HD video camera (Polycom High Definition Group 700) and a high-quality microphone (Polycom RealPresence), enabling two-way teleconsultations. Dedicated mobile workstations (with an integrated HD videocamera and microphone) allow teleconsultations from home with the same quality as video conferencing from the TeleKap center. Spokes are equipped with a mobile telemedical cart consisting of a fully remotely controlled HD videocamera, HD display, and microphone (Polycom Real Presence). Videoconferencing is enabled at a resolution of at least 720 p HD at 60 fps. The servers and software used in the network are part of the Slovenian medical network "eZdravje" (English: "eHealth"). Safety and confidentiality are ensured through safe connections over the closed Z-NET network, which is an integral part of eHealth. There is a platform for reviewing neuroradiological images in DICOM_ format (XERO Viewer; Agfa HealthCare) [5].

Workflow

The TeleKap network operates on a 24/7/365 schedule. Telemedicine comprises telephone contact between a physician at a spoke centre and a consultant vascular neurologist, transmission of digital images, and real-time clinical examination of patients via videoconferencing. It is not mandatory for a treating physician to consult TeleKap for every patient suspected with AIS and other cerebrovascular diseases. A spoke physician asks for a teleconsultation at his/her own judgment. A dedicated stroke team assists in the workflow at spoke hospitals, including a nurse. Every spoke is able to provide hyperacute care as well as further medical care of AIS patients, which is not the case for patients with intracranial haemorrhages. A telemedical follow-up can be requested at any time during the in-hospital stay for care-related issues of patients who remain in the spoke hospital.

Clinical Application of Telekap Network

Acute Ischaemic Stroke

Intravenous Thrombolysis in Acute Ischaemic Stroke

In the TeleKap network, we establish the clinical diagnosis of AIS on the recognition of typical stroke symptoms and consistent findings on neurologic examination, including the National Institutes of Health Stroke Scale (NIHSS) score. This is corroborated by multimodal computed tomography (CT) imaging, including noncontrast brain CT and aortocervical CT angiography in every patient with suspected AIS. CT perfusion is being used increasingly but is limited to both UMCs and three spoke hospitals. MRI is not part of the standard imaging protocol, and we perform it only in exceptional cases predominantly in both UMCs (e.g., pregnant women, very rarely in wake-up AIS). Standard laboratory work-up is simultaneously performed.

Data Collection, Statistical Analysis and Results for Acute Ischemic Stroke

We analysed the data of patients treated in the TeleKap network from January 1, 2015, to December 31, 2017 (observation period). We used linear regression to determine the trends. The IVT rate was calculated as the ratio between the absolute number of IVT patients and the absolute number of AIS patients in a given year. We tested the differences between the groups with Student's t test. For every statistical test, $p < 0.05$ was regarded as statistically significant.

During the observation period, we treated a total of 2224 patients. The demographic data for a subgroup of 2165 patients are available, of which there were 1094 (50.5%) males. A total of 1219 (56.3%) patients were aged >70 years. We treated a total of 1316 patients (59.2%) with AIS, of which 508 (38.6%) received IVT. There were 142 IVT patients in 2015, 158 in 2016, and 208 in 2017. The absolute numbers of AIS patients and IVT patients increased, as depicted in Fig. 18.2. Linear regression showed statistically significant positive trends in the absolute number of IVT patients (B = 4.39, standard error (SE) = 1.59, $p = 0.011$) and the absolute number of AIS patients (B = 14.42, SE = 5.19, $p = 0.010$).

The trend of IVT rate was numerically negative (43.5% in 2015, 37.3% in 2016, and 36.7% in 2017), although not statistically significant ($p = 0.302$). We treated a total of 591 (26.5%) patients with nonvascular neurologic conditions, of which we identified 261 (16.6% of 1577 patients with suspected AIS) patients with stroke mimics (almost a half of them due to postictal neurologic deficit or encephalopathy).

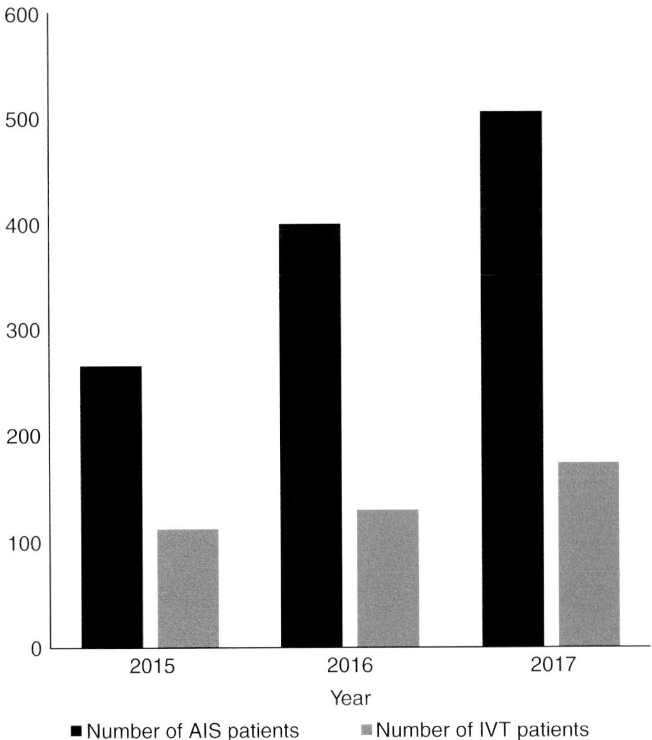

Fig. 18.2 Absolute numbers of patients with AIS and patients treated with IVT from 2015 to 2017. AIS, acute ischemic stroke; IVT, intravenous thrombolysis

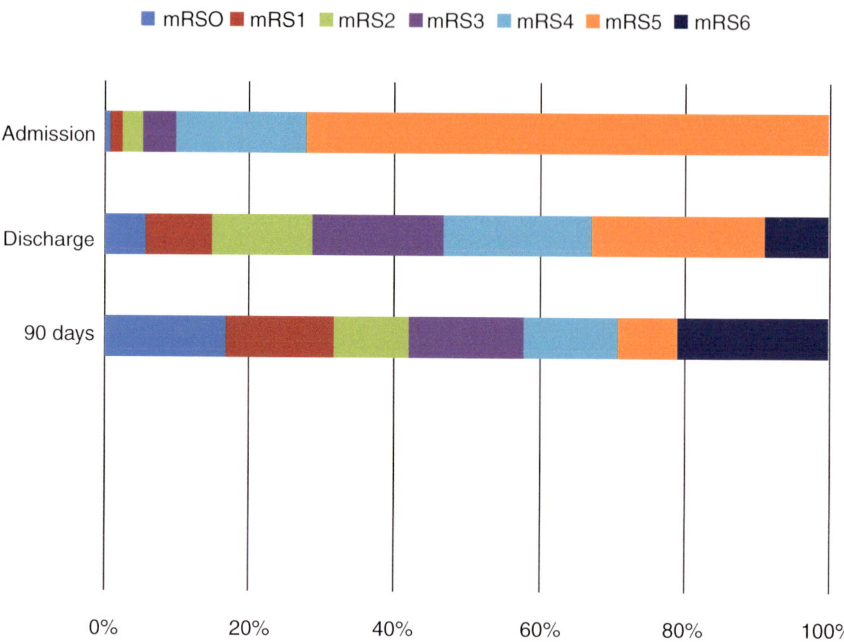

Fig. 18.3 Functional outcomes in patients treated with MeR in the TeleKap network according to the modified Rankin Score on admission, at discharge and 90 days post-admission

We analysed the trend of IVT patients in different types of spokes (Fig. 18.3). Regarding the whole network, our analysis showed significant positive trends in the absolute number of AIS patients (B = 7.66, SE = 0.59. $p < 0.001$) and in the absolute number of IVT patients (B = 2.16, SE = 0.56, $p = 0.003$). Furthermore, there was a significant positive trend in other diagnoses, principally nonvascular stroke mimics (B = 3.58, SE = 0.90, $p = 0.003$).

The main finding of our study is a constant, progressive increase in the absolute number of IVT patients in the TeleKap network. In addition, we found a statistically significant progressive increase in the absolute number of AIS patients. At the same time, we did not observe significant changes in the IVT rate. The analysis showed a significant relationship between teleconsultations and IVT use. The IVT rates were 57.1% in 2015, 40.0% in 2016, and 47.4% in 2017. We believe that AIS patients treated in TeleKap are primarily highly selected patients often eligible for IVT. This consideration is the main reason the IVT rate is so high, which is in stark contrast with the Telemedical Project for integrative stroke care in the region of southeastern Bavaria (TEMPiS [in German]) network (IVT rate of 16.7% according to contemporary data), where virtually every patient with suspected AIS is treated inside the network [9].

We expect that with further years of TeleKap activity, internists from spoke hospitals will increasingly opt for teleconsultations, which will further decrease the IVT rate inside TeleKap and bring it closer to the rate of most other telestroke networks.

Mechanical Recanalization in Acute Ischaemic Stroke

Mechanical recanalization (MeR) in the case of AIS due to large vessel occlusion (LVO) referred to the TeleKap network has been provided from the very beginning of the network. Patients are transferred from network hospitals to the University Medical Centre in Ljubljana on a drip-and-ship principle.

In a retrospective analysis, we included patients with AIS due to LVO from 9/15/2014 to 8/31/2019 who were treated with MeR. The primary outcome event was the functional outcome according to the modified Rankin scale (mRS) at discharge and 90 days after admission. The mRS shares were dichotomized into groups with good functional state (GFS; mRS 0-2), and poor functional state (PFS; mRS 3-6). Trends were analysed using cubic, exponential and linear functions. A logistic univariate regression was used to analyse the association between binary variables.

We included 161 patients aged 68.9 ± 11.9 years. The trend in the number of interventions performed monthly increased with a significant fit to the cubic function ($p = 0.047$). 90 days after admission, 42 patients (43.8%) had GFS (at discharge 36 patients (28.8%), $p = 0.041$), and 33 patients (53.2%) had PFS (at discharge 44 patients (53%), $p < 0.001$). 90 days after admission, 20 patients (21.1%) died and 11 patients (8.8%) died at discharge, $p < 0.001$. The median mRS on admission was 5, at discharge 4, and 90 days after admission 3. Figure 18.3 shows further details.

Based on the analysis, the number of MeR interventions performed in the TeleKap network has been growing exponentially. Treatment outcomes are favourable and comparable to data from randomised international studies with more than 40% of patients achieving GFS and about 20% mortality 90 days after admission. In the TeleKap network, we triage patients for MeR in network hospitals based on clinical picture and bimodal CT imaging (non-contrast CT and aortocervical CTA) As the results show, this is a fairly effective way for our country. The main shortcoming is inconsistent recording of data in the TeleKap network. For patients from northeastern Slovenia we would like to ensure that MeR would be available at the University Medical Centre Maribor on a 24/7 schedule as soon as possible.

Intracranial Haemorrhage

As many as 30% of the 4400 people with stroke in Slovenia yearly experience it in an active working period. In AIS, the mortality rate is lowest, about 14%, while in haemorrhagic stroke it is between 30% and 40%. Like elsewhere in Europe, Slovenia is experiencing a gradual decline in the incidence of stroke. Contrary to the assumption that absolute numbers of new stroke cases will increase in the coming decades, this is currently not observed in Slovenia. In recent years, there has been a trend of the increasing average age of patients when they experience cerebral infarction and cerebral haemorrhage (CH). Data show that the patients' care, especially those with severe strokes, is gradually improving [10].

The analysis of the TeleKap data showed that the number of patients with intracerebral haemorrhage (ICH) and subarachnoid haemorrhage (SAH) treated in the TeleKap network did not change between 2016 and 2019. The trend of patients with SAH shows an increase, but the trend of patients with ICH does not change significantly. Transfer of patients with SAH to a tertiary centre indicates a positive trend that is statistically borderline significant. The decision to transfer the patient to a tertiary centre indicates a negatively significant association with age, which means that we decide to relocate primarily younger patients with intracranial bleeding. This also applies to the subgroup of patients with ICH, but not to SAH, as most patients with SAH are transferred to the UMCL.

The best decision on how to treat patients with SAH and ICH requires the close cooperation of a multidisciplinary team consisting of a vascular neurologist, a neurosurgeon, an interventional neuroradiologist, and often an intensive care specialist. A physician in the emergency centre in the network hospital and a neurologist at the emergency neurology department at the tertiary centre are also of key importance in the treatment chain of these patients.

In our opinion, most patients with SAH, and in particular younger and more at-risk patients with ICH who do not meet the criteria for palliative care, should be transferred to a tertiary institution. The method of treatment of patients with ICH or SAH is determined by a multidisciplinary team, which gets acquainted with the clinical picture and imaging examinations already in the TeleKap network. The multidisciplinary team plays an important role in deciding whether a patient with SAH or ICH needs palliative care. Last but not least, the will of the patient and his relatives must always be taken into account in the treatment.

Conclusion

Today, it is quite clear that telestroke benefits are becoming more and more extensive as it develops at lightning speed. Slovenian experience has shown that TeleStroke is a versatile tool for improving acute stroke care of inhabitants in neurologically underserved regions. The telemedicine services can also be used in other areas of vascular neurology and not only for patients with AIS, for example in patients with SAH and ICH. Undoubtedly, the TeleKap system has enabled more patients in the country to be treated with reperfusion therapy in AIS and appropriately address the intracranial haemorrhage patients` needs. The possibility of consultations in the TeleKap network is also important from the point of view of equal treatment of all patients with stroke and with intracranial haemorrhage regardless of the admitting hospital.

We can conclude that the reduced mortality rate of patients with severe stroke in Slovenia, including ICH and SAH, is the result of improved care, which is certainly influenced by the national TeleKap network.

Unfortunately, the TeleKap system does not yet have a health and economic analysis that would show cost-effectiveness and long-term savings in the costs of hospitals and society, as it still does not have a register for patients with cerebrovascular diseases.

From a global perspective, telestroke has now become a reality in the treatment of patients with cerebrovascular diseases worldwide and not just an illusion.

References

1. Levine SR, Gorman M. Telestroke : the application of telemedicine for stroke. Stroke. 1999;30:464–9.
2. Lees KR, Bluhmki E, von Kummer R, et al. Time to treatment with intravenous alteplase and outcome in stroke: an updated pooled analysis of ECASS, ATLANTIS, NINDS, and EPITHET trials. Lancet. 2010;375:1695–703.
3. Audebert HJ, Kukla C, Vatankhah B, et al. Comparison of tissue plasminogen activator administration management between Telestroke Network hospitals and academic stroke centers: the Telemedical Pilot Project for Integrative Stroke Care in Bavaria/Germany. Stroke. 2006;37:1822–7.
4. Carmen Jime'nez M, Tur S, Legarda I, et al. The application of telemedicine for stroke in the Balearic Islands: the Balearic Telestroke project. Rev Neurol. 2012;154:31–40.
5. Zupan M, Zaletel M, Žvan B. Enhancement of intravenous thrombolysis by Nationwide Telestroke Care in Slovenia: a model of care for middle-income countries. Telemed J E Health. 2019; https://doi.org/10.1089/tmj.2019.0046. [Epub ahead of print]
6. Ali SF, Hubert GJ, Switzer JA, et al. Validating the TeleStroke mimic score: a prediction rule for identifying stroke mimics evaluated over telestroke networks. Stroke. 2018;49:688–92.
7. Tatlisumak T, Soinila S, Kaste M. Telestroke networking offers multiple benefits beyond thrombolysis. Cerebrovasc Dis. 2009;27(Suppl 4):21 7.
8. Backhaus R, Schlachetzki F, Rackl W, Baldaranov D, Leitzmann M, Hubert GJ, et al. Intracranial hemorrhage: frequency, location, and risk factors identified in a TeleStroke network. Neuroreport. 2015;26(2):81–7.
9. Hubert G. TEMPiS. Telemedizinisches Projekt zur integrierten Schlaganfallversorgung in der Region Su" d-Ost-Bayern. 2016. Available at www.tempis.de/index.php/component/jdownloads/finish/3-jahresberichte/1070-jahresbericht-2016.html (last accessed November 2, 2017).
10. Perko D, Korošec A. Epidemiologija možganske kapi v Sloveniji. In: Žvan B, Zaletel M, Zupan M, editors. Akutna možganska kap XIII. [Elektronski vir]. Ljubljana: Društvo za preprečevanje možganskih in žilnih bolezni; 2019. p. 33–52.

Chapter 19
Hemodynamic Changes on Transcranial Duplex Sonography to Predict Poor Outcome after Mechanical Thrombectomy

Markus Kneihsl, Kurt Niederkorn, Hannes Deutschmann, and Thomas Gattringer

Abbreviations

ACA	Anterior Cerebral Artery
DSA	Digital Subtraction Angiography
ICA	Internal Carotid Artery
ICH	Intracranial Hemorrhage
IVT	Intravenous Thrombolysis
LVO	Large Vessel Occlusion
MBF	Mean Blood Flow
MCA	Middle Cerebral Artery
MRA	Magnetic Resonance Angiography
MRI	Magnetic Resonance Imaging
MT	Mechanical Thrombectomy
OR	Odds Ratio
TCD	Transcranial Duplex Sonography
TIBI	Thrombolysis In Brain Ischemia classification

M. Kneihsl · T. Gattringer (✉)
Department of Neurology, Vascular and Interventional Radiology, Medical University of Graz, Graz, Austria

Department of Neuroradiology, Vascular and Interventional Radiology, Medical University of Graz, Graz, Austria
e-mail: thomas.gattringer@medunigraz.at

K. Niederkorn
Department of Neurology, Vascular and Interventional Radiology, Medical University of Graz, Graz, Austria

H. Deutschmann
Department of Neuroradiology, Vascular and Interventional Radiology, Medical University of Graz, Graz, Austria

Endovascular recanalization with **mechanical thrombectomy (MT)** has become the gold-standard treatment option of acute ischemic stroke due to large vessel occlusion (LVO) of the anterior cerebral circulation. Recent randomized-controlled clinical trials have demonstrated that MT combined with intravenous thrombolysis (IVT) results in good functional outcome in up to 50% of patients with LVO stroke as compared to only 25% who receive standard care including IVT [1].

With increasing experience and technical advances, **successful recanalization** can nowadays be achieved in 80–90% of all thrombectomy cases [2, 3]. This is crucial as early successful cerebrovascular reperfusion has been identified as the strongest independent predictor of favorable short-term prognosis in stroke/ MT patients [4].

Despite high recanalization rates of up to 90%, every second ischemic stroke patient with LVO remains functionally dependent. **Complications** in the peri−/postinterventional period are important predictors for poor clinical outcome after MT [5, 6]. While MT is overall relatively safe, the most frequent and feared complication in this setting is **secondary brain hemorrhage**, which affects **20-25%** of patients and consequently often leads to **worse clinical outcome** (rate of immediately symptomatic intracranial hemorrhage: approximately **5%**). Moreover, intracranial hemorrhage (ICH) can prolong or prohibit the use of anticoagulants/ antithrombotics which would be necessary for early prevention of recurrent ischemic stroke. Severe clinical stroke syndromes at baseline, large infarct size, arterial hypertension, elevated blood glucose, diabetes mellitus and additional thrombolytic therapy have been identified as risk factors for ICH after MT [3, 7–10]. However, the **exact mechanisms still remain unclear** and, in clinical practice, ICH after MT is also commonly observed in patients without these known risk factors.

ICH and **vasogenic brain edema** are the main morphological features of **reperfusion injury** – a pathophysiological term describing the complex biochemical mechanisms underlying further damage of the ischemic tissue after successful recanalization [11]. While reperfusion injury has been well studied in experimental **animal stroke models**, comprehensive **prospective data in humans** in the setting of stroke thrombectomy is **lacking**.

Post-Thrombectomy Hyperperfusion as a Possible Mechanism of Secondary Brain Hemorrhage and Reperfusion Injury

Reactive hyperemia/hyperperfusion is a well-known condition after rapid revascularization therapy of acute arterial occlusion in general [12]. In patients with longstanding stenosis, cerebral hyperperfusion syndrome has been identified as an important complication after extracranial carotid artery endarterectomy and stenting, as well as stenting of intracranial arterial stenosis. Cerebral hyperperfusion syndrome can be complicated by **vasogenic brain edema**, but may also lead to **ICH** and poor outcome if not treated properly (i.e. aggressive lowering and tight control of blood pressure) [13].

More recently, focal increase in cerebral perfusion in and surrounding ischemic brain lesions has been found on perfusion MRI after recanalization therapy for acute LVO stroke of the anterior circulation and has also been related to hemorrhagic transformation of brain infarction [14, 15]. However, besides technical limitations, perfusion MRI is not yet widely implemented and its use in patients who are unwell/unstable or have contraindications for MRI scanning may be difficult or prohibited. Such restrictions do not apply for **transcranial Duplex sonography (TCD)**, which allows investigating severely affected stroke patients at bedside and has other advantages, such as easy repeatability [16].

Therefore, **TCD could be a good candidate diagnostic tool to assess hemodynamic status and thereby predicting complications and prognosis in patients after MT**, which is supported by pilot studies from our group [16, 17] *and has been confirmed recently by Baracchini and colleagues* [18].

Pilot Studies: TCD as a Predictor of ICH after Thrombectomy

In a single-center pilot study [17], we retrospectively investigated hemodynamic changes on bedside TCD after MT for acute LVO of the anterior cerebral circulation and their association with postinterventional ICH and short-term prognosis. For this purpose, we identified all consecutive ischemic stroke patients with successful endovascular recanalization for anterior circulation LVO (defined by Thrombolysis in Cerebral Infarction grading of 2b–3) at our department between the years 2010 and 2017, and reviewed their postinterventional TCD examinations for mean blood flow (MBF) velocities of the ipsilateral (recanalized) and contralateral middle cerebral artery (MCA). Furthermore, we calculated an **MCA MBF velocity index** to account for inter-individual physiological differences in blood flow velocities and factors that may influence individual MBF (e.g. hematocrit, blood pressure, heart rate). The MCA MBF velocity index was calculated as follows:

$$\left(MCA\ MBF\ velocity\ index = \frac{recanalized\ MCA\ MBF\ velocity}{contralateral\ MCA\ MBF\ velocity} \right)$$

We included **123 stroke patients** (mean age 63 ± 14 years, 40% women) with successful MT. Of those, 18 patients had postinterventional ICH, which was symptomatic in 6 patients and all bleeds occurred at the side where MT had been performed.

TCD was performed at a median of 6.6 ± 2.3 h after MT. Notably, **ICH patients had an increased MCA MBF velocity index** compared with non-ICH patients (1.32 ± 0.39 versus 1.02 ± 0.32, $P < 0.001$). In a logistic regression model adjusting for age, sex, and traditional outcome predictors (arterial hypertension, IVT, and stroke severity at presentation), **higher MCA MBF velocity index levels** (>1.25; quartile 4) remained independently associated with **postinterventional ICH** (OR 3.6, 95% CI 1.1 to 13.2) and **poor outcome at 90 days** after stroke (modified Rankin Score 3–6; OR 3.2, CI 1.1 to 9.7).

Figure 19.1 shows a representative case of a patient who had an increased MCA MBF velocity index on bedside TCD early after successful MT and developed **secondary brain hemorrhage and vasogenic brain edema (reperfusion injury)** one day later.

Recently, we conducted a larger retrospective analysis of consecutive stroke patients [16] who had undergone MT for acute anterior LVO over a 9-year study period. All included patients received TCD within 72 hours after MT and MCA blood flow was graded according to the Thrombolysis in brain ischemia (TIBI) classification, which is a well-established tool to standardize cerebral blood flow

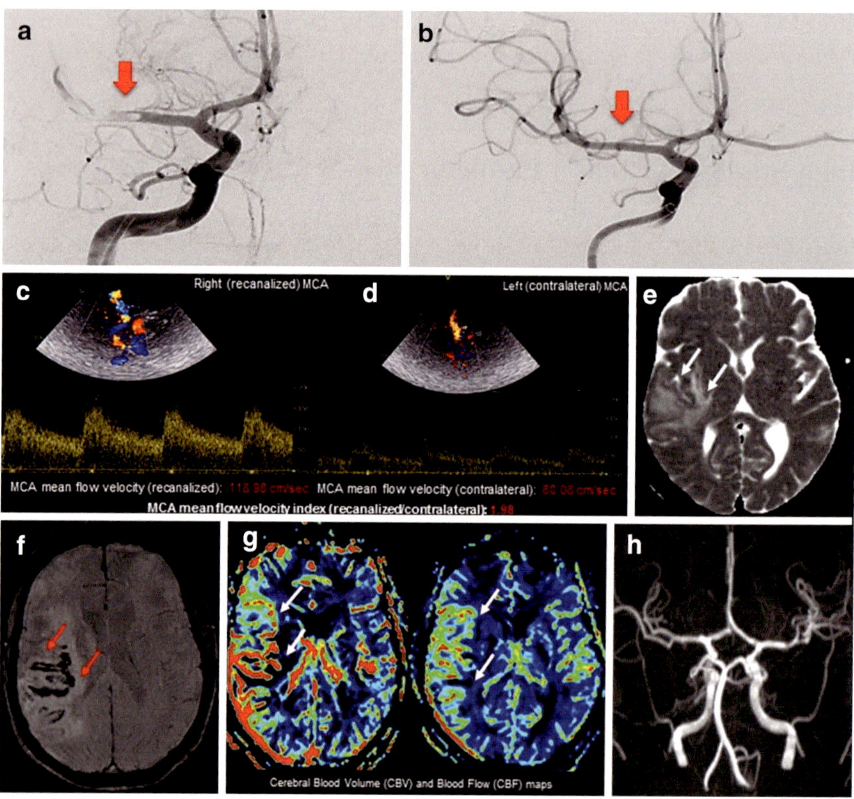

Fig. 19.1 Post-thrombectomy reperfusion injury. Representative images of a patient with acute right middle cerebral artery (MCA) occlusion. Pre-interventional digital subtraction angiography (DSA) confirms proximal MCA occlusion (**a**) and successful recanalization with mechanical thrombectomy (**b**). Postinterventional transcranial Duplex sonography 2 h after thrombectomy shows a twofold higher mean blood flow velocity in the ipsilateral (recanalized) compared to the contralateral MCA (**c, d**). Brain MRI on day one after mechanical thrombectomy shows vasogenic brain edema (**e**, white arrows) and hemorrhagic infarct transformation (**f**, red arrows), two indicators of reperfusion injury. MRI perfusion maps demonstrate underlying post-reperfusion hyperperfusion (**g**). Time-of-flight magnetic resonance angiography confirms normal intracranial vessel status (**h**)

alterations in acute ischemic stroke patients [19]. We studied 193 patients at a mean age of 66 years (42% female) and found that, despite successful vessel recanalization according to digital subtraction angiography (DSA) at the end of the intervention, only two thirds (64%) had normal MCA blood flow (TIBI Grade of 5) on postinterventional TCD. Residual 69 patients showed abnormal cerebral blood flow according to TIBI Grades 0–4, which was significantly associated with a poor outcome at 90 days in multivariable analyses (odds ratio: 2.4, confidence interval 1.3–4.6).

- These studies support the **clinical potential of TCD** to investigate cerebral hemodynamics in the early phase after MT, which could be a valuable tool **to predict postinterventional reperfusion injury including ICH as well as short-term prognosis.**
- Regional post-reperfusion hyperperfusion might be the underlying pathophysiological mechanism and could be identified and monitored using repeated bedside TCD, which is **readily available** in most stroke units and (neuro)intensive care units.
- If confirmed, this might help to identify patients at risk for postinterventional intracranial bleeding complications and could **guide individualized treatment strategies (e.g. blood pressure management) after MT.**

Limitations of the Pilot Studies

Important limitations of these pilot studies have to be acknowledged. Due to the retrospective design, it was not possible to account for **important peri-/postprocedural parameters** possibly influencing MCA velocities (e.g., vasoactive drugs, peri-/postprocedural blood pressure). Most importantly, it cannot be excluded that hemodynamic changes in the MCA, which were identified on TCD were induced by **local endothelial reaction/vasospasm,** for example due to a local stimulus by the interventional catheter. To confirm hyperperfusion and exclude vessel stenosis as the cause of an increased MCA MBF velocity index, concomitant information from **neuroimaging methods to determine vessel patency and perfusion status** (e.g. MRI with angiography and perfusion sequences) is needed. From the data of the pilot studies, it also remains unclear how cerebral hemodynamics change within the first days after the MT as we had no prespecified time points for TCD examinations. It is therefore also unknown how long MCA MBF velocity increases preceded a subsequent ICH as intracerebral bleeding did not lead to clinical deterioration in two thirds of patients and was just detected on routine imaging follow-up. For these reasons serial TCD examinations are needed, which could further improve the pathophysiological understanding of cerebral hemodynamics and how it impacts patient's clinical course in the early phase after MT.

Furthermore, our first results should be replicated in a larger cohort with a higher a number of (symptomatic) ICH patients, and in a **multicenter setting**. For this purpose, we initiated the prospective DYNASTROKE study, an Austrian multicenter project to investigate reperfusion injury after endovascular stroke treatment (for details see https://clinicaltrials.gov/ct2/show/NCT05273216 and the study flow chart, Fig. 19.2).

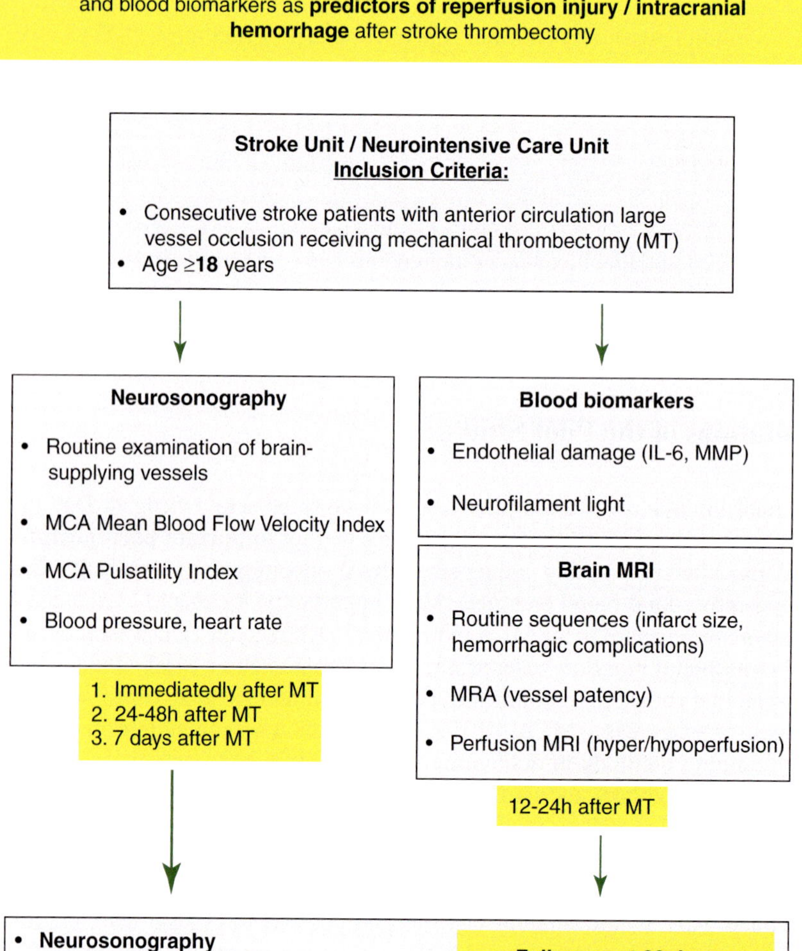

Fig. 19.2 Study flow chart. *ICA* internal carotid artery, *MCA* middle cerebral artery, *MT* mechanical thrombectomy, *MRA* magnetic resonance angiography, *IL-6* interleukin-6, *MMP* Matrix Metalloproteinases

While TCD has been employed for long to assess intracranial vessel status as a cause of stroke, it is **novel to use its possibly predictive capacity for complications and outcome in patients with LVO following MT**.

The **clinical importance** of further studies on this topic comes from the fact that TCD examinations thus may lead the way to optimize patient treatment after MT for avoiding complications and improving their outcome. Thus, there is rather conclusive evidence that patients at risk for cerebral hyperperfusion syndrome benefit from more aggressive **blood pressure lowering with target values < 140/90 mmHg or in high risk even < 120/80 mmHg** [20]. However, such blood pressure lowering could have deleterious effects in general stroke patients, especially in those with residual vessel stenosis, vasospasm or early re-occlusion.

In this context, it seems very likely that patients with post-reperfusion hyperperfusion would benefit from tight blood pressure control to reduce the risk for secondary brain hemorrhage / vasogenic brain edema formation [21].

Potentially, **blood biomarkers indicating blood-brain barrier disruption / endothelial dysfunction and neuroaxonal injury** could also help in the risk stratification of bleeding complications / reperfusion injury, and functional neurological outcome in LVO stroke patients undergoing MT [22, 23].

References

1. Goyal M, Menon BK, van Zwam WH, Dippel DW, Mitchell PJ, Demchuk AM, Dávalos A, Majoie CB, van der Lugt A, de Miquel MA, Donnan GA, Roos YB, Bonafe A, Jahan R, Diener HC, van den Berg LA, Levy EI, Berkhemer OA, Pereira VM, Rempel J, Millán M, Davis SM, Roy D, Thornton J, Román LS, Ribó M, Beumer D, Stouch B, Brown S, Campbell BC, van Oostenbrugge RJ, Saver JL, Hill MD, Jovin TG, HERMES collaborators. Endovascular thrombectomy after large-vessel ischaemic stroke: a meta-analysis of individual patient data from five randomised trials. Lancet. 2016;387(10029):1723–31. https://doi.org/10.1016/S0140-6736(16)00163-X.

2. Saver JL, Goyal M, van der Lugt A, Menon BK, Majoie CB, Dippel DW, Campbell BC, Nogueira RG, Demchuk AM, Tomasello A, Cardona P, Devlin TG, Frei DF, du Mesnil de Rochemont R, Berkhemer OA, Jovin TG, Siddiqui AH, van Zwam WH, Davis SM, Castaño C, Sapkota BL, Fransen PS, Molina C, van Oostenbrugge RJ, Chamorro Á, Lingsma H, Silver FL, Donnan GA, Shuaib A, Brown S, Stouch B, Mitchell PJ, Davalos A, Roos YB, Hill MD, HERMES Collaborators. Time to treatment with endovascular thrombectomy and outcomes from ischemic stroke: a meta-analysis. JAMA. 2016;316(12):1279–88. https://doi.org/10.1001/jama.2016.13647.

3. Serles W, Gattringer T, Mutzenbach S, Seyfang L, Trenkler J, Killer-Oberpfalzer M, Deutschmann H, Niederkorn K, Wolf F, Gruber A, Hausegger K, Weber J, Thurnher S, Gizewski E, Willeit J, Karaic R, Fertl E, Našel C, Brainin M, Erian J, Oberndorfer S, Karnel F, Grisold W, Auff E, Fazekas F, Haring HP, Lang W, Austrian Stroke Unit Registry Collaborators. Endovascular stroke therapy in Austria: a nationwide 1-year experience. Eur J Neurol. 2016;23(5):906–11. https://doi.org/10.1111/ene.12958.

4. Zaidat OO, Yoo AJ, Khatri P, Tomsick TA, von Kummer R, Saver JL, Marks MP, Prabhakaran S, Kallmes DF, Fitzsimmons BF, Mocco J, Wardlaw JM, Barnwell SL, Jovin TG, Linfante I, Siddiqui AH, Alexander MJ, Hirsch JA, Wintermark M, Albers G, Woo HH, Heck DV, Lev M, Aviv R, Hacke W, Warach S, Broderick J, Derdeyn CP, Furlan A, Nogueira RG,

Yavagal DR, Goyal M, Demchuk AM, Bendszus M, Liebeskind DS; Cerebral Angiographic Revascularization Grading (CARG) Collaborators; STIR Revascularization working group; STIR Thrombolysis in Cerebral Infarction (TICI) Task Force. Recommendations on angiographic revascularization grading standards for acute ischemic stroke: a consensus statement. Stroke. 2013;44(9):2650-2663. doi: https://doi.org/10.1161/STROKEAHA.113.001972.

5. Molina CA. Reperfusion therapies for acute ischemic stroke: current pharmacological and mechanical approaches. Stroke. 2011;42(1 Suppl):S16–19. https://doi.org/10.1161/STROKEAHA.110.598763.

6. Hussein HM, Georgiadis AL, Vazquez G, Miley JT, Memon MZ, Mohammad YM, Christoforidis GA, Tariq N, Qureshi AI. Occurrence and predictors of futile recanalization following endovascular treatment among patients with acute ischemic stroke: a multicenter study. AJNR Am J Neuroradiol. 2010;31(3):454–458. https://doi.org/10.3174/ajnr.A2006.

7. Balami JS, White PM, McMeekin PJ, Ford GA, Buchan AM. Complications of endovascular treatment for acute ischemic stroke: prevention and management. Int J Stroke. 2018;13(4):348–61. https://doi.org/10.1177/1747493017743051.

8. Goyal N, Tsivgoulis G, Pandhi A, Chang JJ, Dillard K, Ishfaq MF, Nearing K, Choudhri AF, Hoit D, Alexandrov AW, Arthur AS, Elijovich L, Alexandrov AV. Blood pressure levels post mechanical thrombectomy and outcomes in large vessel occlusion strokes. Neurology. 2017;89(6):540–7. https://doi.org/10.1212/WNL.0000000000004184.

9. Montalvo M, Mistry E, Chang AD, Yakhkind A, Dakay K, Azher I, Kaushal A, Mistry A, Chitale R, Cutting S, Burton T, Mac Grory B, Reznik M, Mahta A, Thompson BB, Ishida K, Frontera J, Riina HA, Gordon D, Parella D, Scher E, Farkas J, McTaggart R, Khatri P, Furie KL, Jayaraman M, Yaghi S. Predicting symptomatic intracranial haemorrhage after mechanical thrombectomy: the TAG score. J Neurol Neurosurg Psychiatry. 2019;90(12):1370–1374. https://doi.org/10.1136/jnnp-2019-321184.

10. Fandler-Höfler S, Odler B, Kneihsl M, Wünsch G, Haidegger M, Poltrum B, Beitzke M, Deutschmann H, Enzinger C, Rosenkranz AR, Gattringer T. Acute and chronic kidney dysfunction and outcome after stroke thrombectomy. Transl Stroke Res. 2021; https://doi.org/10.1007/s12975-020-00881-2.

11. Bai J, Lyden PD. Revisiting cerebral postischemic reperfusion injury: new insights in understanding reperfusion failure, hemorrhage, and edema. Int J Stroke. 2015;10(2):143–52. https://doi.org/10.1111/ijs.12434.

12. Clement DL, Brugmans J. Reactive hyperemia in patients with intermittent claudication, and correlation with other diagnostic methods. Angiology. 1980;31(3):189–197. https://doi.org/10.1177/000331978003100306.

13. van Mook WN, Rennenberg RJ, Schurink GW, van Oostenbrugge RJ, Mess WH, Hofman PA, de Leeuw PW. Cerebral hyperperfusion syndrome. Lancet Neurol. 2005;4(12):877–88. https://doi.org/10.1016/S1474-4422(05)70251-9.

14. Okazaki S, Yamagami H, Yoshimoto T, Morita Y, Yamamoto H, Toyoda K, Ihara M. Cerebral hyperperfusion on arterial spin labeling MRI after reperfusion therapy is related to hemorrhagic transformation. J Cereb Blood Flow Metab. 2017;37(9):3087–90. https://doi.org/10.1177/0271678X17718099.

15. Yu S, Liebeskind DS, Dua S, Wilhalme H, Elashoff D, Qiao XJ, Alger JR, Sanossian N, Starkman S, Ali LK, Scalzo F, Lou X, Yoo B, Saver JL, Salamon N, Wang DJ, UCLA Stroke Investigators. Postischemic hyperperfusion on arterial spin labeled perfusion MRI is linked to hemorrhagic transformation in stroke. J Cereb Blood Flow Metab. 2015;35(4):630–7. https://doi.org/10.1038/jcbfm.2014.238.

16. Kneihsl M, Niederkorn K, Deutschmann H, Enzinger C, Poltrum B, Horner S, Thaler D, Kraner J, Fandler S, Colonna I, Fazekas F, Gattringer T. Abnormal blood flow on transcranial duplex sonography predicts poor outcome after stroke thrombectomy. Stroke. 2018a;49(11):2780–2. https://doi.org/10.1161/STROKEAHA.118.023213.

17. Kneihsl M, Niederkorn K, Deutschmann H, Enzinger C, Poltrum B, Fischer R, Thaler D, Hermetter C, Wünsch G, Fazekas F, Gattringer T. Increased middle cerebral artery mean

blood flow velocity index after stroke thrombectomy indicates increased risk for intracranial hemorrhage. J Neurointerv Surg. 2018b;10(9):882–7. https://doi.org/10.1136/neurintsurg-2017-013617.

18. Baracchini C, Farina F, Palmieri A, Kulyk C, Pieroni A, Viaro F, Cester G, Causin F, Manara R. Early hemodynamic predictors of good outcome and reperfusion injury after endovascular treatment. Neurology. 2019;92(24):e2774–83. https://doi.org/10.1212/WNL.0000000000007646.

19. Demchuk AM, Burgin WS, Christou I, Felberg RA, Barber PA, Hill MD, Alexandrov AV. Thrombolysis in brain ischemia (TIBI) transcranial Doppler flow grades predict clinical severity, early recovery, and mortality in patients treated with intravenous tissue plasminogen activator. Stroke. 2001;32(1):89–93. https://doi.org/10.1161/01.str.32.1.89. PMID: 11136920.

20. Lin YH, Liu HM. Update on cerebral hyperperfusion syndrome. J Neurointerv Surg. 2020;12(8):788–93. https://doi.org/10.1136/neurintsurg-2019-015621.

21. Jadhav AP, Molyneaux BJ, Hill MD, Jovin TG. Care of the Post-Thrombectomy Patient. Stroke. 2018;49(11):2801–7. https://doi.org/10.1161/STROKEAHA.118.021640.

22. Pujol-Calderón F, Zetterberg H, Portelius E, Löwhagen Hendén P, Rentzos A, Karlsson JE, Höglund K, Blennow K, Rosengren LE. Prediction of outcome after endovascular embolectomy in anterior circulation stroke using biomarkers. Transl Stroke Res. 2021; https://doi.org/10.1007/s12975-021-00905-5.

23. Turner RJ, Sharp FR. Implications of MMP9 for blood brain barrier disruption and Hemorrhagic transformation following ischemic stroke. Front Cell Neurosci. 2016;10:56. https://doi.org/10.3389/fncel.2016.00056.

Chapter 20
Extra- and Intracranial Arterial Dissections in Young Adults: Endovascular Treatment with Stent Implantation

Guenther E. Klein, Claudia Wallner, and Kurt Niederkorn

Introduction

Artery dissection is defined as separation between the layers of the artery wall caused by a tear of the inner layer of the artery wall. Mural hemorrhage between the inner and medial layers of the artery wall can occur secondary to an intimal tear or by rupture of vasa vasorum, therefore bleeding can be in the medial layer or in the outer layer [1, 2]. The tear in the intimal layer may allow blood to split the layers, resulting in a false lumen, causing stenosis, pseudoaneurysm (if bleeding involves the outer layer of the artery wall), or both [3–5].

Dissection may be followed by three pathomorphological changes [6]:

- The integrity of the intima is compromised, associated with disruption of the endothelial layer, resulting in favoring local thrombus formation which may lead to thromboembolic events [7].
- Mural hemorrhage may cause stenosis or complete occlusion of the vessel and impairment of cerebral perfusion [4, 8].
- Presence of a pseudoaneurysm with the potential risk of rupture and bleeding [5, 9].

The most important point of these pathomorphological changes is the local thrombus formation which may embolize distally and result in brain infarction [10, 11].

G. E. Klein (✉)
Department of Radiology, Medical University Graz, Graz, Austria
e-mail: guenther.klein@medunigraz.at

C. Wallner · K. Niederkorn
Department of Neurology, Medical University Graz, Graz, Austria

© The Author(s), under exclusive license to Springer Nature Switzerland AG 2023
V. Demarin et al. (eds.), *Mind, Brain and Education*,
https://doi.org/10.1007/978-3-031-33013-1_20

Stenotic lesions of brain supplying vessels followed by dissection usually do not cause ischemic symptoms due to hypoperfusion. High-grade stenosis or occlusion of the vessel is usually compensated by collateralisation at the level of the circle of willis [12].

Depending on the size and the location of the pseudoaneurysm after dissection it may compromise adjacent structures, leading to neurological symptoms. The risk of rupture and bleeding is higher in intradural vertebro-basilar pseudoaneurysms than in extracranial carotid pseudoaneurysms. Subarachnoid bleeding is more common in children.

Symptomatic dissections of the cervical arteries can be spontaneous, traumatic or iatrogenic in origin. Traumatic dissections occur with massive trauma as well as with minor neck injuries. In most cases blunt traumas of the cervical spine with hyperextension cause dissection, also trivial incidents such as chiropractic manipulation such as hair washing, body-work-out and so on, associated with overtension of the cervical vessels may be responsible for dissection [13–17].

The main causes for a blunt head or neck trauma, followed by dissection, are car accidents, especially when hyperextension is associated with rotation of the neck [17–19].

Connective tissue disorders such as fibromuscular dysplasia Ehlers-Danlos Syndrome Type 4 and Marfan's syndrome are also associated with dissection.

Patients with carotid or vertebral dissection, commonly present with head and neck pain in 90%, 75% of patients presenting with stroke or TIA, in 20% of patients tinnitus is observed [1, 20]. Less than 5% of patients present with symptomatic subarachnoidal bleeding. Patients with carotid dissection may present with Horner syndrome associated with pain. The classical clinical presentation in patients with vertebral artery dissection is occipital headache, posterior neck pain or both, usually unilateral, associated with posterior circulation ischemia [21]. Mechanical irritation of vascular nociceptors due to dilatation of the vessel wall at the level of the dissection is responsible for the pain [22].

Dissection of the brain supplying arteries is the second common cause of stroke in young adults, accounting for approximately 20% [23–25]. There is a slight male predilection, 70% occur in the anterior circulation, 30% in the posterior circulation, extracranial dissections are more common than intracranial ones.

Dissection used to be diagnosed by angiography; today non-invasive techniques, especially CT-angiography and MR-angiography, serve to confirm the diagnosis as well as visualizing mural hematoma in the wall of the dissected artery [17, 26–30].

Sonography is usually the tool of initial diagnosis.

Therapeutic options to treat dissections include medical therapy, endovascular treatment and surgery. Conservative treatment with medical therapy may be used in asymptomatic patients and in symptomatic patients without contraindication.

Medical therapy generally consists of anticoagulant or antiplatelet agents [31–33]. Symptomatic patients are usually systemically anticoagulated with heparin.

Anticoagulation is commonly used to prevent thromboembolic strokes and may restore the anatomy of the vessel. Persistent pseudoaneurysms can also constitute a continual risk of embolization and bleeding. If symptoms persist or progress despite medical therapy, intervention is indicated.

Surgical intervention is possible in cervical dissection but is difficult because of the distal location of the lesions. Surgery is associated with a high complication rate and usually ligation or bypass is performed [1, 29–34]. Endovascular treatment with angioplasty and stent placement has been reported in selected cases of spontaneous or traumatic dissection. Endovascular treatment has supplanted surgery when intervention is indicated and is now the treatment of choice in symptomatic dissections not responding to anticoagulants, also permit treatment of associated pseudoaneurysms. Additional indications include the progredient arterial stenosis, a hemodynamic impairment of flow such as with multiple dissections [35, 36].

The endovascular treatment consists of the implantation of self-expandable stents or micro-stents without a balloon dilatation. The stents have proven to provide the necessary centrifugal force to enable apposition of the dissected segment to the vessel wall and thereby to obliterate the false lumen and resolve the stenosis [37, 38].

The goal of stenting is to eliminate the three pathomorphological findings associated with arterial dissection [36, 39–44].

1. So called "Jailing of clots" between the struts of the stent and the intimal layer of the vessel wall. Thrombus formation is jailed in between the stent and the vessel wall to prevent thrombus embolize distally and prevent stroke. This intention is the most important part of the treatment with stents.
2. Recanalization of the vessel is achieved by obliteration of the false lumen and treatment of the stenosis. An angioplasty by the use of a balloon dilatation is not necessary and should be avoided since the self-expansion of the stent provides an opening of the stenosis, not immediately but later on, proved in follow-up controls. An immediate balloon dilatation followed after stent deployment to improve cerebral perfusion is not indicated, also high grade stenosis usually are not associated with hypoperfusion of the brain when collateral circulation is provided.
3. The pseudoaneurysm will be obliterated after stent implantation by reconstruction of the true lumen created by inflow reduction and flow-diversion of the blood flow, passing by the aneurysm. In addition the stent can be used as a scaffold through which coils can be positioned in the pseudoaneurysm [41, 45, 46].

In case of occlusion of the vessel, initial recanalization of the dissected artery can be achieved using a micro-catheter/micro-wire-system, positioned distally, followed by deployment of the stent.

Illustrative Cases and Techniques

Extracranial Aneurysm of the Internal Carotid Artery: Endovascular Treatment with Stent and Coils

A 30-year-old woman (patient 1, Fig. 20.1) with a history of a blunt trauma to the neck after a car accident presented with a pseudoaneurysm and stenosis of the left internal carotid artery (ICA) (Fig. 20.1). Three-dimensional computed tomography angiography (3dCTA) showed the pseudoaneurysm below the skull base with a large neck and narrowing of the ICA at the level of the aneurysm. When surgical repair failed, the patient was admitted to our department for endovascular treatment of the pseudoaneurysm. First part of the treatment was to occlude the aneurysm by implantation of Guglielmi detachable coils (GDC). Second part of the treatment consisted in placement of a self- expandable stent, covering the entire narrowed segment of the ICA and the wide neck of the aneurysm. Control angiograms showed that the normal calibre and patency of the ICA had been reconstituted and that the aneurysm remained occluded. The implanted stent prevented a coil migration of protrusion of coils into the parent artery.

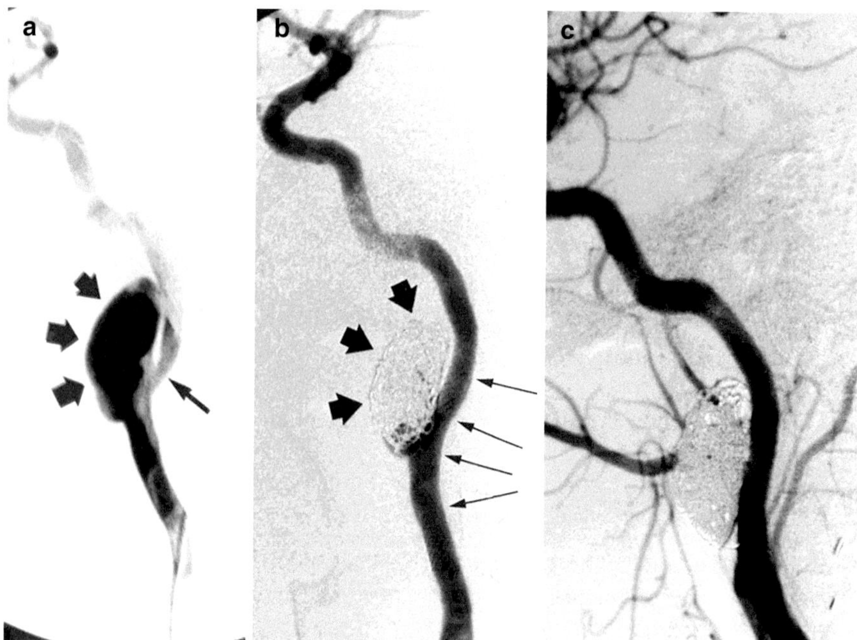

Fig. 20.1 Selective digital subtraction angiography of the left ICA, revealing posttraumatic dissection with pseudoaneurysm (short thick arrows) and narrowing of the ICA (thin arrow) (**a**). Immediately after coilembolization of the pseudoaneurysm (short thick arrows) and stentplacement (Wallstent) showing the remodelled ICA (thin arrows) and a subtotal occlusion of the pseudoaneurysm (**b**). Left common carotid angiogram at six- month-follow-up shows total occlusion of the pseudoaneurysm and normal width and patency of the stented segment of the ICA (**c**)

Symptomatic Spontaneous Carotid Dissection

A 44-year-old male patient (patient 2, Fig. 20.2) presented with focal neurological symptoms due to spontaneous carotid dissection. The indication for endovascular treatment was the progression of symptoms despite medical therapy.

Angiographic findings showing a long dissected segment of the internal carotid artery (ICA) with a pseudoaneurysm at the mid portion of the dissection and the false lumen of the dissection, narrowing the ICA.

The first part of the endovascular treatment consisted of navigation of the true lumen of the ICA with micro-catheter/micro-guidewire technique. After placement

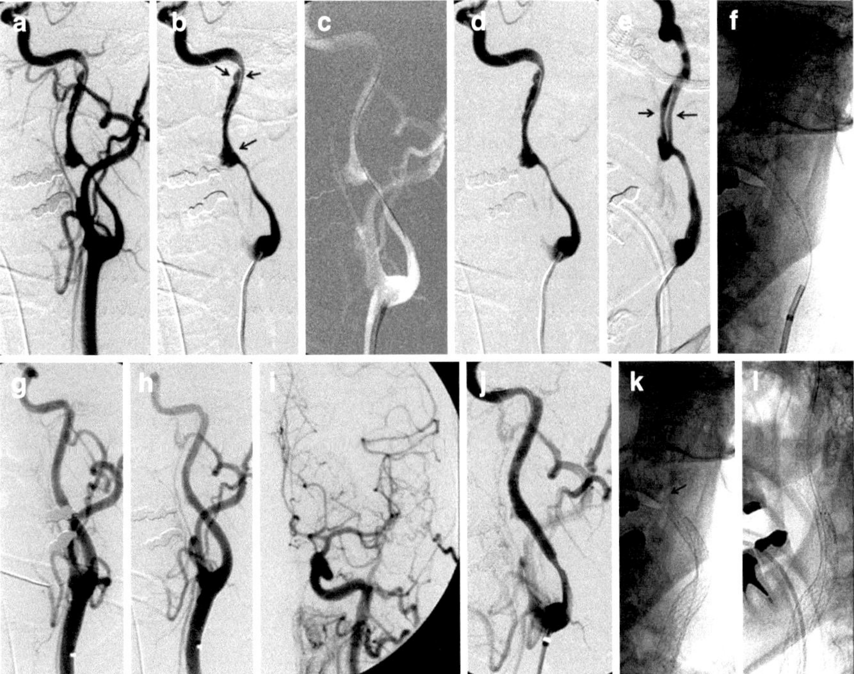

Fig. 20.2 Left common- and internal carotid angiogram (posterior-anterior and lateral projections) showing a long dissected segment of the cervical ICA with narrowing of the entire segment plus a pseudoaneurysm (arrow) at the mid portion of the dissection as well as the demarcation of a false lumen (arrow) adjacent to the true stenosed lumen(arrow) of the distal segment of the cervical ICA (**a, b, d, e**). Roadmapping technique is used to navigate the true lumen of the ICA ending in the normal intracranial portion by micro-wire/micro-catheter-system (**c**). After exchangement of the micro- wire for a long stiffer micro-wire, deployment of the first carotid stent, covering the distal part of the dissection, has been done (**f**). Control angiogram shows the reconstitution of the distal dissected segment after stent placement (**j**). The second step was to place a second stent proximally and overlapping the first one and ending in the common carotid artery (**k, l**). Final post-treatment angiograms of the cervical-(**g, h**) and intracranial ICA (**i**) show remodelling of the cervical ICA (**g, h**) with normal width of the lumen and complete separation of the pseudoaneurysm from the circulation. Normal appearance of the intracranial ICA and branches (**i**)

of the guidewire in the normal distal ICA at the level of the petrous segment, the first expandable stent was implanted, covering the distal portion of the dissected ICA.

Part two of the procedure was the deployment of a second carotid self-expandable stent coaxially, overlapping the first one and covering the proximal part of the dissected ICA, ending in the common carotid artery. The control angiography immediately after stent implantation showed the complete obliteration of the pseudoaneurysm and of the false lumen with normal appearance and calibre of the cervical ICA. Normal appearance of the intracranial distribution of the ICA and branches.

Symptomatic Intracranial Vertebral Dissection

A 48-year-old man (patient 3, Fig. 20.3) with a history of trauma presented with mild neurological symptoms. Non-invasive imaging detected a dissected right vertebral artery with a high grade circumscript stenosis. Despite medical therapy symptoms persisted, therefore endovascular therapy was recommended.

Selective digital subtraction angiography (DSA) of the right vertebral artery showed a short high grade stenosis in a typical location at the level of the V4 segment. Navigation of the stenosis was performed with micro-catheter/micro-wire-system and a self- expandable micro-neurostent (Neuroform stent) was deployed at the stenotic segment, overlapping the stenosis proximally and distally to the normal vessel. Although the immediate control angiogram showed only a minimal opening of the stenosis, a dilatation was not performed. At the follow-up control angiogram after 3 months, a complete remodelling with normal calibre of the artery was achieved by the self- expanding force of the stent.

Symptomatic Basilar Artery Dissection

A 23-year-old woman (patient 4, Fig. 20.4) presented with a symptomatic high grade stenosis of the basilar artery caused by dissection. Despite medical treatment neurological symptoms persist, a control angiogram 10 days after the initial angiographic examination demonstrated the progression of the stenosis as well as unchanged thrombus formation within the stenotic segment at the mid portion of the basilar artery. Progression of the stenosis and persistence of neurological symptoms under medical therapy were the criteria for endovascular therapy.

Using micro-catheter/micro-wire-technique after negotiation of the stenosis with the micro-wire, a self-expanding micro-neurostent was guided over the wire and deployed at the level of the stenotic lesion. Since balloon dilatation is avoided in stent implantation for treatment of dissections, immediate control angiogram after stent placement showed a residual stenosis. However at the six-week-follow-up, a complete reconstitution of the lumen of the basilar artery with normal calibre and no signs of thrombotic material.

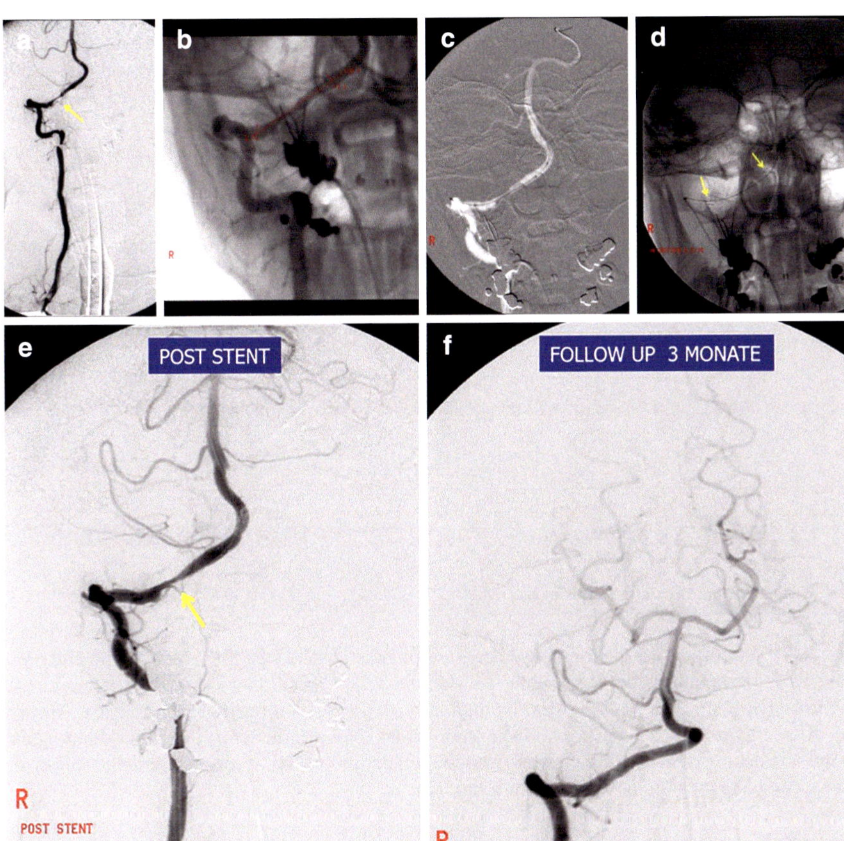

Fig. 20.3 DSA (**a**) and conventional angiography (**b**) of the right vertebral artery, showing a circumscript, highgrade stenosis (arrow) at the V4-segment. Measurement is made to choose the adequate size of the stent (**b**). Roadmapping-technique enables to navigate a micro-wire/micro-catheter-system through the stenosis as well to deploy the micro- stent (Neuroform stent) exactly covering the stenosis (**c**, **d**). Only the markers (arrows) at the ends of the stent are visible (**d**). Immediate control-angiogram after stent placement shows a residual stenosis (arrow), since no dilatation was performed (**e**). The three-month angiographic follow-up shows a normal calibre and regular shape of the artery (**f**)

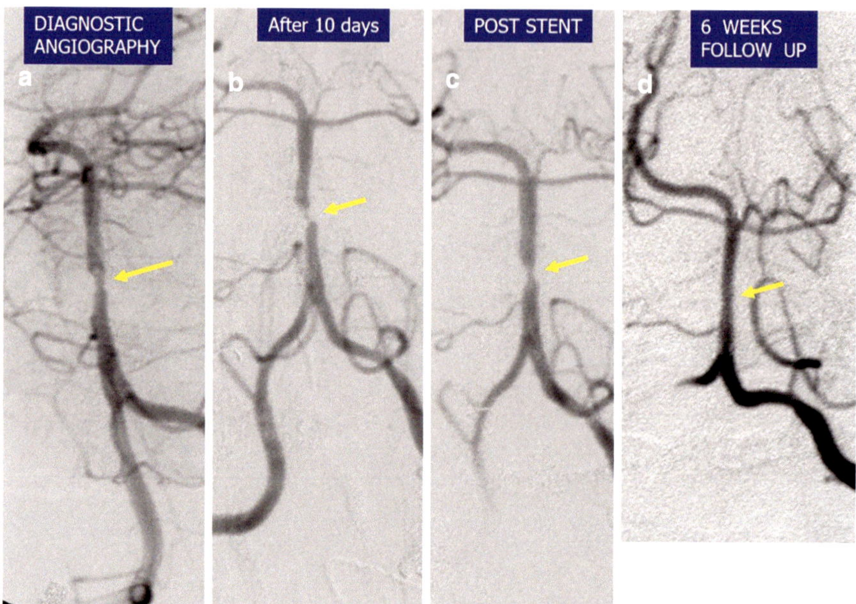

Fig. 20.4 Vertebrobasilar angiography shows an irregular shaped stenosis (arrow) at the mid basilar artery, with contrast defects within the stenotic lesion, according thrombus formation (**a**). Control angiogram after 10 days shows a progression of the stenosis (arrow) and persistent thrombi (**b**). After placement of a self-expandable micro-stent (Neuroform stent) the control angiogram shows a widening of the stenosis (arrow), without dilatation (**c**). Six-week angiographic follow-up shows complete restitution of the basilar artery (**d**)

Symptomatic Intracranial Dissection of the Internal Carotid Artery (ICA)

A 21-year-old woman (patient 5, Fig. 20.5) with symptomatic posttraumatic dissection of the right intracranial ICA at the level of the C1 segment.

Since medical therapy has failed, the patient was referred to endovascular therapy. Digital subtraction angiography (DSA) showed a high grade stenosis of the dissected C1 segment and thrombus formation at the distal part of the dissection, just below the origin of the middle cerebral artery. Road mapping guidance was used to navigate the micro-catheter/micro-wire-system, through the stenotic segment. Care was taken to pass the thrombus formation with the micro-wire without touching or perforating the thrombus.

After placement of the micro-wire with its tip distally in the middle cerebral artery, a self-expandable micro-neurostent was deployed, covering the entire dissection from the C1 segment to the M1 segment of the middle cerebral artery. The

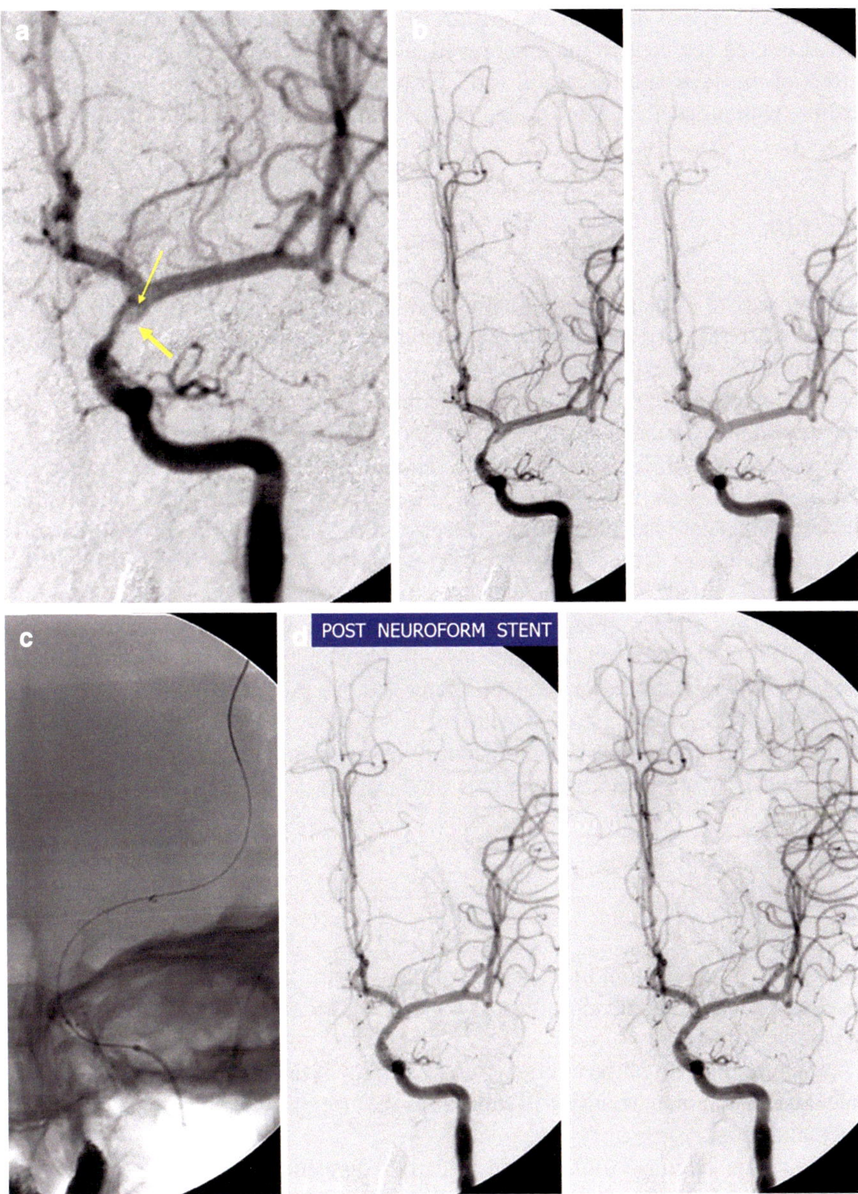

Fig. 20.5 Left internal carotid angiogram shows a high grade intracranial stenosis at the level of C1, associated with a thrombus formation (arrows) corresponding to the contrast defect at the distal part of the dissection (**a**, **b**). Navigation of the true lumen and positioning the micro-wire in the middle cerebral artery (**c**). After micro-stent placement remodelling of the lumen with complete homogenisation (**d**)

centrifugal force of the stent enabled to obliterate the false lumen by apposition of the dissected segment to the vessel wall and thereby jailing the clot between the struts of the stent and the vessel wall. Post-treatment angiograms showed normal calibre with no residual stenosis and elimination of the thrombus formation.

Results

The records of 77 patients with extra- and intracranial dissections at the Medical University Graz, Department of Neurology in Graz from December 1996 to October 2011 were retrospectively reviewed.

The focus was set on stent-associated complications, the rate of recanalization of the dissected vessel as well as the rate of recurrent cerebral ischemic events of dissections. A total of 77 patients with a minimum age of 42.7 years (59% men) with the diagnosis of cerebral artery dissection were identified. 31 of these patients were treated with stent placements at the Department of Radiology in Graz of whom 25 showed according to the modified Rankin-Scale (mRS) a "good" (mRS 0–1), five a "moderate" (mRS 2–3) and one a "worse" (mRS 4–6) functional neurological outcome. During a mean clinical follow-up period of 1 year, one recurrent cerebral ischemic infarct was noted. Two patients suffered from transient neurological symptoms without detection of a cerebral infarct and one presented with reperfusion of an existing pseudoaneurysm. No recurrent dissection was noted, 95% of the patients developed a complete resolution of stenosis or occlusion while one patient showed an in-stent restenosis after endovascular treatment.

Discussion

Dissection of cervical and intracranial portions of the carotid and vertebro-basilar arteries can occur spontaneously, post-traumatic or in association with other underlying risk factors.

A tear in the wall of the arteries causes splitting of the layers of the arteries, producing a false lumen, resulting in stenosis of the arteries or developing a pseudoaneurysm or both.

The annual incidence of spontaneous carotid dissection is estimated to be 2,7 per 100.000 [46–48]. In vertebro-basilar dissections the incidence is about 1,3 per 100.000 per year. Although dissections account only 2% of all ischemic strokes, they are an important cause in young adults [49, 50].

Dissections have been estimated to be the second most common cause for ischemic stroke in young adults, accounting up to 25% [24, 49]. Cardio-embolic events account up to 35%, causing stroke in young adults. 9–14 of 100.000 are estimated to suffer from ischemic stroke in young adults. Most infarcts with dissection have been shown to be embolic in nature.

Clinical presentation in carotid dissection commonly include head/neck/face pain and Horner syndrome, followed by neurological deficits within hours or weeks after the onset. Two thirds of the patients suffer from a TIA or stroke.

In our study, 81% presented with neurological deficits, 68% with stroke and 13% with TIA. Typical symptoms of vertebro-basilar dissection presenting with neck and head pain, especially occipitally, usually homolaterally to the side of the dissection lesion, followed by ischemic symptoms within hours. Pain as single symptom is also being reported [20].

Optimal diagnosis and management of stroke in young adults benefit from a multidisciplinary team including a vascular neurology specialist. If dissection is suspected, color doppler sonography is used in the acute phase. Even when the dissection itself cannot be visualized, abnormal flow signal may prove a dissection. Invasive diagnostic procedure such as digital subtraction angiography (DSA) is today replaced by MR-angiography and CT-angiography with the advantage to detect intramural hematoma in the wall of the dissected artery.

Conservative treatment with medical therapy consists of anticoagulant or antiplatelet agents. Since most strokes or TIAs are embolic in nature, anticoagulation is commonly used to prevent thromboembolic strokes. The ASA/AHA guidelines recommend medical therapy which is successful in the majority of cases. Further the use of either anticoagulation or antiplatelet therapy is recommended [35, 51]. Despite anticoagulation, some dissections may progress or develop pseudoaneurysms leading to a danger of distal embolization, or constitute a risk of bleeding.

Surgery has been performed for direct repair of carotid dissections, the risk of complications is significantly greater compared to surgery for atherosclerotic carotid lesions. Because of high mortality and morbidity rates, indications for surgical interventions such as thrombectomy and arterectomy or extracranial-intracranial bypass are given only in rare cases of dissection with complete vessel occlusion [52, 53].

The use of carotid stents for atherosclerotic diseases has been widely studied and shown to be an effective and safe therapeutic alternative to carotid endarterectomy. The risks associated with stent placement are not the same as the risks with stent placement for dissection. Endovascular treatment by placement of metallic stents has a lower risk than surgical treatment and in most cases, stents have replaced surgery as the initial therapy of choice when medical therapy failed. Endovascular stent/assisted angioplasty has been reported in selected cases of spontaneous, traumatic or iatrogenic dissection. One of the drawbacks of stent supported balloon angioplasty of the carotid artery for atherosclerosis has been the risk of distal embolization of atheromatous debris during the balloon dilatation phase of the procedure [26].

In almost all publications/studies reporting on carotid stent angioplasty, a more or less aggressive balloon dilatation (over-dilatation) of the implanted stent has been done, leading to complications such as distal embolization of debris. The reason behind this procedure was a misconception about the goal of stent PTA for treatment of carotid atherosclerotic disease. The primary goal is not to improve the hemodynamic situation in a patient with embolic stroke but to eliminate the source of embolization, that means to cover the dangerous plaque by stent implantation in order to prevent a distal migration of plaque debris. Most common type of stroke is a thromboembolic one, one should not forget that the brain is supplied by four arteries, in case of occlusion of one, the perfusion of the brain will be maintained by anastomotic supply with communicating arteries at the level of the circle of willis. Contrary to the stent angioplasty in peripheral and coronary arteries, a balloon dilatation of the implanted carotid stent should only performed minimally or should be avoided.

Self-expandable stents have advantages compared to balloon-expandable stents, including a higher flexibility as well as an atraumatic deployment procedure. Due to the selfexpanding force of the stent, there is no risk of overdilatation or perforation of the artery. Therefore balloon-expandable stents were replaced by self-expandable stents for treatment of atherosclerotic lesions, even more for treatment of dissections.

The technique of stent implantation in dissections differs from the technique in atherosclerotic stenosis since dissected stenosis is usually not atheromatous or calcified. Therefore one should definitely not dilate the implanted stent neither extracranially nor intracranially. A self-expandable stent/micro-stent provides the necessary centrifugal force to permit apposition of the dissected segment to the vessel wall, usually not immediately after the deployment of the stent but after a few weeks at the control follow-up. Additional dilatation of the deployed stent is not necessary, it could be a risk for complications [54].

The additional advantage of endovascular therapy of cervical dissections is that it enables the identification of the true and false lumen as well as thrombotic formation by superselective catheterisation and angiography. Even in case of occlusion, initial recanalization of the dissected artery can be achieved, using micro-catheter/micro-wire- system for navigation, followed by stent placement. In case of an extensive dissection, multiple stents can be deployed in an overlapping fashion for a vessel reconstruction.

Common indications for stent therapy include:

Clinical failure of medical therapy, contraindication for medical therapy, therapyresistent or persistent symptomatic patients, progredient stenosis, pseudoaneurysm and hemodynamic impairment.

Conclusion

The incidence of ischemic stroke in young adults has risen over the past years. Increased public awareness ("Time is brain"), increased use of imaging and more accurate diagnosis may in part explain the increased detection of stroke in young adults. With refinements of catheter devices, including micro-catheter/micro-wire-systems and improvement in technologies of implantations such as self-expandable stents and micro- stents, interventional treatment has gained more acceptance, especially in patients remaining symptomatic by recurrent strokes despite medication.

It is important to avoid additional balloon angioplasty of the stent in order to minimize the risk of stenting such as embolization of thrombotic fragments during stent deployment.

The navigation of the true lumen can be difficult, experience in the field of interventional neuroradiology is mandatory to perform the procedure successfully. The use of appropriate micro-catheter/micro-wire-systems provides a safe placement of the guidewire distally to the dissection and deployment of appropriate self-expandable stents/micro-stents.

References

1. Schievink WI. Spontaneous dissection of the carotid and vertebral arteries. N Engl J Med. 2001;344:898–906.
2. Anson J, Crowell AM. Cervicocranial arterial dissection. Neurosurgery. 1991;29:89–96.
3. Ehrenfeld WK, Wylie EJ. Spontaneous dissection of the internal carotid artery. Arch Surg. 1976;111:1294–301.
4. Mokry B, et al. Spontaneous dissection of the cervical carotid artery. Ann Neurol. 1986;19:126–38.
5. Ringelstein E, Dittrich R et al. *S1 Leitlinie Spontane Dissektionen der extra-und intrakraniellen hirnversorgenden Arterien*, in: Deutsche Gesellschaft für Neurologie, Hrsg. Leitlinien für Diagnostik und Therapie in der Neurologie. 2016.
6. Saver JL, Easton JD, Hart GH. Dissection and trauma of cervicocerebral arteries. In: Barnett HJ, Mohr JP, Stein BM, editors. Stroke: pathophysiology, diagnosis and management. New York Churchill Livingstone; 1992. p. 671–88.
7. Lucas C, et al. Stroke patterns of internal carotid artery dissection in 40 patients. Stroke. 1998;29:2646–8.
8. Malek AM, et al. Endovascular management of extracranial carotid artery dissection achieved using stent angioplasty. AJNR Am J Neuroradiol. 2000;21:1280–92.
9. Ahlhelm F, et al. Endovasculat treatment of cervical artery dissection:ten case reports and review of the literature. Intervent Neurol. 2012;1:143–50.
10. Benninger DH, et al. Mechanism of ischemic infarct in spontaneous carotid dissections. Stroke. 2004;35(2):482–5.
11. Bertram M, et al. Spectrum of neurological symptoms in dissections of brainsupplying arteries. Dtsch Med Wochenschr. 1999;124:273–8.
12. Steinke W, et al. Topography of cerebral infarction associated with carotid artery dissection. J Neurol. 1996;243:323–8.
13. Oehler J, et al. Beidseitige A. vertebralis-Dissektion nach chiropraktischer Behandlungen konservativer Behandlungsmethoden bei Patienten mit Claudicatio spinalis. Orthopade. 2003;32:911–3.
14. Marx P, et al. Manipulationsbehandlung der HWS und Schlaganfall. Fortschr Neurol Psychiatr. 2009;77(2):83–90.
15. Biller J, et al. Cervical arterial dissections and association with cervical manipulative therapy: a statement for healthcare professionals from the american heart association/american stroke association. Stroke. 2014;45(10):3155–74.
16. Rubinstein SM, et al. A systematic review of the risk factors for cervical artery dissection. Stroke. 2005;36:1575–80.
17. Hart RG, Easton JD. Dissection of cervical and cerebral arteries. Neurol Clin. 1983;1:155–82.
18. Biffl WL, Moore EE, et al. The unrecognized epidemic of blunt carotid arterial injuries—early diagnosis improves neurological outcome. Ann Surg. 1998;228:462–9.

19. Crissey MM, Bernstein EF. Delayed presentation of carotid intimal tear following blunt craniocervical trauma. Surgery. 1974;75:543–9.
20. Arnold M, et al. Pain as the only symptom of cervical artery dissection. J Neurol Neurosurg Psychiatry. 2006;77:1021–4.
21. Silbert PL, et al. Headache and neck pain in spontaneous internal carotid and vertebral artery dissections. Neurology. 1995;45:1517–22.
22. Sturzenegger M. Headache and neck pain: the warning symptoms of vertebral artery dissection. Headache. 1994;34:187–93.
23. Bogousslavsky J, Pierre P. Ischemic stroke in patients under age 45. Neurol Clin. 1992;10:113–24.
24. Stack CA, Cole JW. A diagnostic approach to stroke in young adults. Curr Treat Options Cardiovasc Med. 2017;19(11):84.
25. Mc Carty JL, et al. Ischemic infarction in young adults: a review for radiologists. Radiographics. 2019;39(6):1629–48.
26. Marciniec M, et al. Non-traumatic cervical artery dissection and ischemic stroke: a narrative review of recent research. Clin Neurol Neurosurg. 2019;187:105561.
27. Sturzenegger M. Spontaneous internal carotid artery dissection: early diagnosis and management in 44 patients. J Neurol. 1995;242:231–8.
28. Mokri B, et al. Spontaneous dissections of the vertebral arteries. Neurology. 1988;38:880–5.
29. Schievink WI, Mokri B, O'Fallon WM. Recurrent spontaneous cervical artery dissection. N Engl J Med. 1994;330:393–7.
30. Schievink WI, et al. Spontaneous dissections of cervicocephalic arteries in childhood and adolescence. Neurology. 1994;44:1607–12.
31. Lyrer P, Engelter S. Antithrombotic drugs for carotid artery dissection. Stroke. 2004;35:613–4.
32. Lyrer PA. Extracranial arterial dissection: anticoagulation is the treatment of choice: against. Stroke. 2005;36:2042–3.
33. Norris JW. Extracranial arterial dissection: anticoagulation is the treatment of choice: for. Stroke. 2005;36:2041–2.
34. Kadkhodayan Y, et al. Angioplasty and stenting in carotid dissection with or without associated pseudoaneurysm. Am J Neuroradiol. 2005;26:2328–35.
35. Writing Committee Members, Brott TG, et al. Guideline on the management of patients with extracranial carotid and vertebral artery disease: executive summary. Stroke. 2011;42:e420–63.
36. Cohen JE, et al. Endovascular management of spontaneous bilateral symptomatic vertebral artery dissections. AJNR Am J Neuroradiol. 2003;24:2052–6.
37. DeOcampo J, et al. Stenting: a new approach to carotid dissection. J Neuroimaging. 1997;7:187–90.
38. Moon K, et al. Stroke prevention by endovascular treatment of carotid and vertebral artery dissections. J Neurointerv Surg. 2017;9(10):952–7.
39. Biondi A, et al. Progressive symptomatic carotid dissection treated with multiple stents. Stroke. 2005;36:e80–2.
40. Komotar RS, Mocco J, et al. Rapidly successive symptomatic bilateral spontaneous vertebral artery dissections. Surg Neurol. 2009;72(3):300–5.
41. Klein GE, et al. Posttraumatic extracranial internal carotid aneurysm: combined endovascular treatment with GDC and stent implantation. Am J Neuroradiol. 1997;18:1261–4.
42. Seifert T, Klein GE, Niederkorn K, et al. Bilateral vertebral artery dissection and infratentorial stroke complicated by stress induced cardiomyopathy. J Neurol Neurosurg Psychiatry. 2008;79(4):480–1.
43. Klein GE, Niederkorn K. Schlaganfall: Diagnostisches und interventionelles Management bei jungen Erwachsenen. RöFo. 2008:89. DRG-Kongress, Abstract
44. Klein GE. Stentimplantation zur Behandlung intrakranieller Stenosen. Z Gefässmed. 2007;4(1):4–10.
45. Mericle RA, et al. Stenting and secondary coiling of intracranial internal carotid artery aneurysm: technial case report. Neurosurgery. 1998;43:1229–34.

46. Higashida RT, et al. Intravascular stent and endovascular coil placement for a ruptured fusiform aneurysm of the basilar artery. Case report and review of the literature. J Neurosurg. 1997;87:944–9.
47. Lee VH, et al. Incidence and outcome of cervical artery dissection: a population- based study. Neurology. 2006;67:1809–12.
48. Schievink WI, et al. Internal carotid artery dissection in a community. Rochester, Minnesota 1987-1992. Stroke. 1993;24:1678–80.
49. Biedermann B, et al. Dissektionen der Arteria carotis interna und vertebralis: Ursachen, Symptome, Diagnostik und Therapie. J Neurol Neurochir Psychiatr. 2007;8(2):7–18.
50. Wolff S, et al. Management von Dissektionen zervikaler hirnzuführender Arterien. J Neurol Neurochir Psychiatr. 2001;12(3):222–6.
51. Nirav B, et al. Stroke in young adults. Five new things. Neurol Clin Pract. 2018;8(6):501–6.
52. Calvet D. Ischemic stroke in the young adult. Rev Med Interne. 2016;37(1):19–24.
53. Kernan WN, et al. Guidelines for the prevention of stroke in patients with stroke and transient ischemic attack: a guideline for healthcare professionals from the American Heart Association/ American Stroke Association. Stroke. 2014;45:2106–236.
54. Pham MH, et al. Endovascular stenting of extracranial carotid and vertebral artery dissections: a systematic review of the literature. Neurosurgery. 2011;68(4):856–66.

Chapter 21
Headaches in Cerebrovascular Diseases

Natasa Stojanovski, Marta Jeremić, Ivana Mitić, Milica Dajević, Vuk Aleksić, Nenad Milošević, Draginja Petković, and Milija Mijajlović

Introduction

Cerebrovascular diseases are one of the leading causes of disability and death worldwide and headache is one of the most common complaints among these patients [1]. In 1964 Willis described the relationship between vascular disease and headache, and in 1968 Fisher carried out the first detailed study on this relationship [2].

In this chapter we will give an extensive review on headache characteristics in various cerebrovascular disorders.

N. Stojanovski
Faculty of Medicine, University of Belgrade, Belgrade, Serbia

M. Jeremić · M. Mijajlović (✉)
Faculty of Medicine, University of Belgrade, Belgrade, Serbia

Neurology Clinic, University Clinical Center of Serbia, Belgrade, Serbia

I. Mitić
Department of Neurology, General Hospital Zaječar, Zaječar, Serbia

M. Dajević
Department of Neurology, Clinical Center of Montenegro, Podgorica, Montenegro

V. Aleksić
Department of Neurosurgery, Clinical Hospital Center Zemun, Belgrade, Serbia

N. Milošević
Faculty of Medicine, University of Pristina, Kosovska Mitrovica, Serbia

Clinical-Hospital Center Pristina, Gracanica, Serbia

D. Petković
Institute for Cardiovascular Diseases "Dedinje", Belgrade, Serbia

© The Author(s), under exclusive license to Springer Nature Switzerland AG 2023
V. Demarin et al. (eds.), *Mind, Brain and Education*,
https://doi.org/10.1007/978-3-031-33013-1_21

Acute Ischemic Stroke and Transitory Ischemic Attack

Acute ischemic stroke (AIS) and transitory ischemic attack (TIA) are most commonly presented with focal neurological deficit, but headache as symptom has been often overlooked by physicians and patients as well [3].

Headache attributed to AIS represents new and most often acute-onset headache caused by ischemic stroke associated with focal neurological deficit and usually has a self-limiting course. TIA associated headache usually begins at the same time as the clinical signs of TIA and it usually resolves with the clinical improvement of TIA or within 24 hours [3, 4].

Finding the relationship between a headache and AIS or TIA is essential for correct diagnosis, optimal treatment, and prevention of complications. However, since tension-type headache has a prevalence of 75% and migraine has a prevalence of 15% in the general population, it is very difficult to decide if headache is caused by the vascular event or is just one of the usual headaches that randomly occurs at the time of the AIS or TIA onset. The closer the onset of headache is to vascular event, the lesser the chance of a usual headache to occur at that time. Also, if the headache characteristics are different from regular the probability of a causal relationship increases [3, 5].

Prevalence of headache associated with AIS varies between 7.4% and 34% and more frequently occurs in younger patients. Both in AIS and TIA headache is more often presented when vertebrobasilar vascular territories are involved. After AIS, headache can persist for more than 3 months. Frequency of stroke associated headache is for sure underestimated because of the substantial amount of patients with aphasia or altered level of consciousness. As regards to TIA associated headaches, prevalence range from 26% to 36%. In such patients it is not uncommon for headache to be prominent symptom and some authors say that in patients with amaurosis fugax an ipsilateral frontal or orbital pain frequently follows [3, 5].

Headache can be of abrupt onset as well as gradually appearing. Quality of pain in AIS is most commonly pressure-like, throbbing, but patients also describe pain as stabbing, burning, "wind", cold-like, as well as pulsating. Intensity of pain is usually described in literature as mild to moderate, sometimes followed with nausea, vomiting, photophobia and phonophobia. Some earlier studies suggested that in most patients the onset headache doesn't have specific features. Headache can occur before, simultaneously, or after the onset of stroke. It can be unilateral or bilateral [3–5]. Tentschert et al. found that in patients with hemispheric stroke and unilateral headache, headache was significantly more often on the side of the lesion, and some other studies also showed that the headache is more commonly ipsilateral to the stroke lesion [6]. Regarding localization, vertebrobasilar stroke has predilection for occipital headache, while carotid stroke associated headache is usually located frontally. However, in some cases bilateral headache is also described [7–10].

Apart of being a symptom of ischemic event, headache can be a risk factor or even a cause of vascular event [11]. Although rare, migraine may be a potential cause of stroke as in migrainous infarction. In the third International Classification of Headache Disorders (ICHD-3) migrainous infarction is defined as stroke associated with persistent aura symptoms [4]. In most of such cases the ischemic lesions are located in the posterior circulation, more often in women. These patients are usually without cerebrovascular risk factors, but some studies showed that foramen ovale apertum can be found in more than half of patients. Also, high prevalence of persistent foramen ovale in patients with migraine with aura is already recognized [12–15]. Recent studies showed that migraine with aura is associated with increased risk of TIA and AIS [15, 16]. Finally, some headaches can mimic stroke, like familial hemiplegic migraine or basilar migraine. Also, migraine with aura can mimic TIA [17].

From therapeutic point of view, it is necessary for physicians to recognize the specific form of the so-called sentinel headache which Lebedeva et al. define as a new type of headache or a previous headache with altered features, such as more severe intensity, increased frequency, prolonged duration, development of new accompanying symptoms and absence of drug effects which arose within 7 days before vascular event [18, 19]. Although the ICHD-3 does not include diagnostic criteria for this type of headache, Lebedeva et al. propose diagnostic criteria after they characterized the prevalence and characteristics of sentinel headache in 550 patients [18]. It is important to distinguish this sentinel headache from sentinel headache associated with impending rupture of an intracranial aneurysm, arterial dissections and cerebral venous sinus thrombosis. According to several recent studies, 18.3% of the patients with TIA had sentinel headache within the 7 days before TIA, and about 15% had sentinel headache before AIS. These patients had arrhythmia attacks and a history of atrial fibrillation associated with headache onset, and these data suggest that emboli can play an essential role in the development of sentinel headache. According to these studies sentinel headaches are relatively frequent and specific diagnostic criteria separate them from previous headaches. Since mentioned studies are prospective case-control studies, and suggested diagnostic criteria are specific and sensitive, they should be used in future editions of the ICHD-3 [18, 19].

Nontraumatic Intracranial Hemorrhage

About 15% of all strokes are hemorrhagic. Headache as presenting symptom at the onset of stroke is more common in subarachnoid hemorrhage (SAH) and in intracerebral hemorrhage (ICH) than in AIS [20, 21].

The hallmark of SAH is sudden, severe headache, or "worst headache of my life". It is often called "thunderclap headache", which in less than 1 min reaches maximum intensity. According to recent studies severe headache is presenting symptom in one third of patients with SAH. Although only 1 in 50 patients with severe, sudden onset headache will have SAH, all patients with such headache should be investigated, regardless of normal neurological status. However, SAH is more likely if headache is accompanied with, photophobia, vomiting, neck stiffness, pain, and meningismus, sudden onset with exertion, altered level of consciousness, or focal neurological deficit. Minority of patients presents with headache as only symptom, but clinicians must remain vigilant for the possibility of SAH, because such mistake can result in catastrophic event. In one study that included 482 SAH patients, the diagnosis was missed by physicians in 12%, usually in neurologically intact patients with headache as presenting symptom [20–23].

It is believed that the headache in SAH patients develops due to the chemical irritation of the blood on the meninges, though many other factors are probably associated, such as infiltration of subarachnoid space with immune cells, and immune system activation with increase of inflammatory cytokines. The high blood pressure that tends to be present after hemorrhage may also contribute to the headache development, as may the evolution of hydrocephalus or cerebral vasospasm as a consequence of SAH. Identifying the true cause of SAH headache influences the treatment choice, however, opioids remain the guideline recommended mainstay of acute therapy despite their significant side effects and potential for addiction and tolerance [22–24].

Premonitory symptoms (also called sentinel hemorrhages or "warning leaks"), typically consisting of an unusually severe sudden onset headache that is sometimes associated with nausea, vomiting, and dizziness, are usually attributed to small hemorrhages from the aneurysm. This headache is actually warning sign during the days or weeks before SAH. Mac Grory et al. performed a prospective, observational study in order to characterize the headache associated with SAH. Most often the SAH associated headache was located in the occipital region (55% in the SAH group *vs* 22% in the non SAH group). Most commonly patient described SAH headache as "stabbing" (35% in the SAH group *vs* 5% in non SAH group), Also, meningismus was twice more often in SAH group. Majority of patients (65%) with SAH reported that the intensity of headache reached its maximum in only one second. Also, headache pain increases during the first week after SAH, even on daily basis [21].

In ICH headache is very frequent symptom, with nausea and vomiting, probably due to increase in intracranial pressure and the direct irritation of the trigeminovascular system at the base of the skull. Also, these patients often have altered level of consciousness. Headache can be unilateral or bilateral, from moderate to severe intensity depending on hemorrhage localization, size and evolution. One population-based study even showed that headache in ICH is independently associated with a higher risk of early death [22].

Unruptured Vascular Malformations

Unruptured intracranial aneurysms (UIA) are not uncommon in general population and approximately one third of them have headache as a symptom. Usually, UIA are found in work up for chronic headaches and often are considered isolated finding. Different types of headache were described in patients with UIA, in some studies migraine was most frequent, and symptoms sometimes improve after treatment of the aneurysm whether in severity or frequency of headache. However, most intracranial aneurysms remain clinically silent. Most common clinical sign of intracranial aneurysm is third nerve palsy. Sentinel headaches are observed as previously described. Clinicians should pay special attention in the presence of a new onset headache without the characteristics of a primary headache, a headache that has changed features in a patient already suffering from a primary headache, headache in a patient suffering from heritable disorders such as Ehlers–Danlos syndrome type IV, Marfan's syndrome and others that have been associated with intracranial aneurysms and of course ones with a positive family history or previously ruptured aneurysm [23, 25, 26].

Headache associated to arteriovenous malformation (AVM) according to ICHD-3 criteria has to develop in close temporal relation to AVM signs and symptoms, have proper localization, and to be synchronized with clinical course of AVM [4]. In unruptured AVM headache could be one of the presenting symptoms [27]. Occipital location of AVM is thought to be predominant localization for headaches and in most cases headaches are described as migraine-like, ipsilateral to the AVM and sometimes associated with visual symptoms. As possible mechanisms of headache in unruptured AVM were proposed increased intracranial pressure, steal phenomenon, and cortical spreading depression [27–29].

Dural arteriovenous fistulas (DAVF) are rare, intracranial vascular malformation whose etiology remains uncertain. Symptoms of DAVF depend of venous drainage and location of DAVF. Carotid-cavernous fistulas are more presented with non-migraine headaches and ocular symptoms due to reversal flow in the ophthalmic veins. Other DAVF are most often presented with migraine like headaches, as well as tinnitus and neurological deficits. Headaches are usually described on the side of fistula. According to one of the criteria headache is worse in the morning or due to a Valsalva's maneuver [30, 31].

Sturge–Weber syndrome (SWS) is a rare congenital neurocutaneous disorder with facial vascular nevus (port-wine stain) and ipsilateral leptomeningeal angioma, most commonly followed by epilepsy and mental retardation. In the study conducted by Klapper with 71 patients, almost half of patients had recurrent headache, with migraine prevalence of 28%. Other headaches were related to glaucoma (8%), chronic tension-type headache (4%), and episodic tension-type headache (1%). Recent studies showed that SWS patients have higher prevalence of migraine than in general population. Also, some patients with SWS have episodic headache with atypical features, such as prolonged aura [32–34].

Cervical Vascular Disorders and Intracranial Arterial Dissection

Pain in the head, face and neck can be associated with artery dissection. Internal carotid arteries are most frequently affected. Pain from carotid arteries dissection can be located in face. Most commonly pain is ipsilateral to the site of dissection, and it is often dull, sometimes throbbing and severe. Unusual and sharp neck pain radiating to mandible and ear can be sign of vertebral arteries dissection, and in up to 50% of patient's pain is located on the dorsal side of the neck and in most cases patients also have occipital headache. Pain can last from few days to weeks, especially in vertebral arteries dissection, but it usually resolves with the resolution of the dissection. Pain can be followed by transient monocular blindness in ICA dissection, pulsatile tinnitus in 10–15% of patients, usually in internal carotid artery dissection, and TIA in 10–20% of patients and most frequent sign, especially in internal carotid artery dissection is Horner's syndrome [35–37]. On the other hand, headache can be the risk factor for cerebral artery dissection. Some studies suggest that migraine can increase the risk for cerebral artery dissection, but the exact mechanism is still unclear, but it may be due to different vessel wall pathologies [37–39].

One of the differential diagnoses for thunderclap headache is intracranial artery dissection. Intracranial dissections are rare, and mostly involve vertebrobasilar system. Matzumoto et al. conducted a study with 37 patients with intracranial vertebral artery dissection that had headache as a first, or one of the presenting symptoms. They found that headache was most often localized in occipital and nuchal region, spreading to the temporal and parietal regions, forehead, and face, had a pulsatile character, acute onset, mainly was unilateral and ipsilateral to the dissection. Some studies showed that migraines are possible risk factor for dissection of intracranial arteries [39].

Special type is post-endarterectomy headache. It is usually mild, ipsilateral headache, with different features, that can occur after this surgical procedure. If the pain is severe it can be a part of more serious condition so-called hyperperfusion syndrome, which can be also seen after angioplasty of the carotid artery, and it requires urgent treatment. Angioplasty with or without stenting can be accompanied with headache that can occur during or after procedure, and according to current criteria, angioplasty related headache must occur within 1 week after treatment and must resolve within 1 month. Some studies showed that headaches after stenting are mostly ipsilateral to procedure, with frontotemporal localization and throbbing-pressing nature with mild to moderate intensity. Sometimes pain can also involve face and neck region [40, 41].

Cerebral Venous Thrombosis

Cerebral venous thrombosis (CVT) represents thrombotic disorder of the brain venous sinuses and also one of the "thunderclap" headaches differential diagnoses. CVT can present with wide spectrum of manifestation that depends of localization of the thrombosis [42, 43]. Most common symptom of CVT is headache. It is often the first symptom, reported by 60–90% of patients, while the second one is epileptic seizure. Also, the headache is rarely isolated, and is almost always associated with other neurological signs such as seizures, disorders of consciousness, papilledema, focal deficits, and cranial nerve palsies. These signs can be grouped into four main syndromes: focal syndrome, isolated intracranial hypertension, diffuse encephalopathy, and cavernous sinus syndrome [44, 45]. However, headache is rarely reported as the only CVT symptom [46]. The pathogenesis of isolated headache in CVT is unknown but probably involves changes in the walls of the occluded sinus. The most affected age group is between 20–50 years of age and it is threefold more common in women due to risk factors such as oral contraceptives, pregnancy and puerperium. Risk factors for CVT are thrombophilia, malignancy, dural fistulae, polycythemia, anemia and vasculitis. There are few theories about mechanism of headache: intracranial hypertension, compression or stretching of the nerve fibers in the walls of occluded venous sinuses, activation of trigeminovascular system which then can cause a migrainous headache. Headache in CVT has a variable presentation and can have characteristics of primary headaches, especially migraine. It is usually subacute, often exacerbated by straining (e.g. Valsalva maneuver) or by lying down. Usually this headache is diffuse, bilateral and clinicians should always rule out ICH with CT. Also, examination of the optic fundus and search for papilledema are highly recommended. Nevertheless, headache can be unilateral, with different localizations. If deep veins are affected, diffuse encephalopathy, impaired mental status and coma can occur [42–46].

Pituitary Apoplexy

Another differential diagnosis for thunderclap headache, and potentially life threatening condition is pituitary apoplexy (PA). PA is considered a relatively rare clinical entity which results from hemorrhage in the normal pituitary gland. Incidence of PA varies from 0.6% to 7% [47, 48]. On the other hand, the incidence of apoplectic presentation of pituitary adenoma varies between 15% and 30%, and it is well known that pituitary tumors are intracranial tumors that have bleeding frequency more than 5 times higher than any other primary intracranial tumor [49]. Acute headache is the most frequent presenting symptom in more than 80% of patients with PA, usually with sudden onset and severe intensity often accompanied with nausea and vomiting [50]. Visual symptoms and oculomotor palsies are seen in up to 50% of cases. Partial or complete hypopituitarism is hallmark of this condition. Mechanism of these headaches is thought to be both because of meningeal irritation due to presence of blood in the subarachnoid space and dural stretching from mass effect. Unfortunately, the history of primary headaches in patients can cause a delay in diagnosis of PA which can lead to devastating outcome [50–52].

Giant Cell Arteritis

Giant cell arteritis (GCA) is a chronic vasculitis affecting large and medium-sized arteries, most commonly the temporal or other cranial arteries. The vasculitis in GCA is thought to be result of an antigen-mediated autoimmune response. GCA cause blindness and occasionally may have life-threatening consequences. GCA is cause of a secondary headache which primarily affects women over 50 years of age [53–55].

The International Headache Society (IHS) in 2018 proposed new criteria for diagnosis of GCA [4]:

A. Any new headache fulfilling criterion C
B. GCA has been diagnosed
C. Evidence of causation demonstrated by at least two of the following:

1. Headache has developed in close temporal relation to other symptoms and/ or clinical or biological signs of onset of GCA, or has led to the diagnosis of GCA
2. either or both of the following:

    (a) headache has significantly worsened in parallel with worsening of GCA
    (b) headache has significantly improved or resolved within 3 days of high-dose steroid treatment

3. headache is associated with scalp tenderness and/or jaw claudication

D. Not better accounted for by another ICHD-3 diagnosis

Headache is one of the cardinal symptoms when the temporal arteries are affected. The artery is tender, thick, and nodular. Artery is pulsating in the early phases of the disease, but later on it may become blocked. New-onset headache is typical in about two third of patients, while jaw claudication's are present in under 50% of patients. Other typical clinical features include scalp tenderness, fever, night sweats, fatigue, malaise, anorexia, weight loss, mood change, and polymyalgia. Ophthalmic symptoms can be temporary or permanent and they include amaurosis fugax, arteritic anterior ischemic optic neuropathy due to choroidal infarction, central retinal artery occlusions or posterior ischemic optic neuropathy. Visual loss occurs in up to 20% of GCA patients and risk factors are older age, male gender, and hypertension. Large vessel manifestations include aortitis, limb claudication, and formation of thoracic or abdominal aneurysms [53, 54].

The diagnosis of GCA is based on medical history, clinical examination and a combination of investigations. There should be a high degree of suspicion in individuals who are over 50 years of age and who are presenting with headache, especially new-onset headache or with visual changes, fever, jaw claudication, and/or muscle aches. Laboratory analysis that can help in achieving definitive diagnosis are erythrocyte sedimentation rate (ESR) and CRP, but in some cases, ESR can be normal or only slightly elevated, however, this should not exclude the diagnosis. Color Doppler ultrasound (CDUS) of the affected vessels shows the typical halo sign, a darkened area around the vascular lumen caused by edema (Picture 21.1). This can turn into the compression sign, in which the area of the vessel remains visible after compression by the ultrasound probe. Definitive diagnosis of GCA is made with histopathologic analysis of a temporal artery biopsy [53–55].

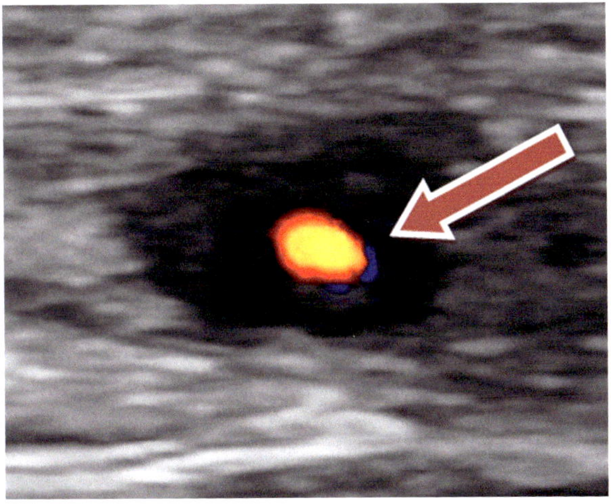

Picture 21.1 Color Doppler ultrasound of the superficial temporal artery in patient with newly diagnosed GCA with typical halo sign in the transverse plane

Early treatment is extremely important, because there is a risk of permanent vision loss or stroke. The treatment includes immediate high dose of glucocorticoid. Second-line therapy includes Methotrexate or Tocilizumab [54, 56]. Better understanding of the disease will help to update treatment guidelines.

Primary Angiitis of the Central Nervous System (PACNS)

Primary angiitis of the central nervous system (PACNS) is an inflammatory disease of the brain and spinal cord, manifesting with different clinical presentation such as headache, stroke, encephalopathy, seizures, and impaired cognition. Around 60% of patients have headache as major symptom. They are slowly progressive, and usually are not severe enough to warrant emergency evaluation for SAH, e.g. beginning of the disease is typically subacute or chronic, starting with headache and impaired cognition, which can be present for few months. The absence of these symptoms dose not excludes the diagnosis of PACNS [57, 58]. If thunderclap headache occurs, a different diagnosis should be taken into consideration, especially reversible cerebral vasoconstriction syndromes (RCVS) in patients younger than 40 years of age. These syndromes include Call-Fleming syndrome, post-partum angiopathy, and migrainous vasospasm. Of course, SAH should be ruled out [58–61]. The second group of disorders which can mimic PACNS are secondary CNS vasculitides, which arising as a complication of numerous infectious, rheumatic, autoimmune inflammatory disorders. The infectious causes of secondary CNS vasculitides include viral (Epstein-Barr or Varicella-zoster virus), bacterial (syphilis or tuberculosis), and fungal (Candida or Aspergillus) pathogens. The rheumatic and autoimmune inflammatory causes include Churg-Strauss disease, systemic lupus erythematosus, and Behcet syndrome. Other disorders which can cause CNS vasculitis include fibromuscular dysplasia, Moyamoya disease, and radiation vasculopathy. A complete workup for these secondary vasculitides should be made before confirming a PACNS diagnosis [62].

Most often, PACNS affects middle-aged men of approximately 50 years of age. Symptoms mimicking PACNS in patients younger than 20 years or older than 70 years should arouse suspicion of other possible diagnoses. The pathogenesis of headaches in PACNS is unknown and most likely occurs as a result of diffuse cortical dysfunction. Neurologic deficits, transient ischemic attacks, aphasia, visual field deficits and seizures may occur, but systemic manifestations such as fever, weight loss, night sweats and elevated ESR are rarely present, and if they are, systemic disorder should be ruled out. Even though headache is not of high diagnostic value, it is important to correlate the duration and characteristics of headache with diagnostic criteria for PACNS. Headache decreases and even diminishes with effective treatment of the disease [60, 61, 63].

Secondary Angiitis of the Central Nervous System (SACNS)

The secondary angiitis of the central nervous system (SACNS) is a central nervous system vasculitis of known etiology, such as lupus vasculitis, rheumatoid vasculitis, Gougerot Sjögren's syndrome and idiopathic hyper eosinophilic syndrome. Additional potential etiologies are cancer, use of certain drugs and infections. The pathogenesis of this headache is multifactorial: inflammation, stroke (ischemic or haemorrhagic), raised intracranial pressure and/or subarachnoid haemorrhage. The brain involvement is rare and usually a late manifestation of the disease, always accompanied with signs of the primary disease (headache, fever, fatigue, myalgia, and arthralgia). Recent studies showed that headache is less severe in SACNS patients than in patients with PACNS. The frequency of headache, neuropathy, cognitive deficit, convulsive crisis, and cerebellar symptoms was similar in both groups. Headache is the most frequent neurological manifestation of SACNS. Other symptoms include neurological deficits such as hemiparesis, aphasia, visual symptoms, ataxia, and numbness. Without headache and cerebrospinal fluid (CSF) pleocytosis, diagnosis of CNS angiitis is unlikely. The treatment of headaches attributed to SACNS should be based on the primary disease [62, 63].

Reversible Cerebral Vasoconstriction Syndrome

Reversible cerebral vasoconstriction syndrome (RCVS) represents a diverse group of disorders caused by acute intracerebral vasoconstriction that is fully reversible within 3 months and characterized by symptoms such as headache, seizures and focal neurological deficits. Headache is the main feature and may be the only manifestation. It is characterized by acute onset and very high intensity, so it has characteristics of thunderclap headache same as headache associated with SAH. The pain is severe, throbbing and reaches its highest point of intensity in about 1 min, after which it subsides [64]. Thunderclap headache in RCVS lasts on average less than 3 h and can be accompanied with a severe emotional response including agitation, screaming, confusion and loss of consciousness [65]. Symptoms such as nausea and vomiting, hypersensitivity to light and sound can be present. Typically, headache in RCVS is located bilaterally, originating from the occipital region and followed by diffuse pain. In some cases, it can be unilateral. RCVS can be spontaneous or have specific triggers, such as sympathetic over-activity, endothelial dysfunction, or oxidative stress, all of which ultimately lead to vasoconstriction. Some of the triggers are vasoactive substances such as cannabis, antidepressants, nasal decongestants, steroids, ergots, triptans, cocaine, nicotine patches, alcohol, epinephrine, interferon alpha and cyclosporine. Other possible triggers are postpartum period, showering, straining (coughing, sneezing, defecation), physical or sexual activity, eclampsia, or surgical operations. An episode of thunderclap headache is usually repeated on average 4 times over a period of 1–4 weeks. In some cases, the headache does not have the typical thunderclap features, and can be less intense with more progressive course. A presentation of RCVS without headache is exceptional. Complications that can follow RCVS are intracranial hemorrhages and posterior reversible encephalopathy syndrome (PRES) that usually occur in the first week, while TIA and ischemic stroke typically occur later, mostly at the end of the second week, during which the headache already subside. Thunderclap headache is of high diagnostic value and indicates further examination and diagnostic procedures. Neuroimaging in many patients with RCVS shows normal findings, but is also necessary to exclude other causes of thunderclap headache. CSF findings are usually normal. Multifocal segmental cerebral artery vasoconstriction can be found with angiography, and complete regression can be seen in up to 12 weeks. A link between migraine and RCVS has been examined. It was found in some studies that 17–27% of patients suffering from migraine also present with RCVS at some point. This can be linked to drugs used in prophylactic (selective serotonin reuptake inhibitors) or acute (triptans and ergots) migraine treatment [64–68]. On the other hand, RCVS has been found to trigger migraine with aura which could be due to cerebral hypoperfusion because of vasoconstriction [69].

Cerebral Endovascular Procedures

Cerebral endovascular procedures, such as angiography, angioplasty or stenting are very important diagnostic and therapeutic methods which are increasingly applied in the treatment of intracranial vascular lesions as they changed the prognosis of neurovascular disease [70, 71]. The ICHD-3 recognizes several headache types connected with endovascular procedures which are grouped under secondary headaches attributed to cranial or cervical vascular disorders [4].

These procedures may trigger headache which may start during or after procedure. The type of headache that occurs includes migraine with aura, migraine without aura, tension-type headache, probable migraine, probable tension-type headache, cluster headache and unclassified headache. There are differences in pain characteristics such as after extracranial procedure the pain is ipsilateral to the treated side, while intracranial procedures determine a more diffuse pain. Peri-procedural headache may start after few seconds or during contrast injection and most often headache is lasting less than 1 h, depending on what procedure is used. Persistence of post-procedural headache depends on how early in the intervention the headache occurred, so the earlier it occurs the shorter it lasts, and the headache that develops after hours to days after the procedure lasts longer. The headache that occurs after endovascular procedure in patients suffering from migraine may mimic their typical migraine pain. According to diagnostic criteria, only post-angiography headache has to resolve during 72 h, while other types of post-procedural headaches may last up to 30 days [4, 71–73]. On the other hand, headache can be a warning sign of a procedure complication that may be serious or even life-threatening, such as arterial perforation or dissection due to vessel injury by micro-catheters or micro-wires. Also, hemorrhagic stroke can occur due to cerebral venous thrombosis induced by the treatment [70–72].

One of the rare, but very important and potentially preventable complication of extracranial or intracranial endovascular procedure is cerebral hyperperfusion syndrome (CHS). Risk factors for CHS are age over 75 years, pre-existing hypertension, high-grade arterial stenosis with poor collateral circulation, decreased cerebrovascular reactivity, increased peak flow velocities, impaired cerebrovascular reserve, intra-operative distal carotid pressure lower than 40 mmHg, intraoperative ischemia, and recent contralateral carotid endarterectomy (CEA). Patients with CHS may develop headache which is unilateral, severe and pulsating, frequently followed with focal neurological deficit, cerebral edema, brain hemorrhage or seizures. The goal of the endovascular procedure is to increase cerebral perfusion, but CHS is the result of the post-procedure increases in cerebral blood flow, impaired cerebral autoregulation and disrupted blood brain barrier (conditions seen in hypertensive patients), sometimes accompanied with baroreceptor dysfunction, which eventually lead to cerebral edema and/or cerebral hemorrhage. Careful control of blood pressure is very important as prevention of CHS in periprocedural period. Also, if a patient develops CHS, the management of blood pressure is the main therapeutic goal [72].

Headache Attributed to Angiography

Angiography headache is defined as a headache caused directly by cerebral angiography. This headache is developed during angiography or within 24 h from the procedure and it is resolved within 72 h after the angiography. Incidence of angiography headache is about 35% during first day after angiography. Typical post-procedure headache features are not identified, probably as a result of its multifactorial etiology. Female gender is more susceptible to post-procedure headache occurrence probably because of the hormonal disturbance which is more common in women. In the study of Gil-Gouveia et al., incidence of post angiography headache within first 24 h was 51%, i.e. 35% if patients with SAH were excluded. Also, 6 months after procedure almost half of patients had frequent headaches, especially women who had a positive headache history before the angiography. In both groups (24 h and 6 months) headaches where mild, most often in the trigeminal area. After 24 h in most cases patients had bilateral headache (55%), while 60% of patients after 6 months reported unilateral headache. In both groups patients didn't have nausea or vomiting, but patients evaluated after 6 months had increased percentages of phonophobia and photophobia accompanying headache [73].

Interestingly, incidence of angiography headaches may vary depending upon the patient's clinical characteristics. The mechanism of headaches induced by angiography is not entirely clear. Post angiography headache with migraine features is explained by endothelial dysfunction and temporary vasodilatation as result of contrast injection. Possible mechanism could also be selective vasodilatation due to entry of contrast agent into the external carotid artery during injections performed through catheters placed in the common carotid artery or in the external carotid artery. Supporting this theory is fact that there are medicaments that can be used as prevention for migraine headache if they are applied before the procedure. One of those medications is indomethacin. Additionally, factors that occur prior to angiography such as psychological stress and fasting as a preparation for the procedure are also triggering factors for migraine headaches. It is important to mention that contrast angiography is contraindicated in patients with hemiplegic migraine because it may provoke a life-threatening attack, with prolonged hemiplegia and coma [74, 75].

Genetic Related Vasculopathy

Cerebral autosomal dominant arteriopathy with subcortical infarcts and leukoencephalopathy (CADASIL) is hereditary ischemic cerebral small vessel disease. Migraine with aura (MA) is found in the earliest stage of the disease, and represents a major symptom of CADASIL. MA may precede decades before stroke, but its frequency decreases at later stage when frequency of ischemic lesions increases. Female CADASIL patients are more likely to develop MA with earlier onset then man. There are some evidence supporting a causal relationship between cortical spreading depression and aura symptoms and, it is showed that the NOTCH3 gene mutations are responsible for early changes in cortical excitability in humans [76].

Mitochondrial encephalopathy, lactic acidosis, and stroke-like episodes (MELAS) syndrome is a rare genetic mitochondrial disease that primarily affects nervous system. Half of the patients have headache as first symptom. Headaches are usually migraine-like with or without aura, but certainly some of patients already have migraine as primary headache disorder. It is postulated that migraine in mitochondrial diseases such as MELAS may be expression of CNS vulnerability to disorders of mitochondrial respiratory chain [77, 78].

Conclusions

Headache is a common finding in different cerebrovascular diseases. Although vast majority of patients with headache will have a benign etiology, treating physicians need to make accurate diagnosis and confirm or rule out serious or even life threatening conditions, since good prognosis is based on clear etiology and precise treatment. Special attention should be paid to the conditions in which headache presents a cardinal symptom, especially if onset of headache is acute, with throbbing or thunderclap nature, and if pain is rapidly reaching the peak of pain intensity, when subarachnoid hemorrhage, cervical artery dissection, cerebral venous thrombosis and certain other conditions and diseases should be immediately ruled out.

References

1. Evans RW, Mitsias PD. Headache at onset of acute cerebral ischemia. Headache. 2009;49(6):902–8.
2. Fisher CM. Headache in cerebrovascular disease. In: Vinken PJ, Bruyn GW, editors. Handbook of clinical neurology. Headache and cranial neuralgias, vol. 5, chapter 11. Amsterdam: North Holland; 1968. p. 124–56.
3. Oliveira FAA, Sampaio Rocha-Filho PA. Headaches attributed to ischemic stroke and transient ischemic attack. Headache. 2019;59(3):469–76.
4. Headache Classification Committee of the International Headache Society (IHS). The international classification of headache disorders. Cephalalgia. 2018;38:1–211.
5. Abadie V, Jacquin A, Daubail B, Vialatte AL, Lainay C, Durier J, Osseby GV, Giroud M, Béjot Y. Prevalence and prognostic value of headache on early mortality in acute stroke: the Dijon Stroke Registry. Cephalalgia. 2014;34(11):887–94.
6. Tentschert S, Wimmer R, Greisenegger S, Lang W, Lalouschek W. Headache at stroke onset in 2196 patients with ischemic stroke or transient ischemic attack. Stroke. 2005;36(2):e1–3.
7. Carolei A, Sacco S. Headache attributed to stroke, TIA, intracerebral haemorrhage, or vascular malformation. Handb Clin Neurol. 2010;97:517–28.
8. Seifert CL, Schönbach EM, Magon S, Gross E, Zimmer C, Förschler A, Tölle TR, Mühlau M, Sprenger T, Poppert H. Headache in acute ischaemic stroke: a lesion mapping study. Brain. 2016;139(Pt1):217–26.
9. Seifert CL, Schonbach EM, Zimmer C, Poppert H, et al. Association of clinical headache features with stroke location: an MRI voxel-based symptom lesion mapping study. Cephalalgia. 2018;38:283–91.
10. Verdelho A, Ferro JM, Melo T, Canhão P, Falcão F. Headache in acute stroke. A prospective study in the first 8 days. Cephalalgia. 2008;28(4):346–54.
11. Magalhães JE, Sampaio Rocha-Filho PA. Migraine and cerebrovascular diseases: epidemiology, pathophysiological, and clinical considerations. Headache. 2018;58(8):1277–86.
12. Viswanathan V, Lucke-Wold B, Jones C, Aiello G, Li Y, Ayala A, Fox WC, Maciel CB, Busl KM. Change in opioid and analgesic use for headaches after aneurysmal subarachnoid hemorrhage over time. Neurochirurgie. 2021;67(5):427–32.
13. Wolf ME, Szabo K, Griebe M, Förster A, Gass A, Hennerici MG, Kern R. Clinical and MRI characteristics of acute migrainous infarction. Neurology. 2011;76(22):1911–7.
14. Laurell K, Artto V, Bendtsen L, Hagen K, Kallela M, Meyer EL, Putaala J, Tronvik E, Zwart JA, Linde M. Migrainous infarction: a Nordic multicenter study. Eur J Neurol. 2011;18(10):1220–6.

15. Androulakis XM, Kodumuri N, Giamberardino LD, Rosamond WD, Gottesman RF, Yim E, Sen S. Ischemic stroke subtypes and migraine with visual aura in the ARIC study. Neurology. 2016;87(24):2527–32.

16. Peng KP, Chen YT, Fuh JL, Tang CH, Wang SJ. Migraine and incidence of ischemic stroke: a nationwide population-based study. Cephalalgia. 2017;37(4):327–35.

17. Schurks M, Rist PM, Bigal ME, Buring JE, Lipton RB, Kurth T. Migraine and cardiovascular disease: systematic review and meta-analysis. BMJ. 2009;339:b3914.

18. Lebedeva ER, Ushenin AV, Gurary NM, Gilev DV, Olesen J. Diagnostic criteria for acute headache attributed to ischemic stroke and for sentinel headache before ischemic stroke. J Headache Pain. 2022;23(1):11.

19. Lebedeva ER, Ushenin AV, Gurary NM, Gilev DV, Olesen J. Sentinel headache as a warning symptom of ischemic stroke. J Headache Pain. 2020;21(1):70.

20. Ogunlaja OI, Cowan R. Subarachnoid hemorrhage and headache. Curr Pain Headache Rep. 2019;23(6):44.

21. Mac Grory B, Vu L, Cutting S, Marcolini E, Gottschalk C, Greer D. Distinguishing characteristics of headache in nontraumatic subarachnoid hemorrhage. Headache. 2018;58(3):364–70.

22. An SJ, Kim TJ, Yoon BW. Epidemiology, risk factors, and clinical features of intracerebral Hemorrhage: an update. J Stroke. 2017;19(1):3–10.

23. Dandurand C, Parhar HS, Naji F, Prakash S, Redekop G, Haw CS, Prisman E, Gooderham PA. Headache outcomes after treatment of unruptured intracranial aneurysm: a systematic review and meta-analysis. Stroke. 2019;50(12):3628–31.

24. Nader ND, Ignatowski TA, Kurek CJ, Knight PR, Spengler RN. Clonidine suppresses plasma and cerebrospinal fluid concentrations of TNF-alpha during the perioperative period. Anesth Analg. 2001;93(2):363–936.

25. Arena JE, Hawkes MA, Farez MF, Pertierra L, Kohler AA, Marrodán M, Benito D, Goicochea MT, Miranda JC, Ameriso SF. Headache and treatment of unruptured intracranial aneurysms. J Stroke Cerebrovasc Dis. 2017;26(5):1098–103.

26. Cumurciuc R, Crassard I, Sarov M, Valade D, Bousser MG. Headache as the only neurological sign of cerebral venous thrombosis: a series of 17 cases. J Neurol Neurosurg Psychiatry. 2005;76:1084–7.

27. Ellis JA, Mejia Munne JC, Lavine SD, Meyers PM, Connolly ES Jr, Solomon RA. Arteriovenous malformations and headache. J Clin Neurosci. 2016;23:38–43.

28. Galletti F, Sarchielli P, Hamam M, Costa C, Cupini LM, Cardaioli G, Belcastro V, Eusebi P, Lunardi P, Calabresi P. Occipital arteriovenous malformations and migraine. Cephalalgia. 2011;31(12):1320–4.

29. Dehdashti AR, Thines L, Willinsky RA, terBrugge KG, Schwartz ML, Tymianski M, Wallace MC. Multidisciplinary care of occipital arteriovenous malformations: effect on nonhemorrhagic headache, vision, and outcome in a series of 135patients. Clinical article. J Neurosurg. 2010;113(4):742–8.

30. Corbelli I, De Maria F, Eusebi P, Romoli M, Cardaioli G, Hamam M, Floridi P, Cupini LM, Sarchielli P, Calabresi P. Dural arteriovenous fistulas and headache features: an observational study. J Headache Pain. 2020;21(1):6.

31. Kim MS, Han DH, Kwon OK, Oh CW, Han MH. Clinical characteristics of dural arteriovenous fistula. J Clin Neurosci. 2002;9(2):147–55.

32. Lisotto C, Mainardi F, Maggioni F, Zanchin G. Headache in Sturge-Weber syndrome: a case report and review of the literature. Cephalalgia. 2004;24(11):1001–4.

33. Klapper J. Headache in Sturge-Weber syndrome. Headache. 1994;34(9):521–2.

34. Iizuka T, Sakai F, Yamakawa K, Suzuki K, Suzuki N. Vasogenic leakage and the mechanism of migraine with prolonged aura in Sturge-Weber syndrome. Cephalalgia. 2004;24(9):767–70.

35. Evers S, Marziniak M. Headache attributed to carotid or vertebral artery pain. Handb Clin Neurol. 2010;97:541–5.

36. Bogousslavsky J, Caplan LR, editors. Stroke syndromes. 2nd ed. Cambridge: Cambridge University Press; 2001.

37. Rist PM, Diener HC, Kurth T, Schürks M. Migraine, migraine aura, and cervical artery dissection: a systematic review and meta-analysis. Cephalalgia. 2011;31(8):886–96.
38. Chaves C, Estol C, Esnaola MM, Gorson K, O'Donoghue M, De Witt LD, Caplan LR. Spontaneous intracranial internal carotid artery dissection: report of 10 patients. Arch Neurol. 2002;59(6):977–81.
39. Matsumoto H, Hanayama H, Sakurai Y, Minami H, Masuda A, Tominaga S, Miyaji K, Yamaura I, Yoshida Y, Hirata Y. Investigation of the characteristics of headache due to unruptured intracranial vertebral artery dissection. Cephalalgia. 2019;39(4):504–14.
40. Suller Marti A, Bellosta Diago E, Velázquez Benito A, TejeroJuste C, Santos LS. Headache after carotid artery stenting. Neurologia. 2019;34(7):445–50.
41. Gündüz A, Göksan B, Koçer N, Karaali-Savrun F. Headache in carotid artery stenting and angiography. Headache. 2012;52(4):544–9.
42. Mehta A, Danesh J, Kuruvilla D. Cerebral venous thrombosis headache. Curr Pain Headache Rep. 2019;23(7):47.
43. deBruijn SF, Stam J, Kappelle LJ. Thunderclap headache as first symptom of cerebral venous sinus thrombosis. CVST Study Group. Lancet. 1996;348(9042):1623–5.
44. Silvis SM, de Sousa DA, Ferro JM, Coutinho JM. Cerebral venous thrombosis. Nat Rev Neurol. 2017;13(9):555–65.
45. Ferro JM, Canhao P, Stam J, Bousser MG, Barinagarrementeria F, ISCVT Investigators. Prognosis of cerebral vein and dural sinus thrombosis: results of the International Study on cerebral vein and dural sinus thrombosis (ISCVT). Stroke. 2004;35(3):664–70.
46. Timóteo Â, Inácio N, Machado S, Pinto AA, Parreira E. Headache as the sole presentation of cerebral venous thrombosis: a prospective study. J Headache Pain. 2012;13:487–90.
47. Johnston PC, Hamrahian AH, Weil RJ, Kennedy L. Pituitary tumor apoplexy. J Clin Neurosci. 2015;22(6):939–44.
48. Satyarthee GD, Sharma BS. Repeated headache as presentation of pituitary apoplexy in the adolescent population: unusual entity with review of literature. J Neurosci Rural Pract. 2017;8(Suppl 1):S143–6.
49. Donovan LE, Welch MR. Headaches in patients with pituitary tumors: a clinical conundrum. Curr Pain Headache Rep. 2018;22(8):57. https://doi.org/10.1007/s11916-018-0709-1. Review
50. Suri H, Dougherty C. Presentation and management of headache in pituitary apoplexy. Curr Pain Headache Rep. 2019;23(9):61.
51. Grangeon L, Moscatelli L, Zanin A, Rouille A, Maltete D, Guegan-Massardier E. Indomethacin-responsive paroxysmal hemicrania in an elderly man: an unusual presentation of pituitary apoplexy. Headache. 2017;57(10):1624–6.
52. Silvestrini M, Matteis M, Cupini LM, Troisi E, Bernardi G, Floris R. Ophthalmoplegic migraine-like syndrome due to pituitary apoplexy. Headache. 1994;34(8):484–6.
53. Klenofsky B, Sheikh UH. Giant cell arteritis. Retrieved from https://practicalneurology.com/articles/2020-may/giant-cell-arteritis
54. Lyons HS, Quick V, Sinclair AJ, Nagaraju S, Mollan SP. A new era for giant cell arteritis. Eye (Lond). 2020;34(6):1013–26.
55. Ing EB, Miller NR, Nguyen A, Su W, Bursztyn LLCD, Poole M, Kansal V, et al. Neural network and logistic regression diagnostic prediction models for giant cell arteritis: development and validation. Clin Ophthalmol. 2019;13:421–30.
56. Mollan SP, Paemeleire K, Versijpt J, Luqmani R, Sinclair AJ. European headache federation recommendations for neurologists managing giant cell arteritis. J Headache Pain. 2020;21(1):28.
57. Beuker C, Schmidt A, Strunk D, Sporns PB, Wiendl H, Meuth SG, Minnerup J. Primary angiitis of the central nervous system: diagnosis and treatment. Ther Adv Neurol Disord. 2018;11:1756286418785071.
58. Bhattacharyya S, Berkowitz AL. Primary angiitis of the central nervous system: avoiding misdiagnosis and missed diagnosis of a rare disease. Pract Neurol. 2016;16(3):195–200.

59. Birnbaum J, Hellmann DB. Primary angiitis of the central nervous system. Arch Neurol. 2009;66(6):704–9.
60. Hajj-Ali RA, Calabrese LH. Primary angiitis of the central nervous system. Autoimmun Rev. 2013;12(4):463–6.
61. Wan C, Su H. A closer look at angiitis of the central nervous system. Neurosciences (Riyadh). 2017;22(4):247–54.
62. Vera-Lastra O, Sepúlveda-Delgado J, Cruz-Domínguez Mdel P, Medina G, Casarrubias-Ramírez M, Molina-Carrión LE, et al. Primary and secondary central nervous system vasculitis: clinical manifestations, laboratory findings, neuroimaging, and treatment analysis. Clin Rheumatol. 2015;34:729–38.
63. Godasi R, Pang G, Chauhan S, Bollu PC. Primary central nervous system vasculitis. 2022 May 10. In: StatPearls [Internet]. Treasure Island (FL): StatPearls Publishing; 2022.
64. Miller TR, Shivashankar R, Mossa-Basha M, Gandhi D. Reversible cerebral vasoconstriction syndrome, part 1: epidemiology, pathogenesis, and clinical course. AJNR Am J Neuroradiol. 2015;36(8):1392–9.
65. Ducros A, Wolff V. The typical thunderclap headache of reversible cerebral vasoconstriction syndrome and its various triggers. Headache. 2016;56(4):657–73. https://doi.org/10.1111/head.12797. Epub 2016 Mar 26
66. Burton TM, Bushnell CD. Reversible cerebral vasoconstriction syndrome. Stroke. 2019;50(8):2253–8.
67. Santos L, Azevedo E. Reversible cerebral vasoconstriction syndrome—a narrative revision of the literature. Porto Biomed J. 2016;1(2):65–71.
68. Ducros A. Reversible cerebral vasoconstriction syndrome. Handb Clin Neurol. 2014;121:1725–41.
69. Mawet J, Debette S, Bousser MG, Ducros A. The link between migraine, reversible cerebral vasoconstriction syndrome and cervical artery dissection. Headache. 2016;56(4):645–56.
70. Khan S, Amin FM, Hauerberg J, Holtmannspötter M, Petersen JF, Fakhril-Din Z, Gaist D, Ashina M. Post procedure headache in patients treated for neurovascular arteriovenous malformations and aneurysms using endovascular therapy. J Headache Pain. 2016;17(1):73.
71. Paolucci M, Longoni M, Agostoni EC. Headache following endovascular procedures. Neurol Sci. 2020;41(Suppl 2):407–8.
72. Farooq MU, Goshgarian C, Min J, Gorelick PB. Pathophysiology and management of reperfusion injury and hyperperfusion syndrome after carotid endarterectomy and carotid artery stenting. Exp Transl Stroke Med. 2016;8(1):7.
73. Gil-Gouveia RS, Sousa RF, Lopes L, Campos J, Martins IP. Post angiography headaches. J Headache Pain. 2008;9(5):327–30.
74. Rasmussen BK, Olesen J. Symptomatic and nonsymptomatic headaches in a general population. Neurology. 1992;42(6):1225–31.
75. Qureshi AI, Naseem N, Saleem MA, Potluri A, Raja F, Wallery SS. Migraine and non-migraine headaches following diagnostic catheter-based cerebral angiography. Headache. 2018;58(8):1219–24.
76. Guey S, Mawet J, Hervé D, Duering M, Godin O, Jouvent E, Opherk C, Alili N, Dichgans M, Chabriat H. Prevalence and characteristics of migraine in CADASIL. Cephalalgia. 2016;36(11):1038–47.
77. Kraya T, Deschauer M, Joshi PR, Zierz S, Gaul C. Prevalence of headache in patients with mitochondrial disease: a cross-sectional study. Headache. 2018;58(1):45–52.
78. Vollono C, Primiano G, Della Marca G, Losurdo A, Servidei S. Migraine in mitochondrial disorders: prevalence and characteristics. Cephalalgia. 2018;38(6):1093–106.

Chapter 22
Calcitonin Gene-Related Peptide Induced Changes of Internal Homeostatic Body Model; Translation from TCD Studies

Marjan Zaletel and Bojana Žvan

Introduction

Migraine is an important neurological disease, affecting more than 10% of the population and causing significant disability. In recent years, it has been found that Calcitonin gene-related peptide (CGRP) is important for migraine development and CGRP antagonism can be used to treat and cope with migraine. Sensitization of trigeminal ganglia appears to be initially even which leads to the most disabling phase of a migraine episode, headache. According to predictive coding and interoception theories, the nociceptive drive after sensitization of trigeminal ganglia is increased leading to error detection in trigemino-cervical complex (TCC) and updating of the internal body model produces a headache that switches mode no fit to purpose. This causes transitory disability of migraincurs. Thus, CGRP could play important role in formatting the internal body homeostatic model.

Migraine and CGRP

Migraine is a common, disabling, neurovascular disorder. In Europe, the prevalence of migraine is estimated at around 10%. It mainly affects the population between the ages of 20 and 50. During a migraine episode, as many as 50% of patients are

M. Zaletel (✉)
University Clinical Center Ljubljana, Department of Vascular Neurology, Ljubljana, Slovenia

University Clinical Center Ljubljana, Pain Clinic, Ljubljana, Slovenia
e-mail: marjan.zaletel@kclj.si

B. Žvan
University Clinical Center Ljubljana, Department of Vascular Neurology, Ljubljana, Slovenia
e-mail: bojana.zvan@kclj.si

© The Author(s), under exclusive license to Springer Nature Switzerland AG 2023
V. Demarin et al. (eds.), *Mind, Brain and Education*,
https://doi.org/10.1007/978-3-031-33013-1_22

unable to work. Migraine has therefore a significant impact on work capacity as well as on the quality of life and as such represents a major socio-economic burden [1]. Migraine is defined by clinically characteristic migraine attacks, among which there are asymptomatic periods of varying lengths. The central symptom of a migraine episode is a migraine headache, which is often unilateral, moderate to severe, throbbing, and lasts from a few hours to a few days. Basically, a migraine episode is a clinical correlate of transient central sensitization of the nervous system.

The pathophysiology of migraine has not been fully elucidated to date. The findings of clinical trials suggest that the neuropeptide CGRP plays an important role in the pathophysiology of migraine. It is involved in the mechanisms of onset, persistence, and deepening of migraine headaches [2, 3]. Elevated serum CGRP levels in the external jugular vein have been found to occur during a migraine attack [4, 5]. In addition, serum interictal CGRP levels are higher in patients with chronic migraine than in patients with episodic migraine [5]. Migraine medications such as onabotualinumtoxin A and sumatriptan reduce serum CGRP levels [6, 7]. The importance of newer anti-CGRP target drugs for both acute and preventive treatment of migraine also speaks in favor of the importance of CGRP. CGRP receptor antagonists have been shown to be effective in the treatment of acute migraine headaches. Monoclonal antibodies against CGRP and CGRP receptors have been shown to be effective in the preventive treatment of episodic and chronic migraines. They significantly reduce the number of headache days and the use of drugs to acutely relieve migraine headaches [8]. Current data indicate that the CGRP target is important in migraine although the exact mechanism of action is still unclear. There are still many questions regarding anti-CGRP monoclonal antibodies since they cannot pass the blood-brain barrier. There is still uncertainty about whether plasma concentration (anti-CGRP mAb) of CGRP is elevated in chronic and episodic migraine. Nevertheless, anti-CGRP mAb could set off these effects by acting as a scavenger of CGRP in human blood.

CGRP in Human

CGRP is a neuropeptide composed of a sequence of 37 amino acids. It is formed in the nervous system in the process of alternative processing of RNA transcripts from the calcitonin gene, resulting in differential production of CGRP mRNA. Two isoforms are known in humans: αCGRP, which is found in the peripheral and central nerves, and ßCGRP, which is found primarily in the enteric nervous system [2]. CGRP is a potent vasodilator. It works by binding to the CGRP receptor, which consists of three components. CGRP and its receptor components are demonstrated in myelinated and unmyelinated nerve fibers involved in pain transmission, the trigeminal ganglion, its satellite glial cells, and many other central nervous system structures such as the cerebral cortex, thalamus, hypothalamus, cerebellum [9–11].

The pharmacokinetics of exogenous CGRP was investigated in animal and human studies, where it was found that CGRP pharmacokinetics follows the first order with a plateau reached within one hour. The elimination of CGRP shows a two-phase, bi-exponential decay [12]. The half-life was found to be 6.9 min for the first phase, and 26.4 min for the second, which supports a modulatory role of CGRP. Indeed, CGRP level in human blood shows high individual variance and it is released by CGRP afferents, majority of which form C-fibers but a minor proportion also Aδ-fibers, have nociceptive functions and are usually co-expressing transient receptor potential (TRP) cation channels of the vanilloid 1 and ankyrin type (TRPV1 and TRPA1) [13]. In peripheral blood, CGRP occurs in pico-gram levels. Despite CGRP being a permanently existing peptide in human circulation, in chronic migraine, it could be increased.

Cerebrovascular Reactivity to CGRP (CVR-CGRP)

CGRP is a potent vasodilator of intracranial arteries. Perivascular administered CGRP induces dose-dependent dilatation of many cerebral arteries. In animal models, it induces vasodilation of the middle cerebral artery (MCA), basilar artery, and cortical arteries [14, 15]. *In vitro* studies of human intracranial arteries' responses to CGRP have shown that CGRP induces dose-dependent vasodilation of human pial arteries [16].

Cerebrovascular reactivity to intravenously administered CGRP (CVR-CGRP) of anterior and posterior cerebral circulation CGRP has been elucidated. Vasodilation of the middle meningeal artery and the superficial temporal artery has been demonstrated [17, 18]. The results of a small number of studies on CVR-CGRP of MCA were not consistent. For instance, in aura-free migraine patients and migraine-free subjects, intravenous infusion of CGRP induces vasodilation of MCA [18, 19]. Contrary, the Asghar study did not confirm that CGRP induces vasodilation of MCA in healthy subjects [18]. Recently, Visočnik et al. using Transcranial Doppler (TCD) unequivocally demonstrated that αCGRP induces vasodilatation of MCA likely via activation of TCC [20]. In addition, they reported a decrease in end-tidal carbon dioxide (Et-CO$_2$) which could reflect a compensatory decrease in arterial partial pressure of carbon dioxide (pCO$_2$), underlying the normalization of cerebral blood flow during CGRP stimulation. In addition, the authors found that an intravenous αCGRP infusion could induce vasodilatation not only in MCA but in the posterior cerebral artery (PCA) as well [21]. Furthermore, CGRP produces systemic effects with a significant decrease in mean arterial pressure (MAP) and an increase in heart rate (HR). They explained that Et-CO2 decreases a compensatory mechanism in preserving cerebral blood flow and intracranial pressure during CGRP stimulation. In patients with migraine, CVR-CGRP is significantly enhanced and is associated with CGRP-induced headache (CGRP-IH) [20]. This might indicate that patients with migraine are more prone to sensitization. Therefore, studies of CVR-CGRP suggest vasodilatory effects of αCGRP in human blood on cerebral and systemic vessels.

Methodology for CVR-CGRP

The investigation of CVR-CGRP in our laboratory [21] is conducted in a quiet room under constant conditions. During the experiment, the participants are resting in a supine position. The experiment consists of a 10 min baseline period, a 20 min period during which an intravenous infusion of αCGRP 1.5mcg/min (Calbiochem, Merck4 Biosciences, Darmstadt, Germany) is given, and a 10 min period after the end of the application of αCGRP. The incidence of CGRP-IH is recorded within 1 h of the experiment (immediate CGRP-induced headache) and within the next 12 h after the experiment (delayed CGRP-induced headache). A visual analog scale is used to measure the intensity of CGRP-IH.

TCD with 2 MHz ultrasound probes is applied to measure mean flow velocity (vm) in left MCA and right PCA through the transtemporal acoustic windows. The signals of the MCA and PCA were defined according to the direction of the blood flow, the typical depth of the signal, and the response to compression. A mechanical probe holder is used to ensure a constant probe position. During the entire experiment MAP and HR are continuously measured using non-invasive plethysmography (Colin 7000, 12 Komaki-City, Japan). The Et-CO_2 is measured by a ventilation mask and an infrared capnograph (Capnograph, Model 9004, Smith medical, USA) using the standard protocol. The capnograph is connected to a computer. Et-CO_2 signals were recorded on the same time scale as other variables.

TCD Multi-Dop X4 software (DWL, Sipplingen, Germany) is used to define mean values of vmMCA, vmPCA, MAP, HR, and Et-CO_2 during 5 min intervals:first one interval during the baseline period (5–10 min of the experiment-measurement), second two intervals during the αCGRP infusion (15–20 min and 25–30 min of the experiment) and last one interval after the αCGRP infusion (35–40 min of the experiment). The mean vm MCA is calculated for each 5-min interval using the following equation:

$$\text{vm} = \int v\,\mathrm{d}t\,/\left(t_0 - t_5\right).$$

The mean values of other variables (MAP, HR, and Et-CO_2) are also calculated for the same time intervals as vmMCA and vmPCA using TCD software. We determine calculated responses to αCGRP as the differences between measuring points and responses are determined.

CGRP-Induced Headache (CGRP-IH)

Cerebral hemodynamics is supposed to be associated with a migraine headache. CGRP seems to be a key signaling molecule in connecting cerebral arteries with the nervous system. The changes in pial cerebral arteries may be a source of nociceptive signals to the central nervous system. Although the exact mechanism is elusive, it is well known that parenteral administration of CGRP induces CGRP-IH and even migraine-like attacks in migraineurs [22]. Recent study found [23] excellent response to erenumab in patients, in whom CGRP-IH with migraine futures has been evoked. Visočnik et al. established relationships between hemodynamic changes of arterial velocity changes in MCA, PCA, and CGRP-IH [24].

According to the concept of distal and proximal segments in cerebral circulation [25], CGRP could dilate both proximal and distal segments of the cerebral circulation and consequently lower overall cerebral arterial resistance. The response to increased cerebral blood flow could be mediated through the lowering of pCO_2 which acts on the distal cerebral segment and increase cerebral resistance due to CGRP provocation [20]. Therefore, the decrease in vm MCA and vm PCA could be solely due to vasodilatation of MCA and PCA during normalization of cerebral blood flow after lowering of pCO_2. This finding suggests that CGRP causation of CGRP-IH is probably due to its CGRP's direct effect on hemodynamics and consequent nociception related to it.

Interoception and Internal Body Homeostatic Model

According to the predictive coding concept, a human forms internal models of the homeostatic states through experiences. This represents priors under the principles of predictive coding and Bayesian inference. It is predicted that the central nervous system builds an internal homeostatic model through the process of learning which represents a template for further incoming homeostatic information. Incoming sensory data from the periphery are adjusted with prediction based on a current internal homeostatic model in prediction error units at different levels of the central nervous system [26]. We presume that the lowest prediction error unit is represented by TCC. The higher levels are located in the modulatory pain system and brainstem including periaqueductal gray (PAG) with descending pathways to TCC. The highest level consists of the cerebral cortex and other brain structures. Thus, the central nervous system including brain, brain stem, and spine represents a material base for the ever-changing internal homeostatic model.

In our described system, the sensory input is represented by trigeminal afferents which contain CGRP. In addition, CGRP is located in peripheral as well in central nervous system. When αCGRP is delivered intravenously, the CGRP in the blood increases in accordance with pharmacokinetic principles. Thus, CGRP in systemic blood could represent nociceptive stimuli, which induce nociceptive sensation when the human central nervous system does not predict a painful expectation. In TCC, the error unit of central nervous system and, the prediction error is detected and sent via ascending pathways to PAG and further to cerebral cortex. In this context, predictive error, carrying painful information, updates the internal model from set the subject to not fit purpose mode. In other words, an increase in CGRP in the systemic blood evokes CGRP-IH and disables the subject. It seems, that a similar situation occurs during a migraine episode leading to disability.

Conclusion

TCD studies of CVR-CGRP suggested that exogenous αCGRP induces vasodilatation of cerebral vessels. This could be induced directly through peripheral access or indirectly via sensitization of TCC. Nevertheless, the CGRP-IH can serve as a clinical model of a migraine episode. In addition, cerebral vascular response to αCGRP can be utilized to distinguish CGRP sensitive from insensitive migraine and other headaches. Therefore, we could predict the efficiency of anti-CGRP therapy in migraineurs.

References

1. Stovner LJ, Andree C. Prevalence of headache in Europe: a review for the Eurolight project. J Headache Pain. 2010;11(4):289–99.
2. Russell FA, King R, Smillie SJ, Kodji X, Brain SD. Calcitonin gene-related peptide: physiology and pathophysiology. Physiol Rev. 2014;94(4):1099–142.
3. Edvinsson L. Role of CGRP in migraine. Handb Exp Pharmacol. 2019;255(2):121–30.
4. Sarchielli P, Alberti A, Codini M, Floridi A, Gallai V. Nitric oxide metabolites, prostaglandins and trigeminal vasoactive peptides in internal jugular vein blood during spontaneous migraine attacks. Cephalalgia. 2000;20(10):907–18.
5. Pérez-Pereda S, Toriello-Suárez M, Ocejo-Vinyals G, Guiral-Foz S, Castillo-Obeso J, Montes-Gómez S, Martínez-Nieto RM, Iglesias F, González-Quintanilla V, Oterino A. Serum CGRP, VIP, and PACAP usefulness in migraine: a case-control study in chronic migraine patients in real clinical practice. Mol Biol Rep. 2020;47(9):7125–38.
6. Cernuda-Morollon E, Martinez-Camblor P, Ramon C, Larrosa D, Serrano-Pertierra E, Pascual J. CGRP and VIP levels as predictors of efficacy of Onabotulinum toxin A in chronic migraine. Headache. 2014;54(6):987–95.
7. Cerunda-Morollon E, Ramon C, Martinez-Camblor P, Serrano-Pertierra E, Larrosa D, Pascual J. Onabotulinum toxin A decreases interictal CGRP plasma levels in patients with chronic migraine. Pain. 2015;156(5):820–4.
8. Tepper SJ. CGRP and headache: a brief review. Neurol Sci. 2019;40:99–105.
9. Miller S, Liu H, Warfvinge K, Shi L, Dovlatyan M, Xu C, Edvinsson L. Immunohistochemical localization of the calcitonin gene-related peptide binding site in the primate trigeminovascular system using functional antagonist antibodies. Neuroscience. 2016;328:165–83.
10. Eftekhari S, Salvatore CA, Calamari A, Kane SA, Tajti J, Edvinsson L. Differential distribution of calcitonin gene-related peptide and its receptor components in the human trigeminal ganglion. Neuroscience. 2010;169(2):683–96.
11. Eftekhari S, Edvinsson L. Calcitonin gene-related peptide (CGRP) and its receptor components in human and rat spinal trigeminal nucleus and spinal cord at C1-level. BMC Neurosci. 2011;12:112.
12. Kraenzlin ME, Ching JL, Mulderry PK, Ghatei MA, Bloom SR. Infusion of a novel peptide, calcitonin generelated peptide (CGRP) in man. Pharmacokinetics and effects on gastric acid secretion and on gastrointestinal hormones. Regul Pept. 1985;10(2–3):189–97.
13. Messlinger K, Vogler B, Kuhn A, Sertel-Nakajima J, Frank F, Broessner G. CGRP measurements in human plasma - a methodological study. Cephalalgia. 2021;41(13):1359–73.
14. McCulloch J, Uddma R, Kingman TA, Edvinsson L. Calcitonin gene-related peptide: Functional role in cerebrovascular regulation. Neurobiology. 1986;83(15):5731–5.

15. Edvinsson L, Ekman R, Jansen I, McCulloch J, Uddman R. Calcitonin gene-related peptide and cerebral blood vessels: distribution and vasomotor effects. J Cereb Blood Flow Metab. 1987;7(6):720–8.
16. Edvinsson L, Fredholm BB, Hamel E, Jansen I, Verrecchia C. Perivascular peptides relax cerebral arteries concomitant with stimulation of cyclic adenosine monophosphate accumulation or release of an endothelium-derived relaxing factor in cat. Neurosci Lett. 1985;58(2):213–7.
17. Petersen KA, Lassen LH, Birk S, Lesko L, Olesen J. BIBN4096BS antagonizes human α-calcitoninj gene related peptide-induced headache and extracerebral artery dilatation. Clin Pharmacol Ther. 2005;77(3):202–13.
18. Asghar MS, Hansen AE, Kapijimpanga T, van der Geest RJ, van der Koning P, Larsson HB. Dilation by CGRP of middle meningeal artery and reversal by sumatriptan in normal volunteers. Neurology. 2010;75(17):1520–6.
19. Petersen KA, Birk S, Lassen LH, Kruuse C, Jonassen O, Lesko L, Olesen J. The CGRP-antagonist, BIBN4096BS does not affect cerebral or systemic hemodynamics in healthy volunteers. Cephalalgia. 2005;25(2):139–47.
20. Visočnik D, Zaletel M, Žvan B, Zupan M. Enhanced hemodynamic and clinical response to αCGRP in migraine patients-a TCD study. Front Neurol. 2021;12:638903.
21. Zupan M, Zaletel M, Visočnik D, Žvan B. The influence of calcitonin gene-related peptide on cerebral hemodynamics in nonmigraine subjects with calcitonin gene-related peptide-induced headaches. Biomed Res Int. 2021;2021:5540254.
22. Edvinsson L. The trigeminovascular pathway: role of CGRP and CGRP receptors in migraine. Headache. 2017;57(Suppl 2):47–55.
23. Christensen CE, Younis S, Deen M, Khan S, Ghanizada H, Ashina M. Migraine induction with calcitonin gene-related peptide in patients from erenumab trials. J Headache Pain. 2018;19(1):105.
24. Visočnik D, Žvan B, Zaletel M, Zupan M. αCGRP-induced changes in cerebral and systemic circulation; a TCD study. Front Neurol. 2020;11:578103.
25. Hoiland RL, Fisher JA, Ainslie PN. Regulation of the cerebral circulation by arterial carbon dioxide. Compr Physiol. 2019;9(3):1101–54.
26. Seth AK, Suzuki K, Critchley HD. An interoceptive predictive coding model of conscious presence. Front Psychol. 2012;2:395.

Chapter 23
Contemporary Cerebrospinal Fluid Analysis in Multiple Sclerosis

Uroš Rot and Andreja Emeršič

Introduction

Multiple sclerosis (MS) is a chronic inflammatory demyelinating disease of the central nervous system, which is the second most common cause of disability in young people. Yet, treatment of MS is one of the great medical successes in the last decades. There are now more than fifteen disease-modifying therapies for relapsing-remitting (RR) MS [1]. Early natural history studies in untreated populations showed that fifty percent of patients needed help with walking after 15 years of the disease. With the modern treatment, this frequency is much lower, approximately 10% [2]. Moreover, anti-B cell treatment ocrelizumab was shown to slow the progression of primary progressive (PP) MS and siponimod (S1P modulator) is effective in active secondary progressive (SP) MS [3, 4]. All medications for MS have anti-inflammatory mechanism of action therefore efficacy of the drugs is highest when given early in disease course.

MS diagnosis is made with the McDonald diagnostic criteria which require evidence of dissemination of symptoms and signs in 'time' and 'space' of the central nervous system either clinically or with the help of paraclinical investigations (almost exclusively MRI). The cerebrospinal fluid (CSF) analysis has an added

U. Rot (✉)
Laboratory for CSF Diagnostics, Department of Neurology, University Medical Centre Ljubljana, Ljubljana, Slovenia

Faculty of Medicine, University of Ljubljana, Ljubljana, Slovenia
e-mail: uros.rot@kclj.si

A. Emeršič
Laboratory for CSF Diagnostics, Department of Neurology, University Medical Centre Ljubljana, Ljubljana, Slovenia
e-mail: andreja.emersic@kclj.si

V. Demarin et al. (eds.), *Mind, Brain and Education*,
https://doi.org/10.1007/978-3-031-33013-1_23

value to MRI findings because it confirms inflammatory nature of the disease and supports accurate MS diagnosis [5–7]. A retrospective multicentre study revealed that patients with first clinical manifestation suggestive of MS and positive CSF oligoclonal bands (OB) were twice as likely to experience second clinical attack as OB negative patients [8]. The most recent, 2017 revisions of the McDonald criteria, thus permit substitution of positive OB result for dissemination in time in order to facilitate early MS diagnosis and treatment. Applying these criteria, diagnosis can be confirmed in a patient with clinically isolated syndrome (CIS) who has MRI evidence of dissemination in space and CSF restricted OB [5].

Another important reason for comprehensive CSF examination is improved differential diagnosis and a possibility to exclude alternative diagnoses, like other neurological and inflammatory diseases which may present with negative or positive OB but substantially higher total protein concentrations and/or cell counts [5, 6]. Studies characterising patients misdiagnosed with MS demonstrate that overreliance on MRI criteria, omission of CSF analysis and inappropriate interpretation of symptoms are main contributors to MS misdiagnosis [9, 10]. Diagnostic errors most often occurred in patients with unspecific neurological symptoms and unspecific MRI abnormalities in whom lumbar puncture was either not done or who had negative CSF OB [10].

Cerebrospinal Fluid in the Diagnosis of MS

Basic CSF findings are normal or mildly abnormal in MS. Total protein concentration is slightly elevated in approximately 30% of patients and mild pleocytosis is observed in about 60% of patients. CSF leukocyte counts are usually higher (but typically less than $50 \cdot 10^6$ cells/l) in younger patients and in patients with early MS (CIS) and correlate with disease duration [6, 11–13]. This simple finding is an important biomarker of inflammation in MS, which may be present when clinical symptoms are not yet fully developed and MRI burden is still low and less specific. CSF cytological examination shows lymphocytic predominance, plasma cells and monocytes may also be found (Fig. 23.1), whereas granulocytes (neutrophils, eosinophils) are extremely rare and should be regarded as a red flag, suggesting the possibility of an alternative diagnosis. The same consideration is needed in case of decreased CSF glucose (CSF/serum glucose ratio) or elevated CSF lactate, indicative of central nervous system infections or neoplastic meningitis [6, 14].

When interpreting CSF results against normal or expected values it is important to keep in mind that establishment of universal or assay-independent reference values is challenging even for the basic CSF parameters due to the limited availability of reference methods, certified reference materials and CSF samples in reference population (6). Upper reference limit for CSF total protein, for example, was shown to be age-dependent [15], but many laboratories still use adopted historical reference values without proper validation, which may lead to false interpretations. Sufficient method validation, internal quality control and participation in external

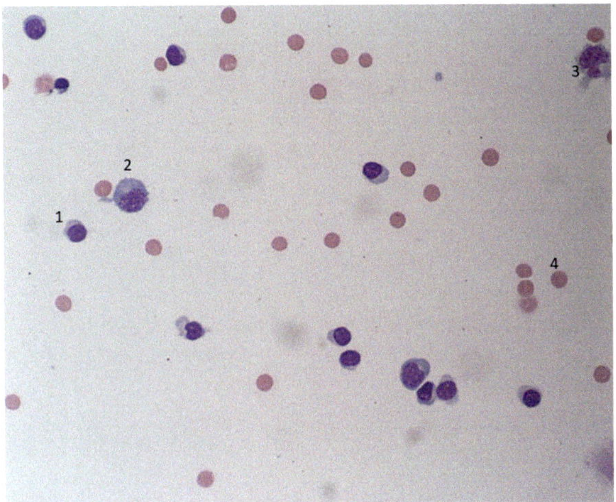

Fig. 23.1 Cerebrospinal fluid cytospin with predominant mononuclear cell pleocytosis (May-Grünwald-Giemsa staining). Lymphomonocytic cell picture with lymphocytes (**1**), plasma cells (**2**) and monocyte (**3**). Erytrocytes (**4**) due to artificial bleeding induced by lumbar puncture may be seen

quality assessments is needed to assure precise and accurate results, especially when standardised and certified CSF assays are not available [6]. Even for OB, laboratory-developed tests are not uncommon despite the availability of commercial assays, likely due to several optimizations of in-house methods and critical expertise gained through their use [16].

Knowledge about the presence of intrathecal IgG synthesis in MS is 80 years old and OB were discovered more than 60 years ago [17]. However, the presence of CSF restricted OB is still one of the most important diagnostic observations in MS [17]. OB bands are detected in more than 90% of our MS patients and in approximately 80% of patients with CIS [12, 16]. Conversely, elevated IgG index and intrathecal IgG synthesis according to Reiber's formula are seen in only 60–70% of our MS patients [11, 12]. Because of superior sensitivity of qualitative OB analysis, quantitative assessment is nowadays rarely performed as the only test for demonstration of intrathecal IgG production [7, 8]. Quantitative IgA and IgM intrathecal synthesis may, however, have a role in differential diagnosis of central nervous system infections [14].

OB analysis should be performed by isoelectric focusing with IgG specific immunodetection (blotting or fixation), which is a recommended standard that commercial and most of the in-house assays follow [14, 18] In order to distinguish between five different OB patterns, paired CSF and serum samples are analysed simultaneously in the same run and intrathecal IgG production is considered when bands occur in CSF only (type 2) or more bands are observed in CSF compared to serum (type 3) (Fig. 23.2). Several modifications of in-house methods have been described over the decades [16, 19]. We compared the sensitivity and specificity of a cheaper, easier to perform and shorter alkaline phosphatase assay with the previous method using double-antibody peroxidase staining. Samples of 160 patients with MS and CIS were analysed using both methods and comparison of blots showed that alkaline phosphatase assay enabled bands detection at 4 times lower IgG concentrations compared to the peroxidase method. This resulted in higher

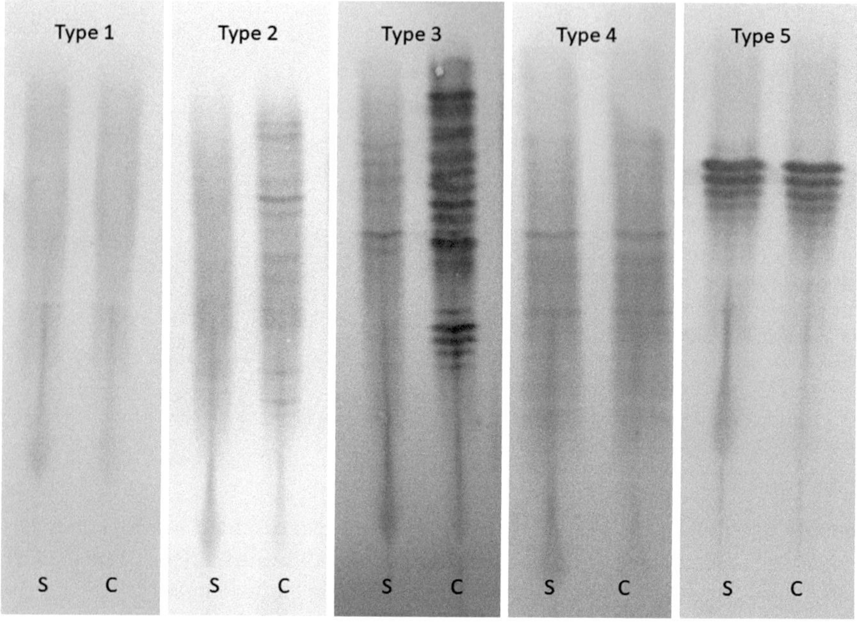

Fig. 23.2 Five consensus patterns of oligoclonal bands (isoelectric focusing with IgG immunodetection, Department of Neurology, Ljubljana). Paired serum (S) and cerebrospinal fluid (C) samples are run simultaneously in an adjacent track. Intrathecal IgG synthesis is considered in type 2 and 3. **Type 1** (no bands in CSF and serum), **type 2** (bands in CSF but not in serum), **type 3** (bands in CSF and serum with additional bands in CSF), **type 4** (identical bands in CSF and serum), **type 5** (monoclonal IgG pattern in CSF and serum)

frequency of OB positive patterns (89% versus 83%), higher number and sharper and stronger bands with alkaline phosphatase assay [16].

OB determination is a time-consuming procedure and the performance of this qualitative method still largely depends on technical skills and subjective interpretation of the results. Therefore, a shorter and easier to standardise assay with similar diagnostic accuracy would be preferable for detection of intrathecal inflammation. Quantification of immunoglobulin free light chains (FLC) rather than intact antibodies was proposed to overcome issues with OB method. As FLC are secreted in excess by plasma cells during the immunoglobulin production they also accumulate in the CSF in the presence of intrathecal B cell activity. Comparable diagnostic accuracy of kappa FLC (KFLC) and OB in MS was reported in several studies [20–23]. We included 80 MS patients and 50 controls in our method comparison, which revealed slightly higher sensitivity of KFLC index and intrathecal KFLC synthesis compared to OB. Ninety-six percent of our MS patients had elevated KFLC index and 95% local KFLC synthesis, whereas OB were positive in 91%. Specificity, on the other hand, was higher for OB determination, namely 100% for OB and 96% for KFLC index and synthesis [23].

Very high accuracy and simple (nephelometric or turbidimetric) measurement procedures with objective read-outs make KFLC determination potentially useful test in clinical routine, especially in laboratories with no or limited expertise in OB analysis. Alternatively, sensitivity of KFLC index (local synthesis) could be further increased by identifying and validating a cut-off below which MS diagnosis would be very unlikely. With greater negative predictive value, KFLC measurements could then be used as a screening test to guide further investigations and confirmatory OB analysis in dedicated, specialized laboratories [23]. Depending on pre-test probability of intrathecal inflammation and material expenses, our calculations show that this approach could reduce the number of OB determinations and manual work by 35% and material costs by 40% (unpublished observation). Clearly, more studies including different inflammatory diseases and control groups are needed before KFLC could be considered a valid alternative to OB [6]. Nevertheless, specialized clinical settings might still benefit from KFLC as an additional test [23]; due to the independent synthesis of immunoglobulin light and heavy chains, an elevated KFLC index is sometimes seen in OB negative CIS patients with very early immune response and intrathecal IgM (but not IgG) synthesis (unpublished observation).

To shortly summarize, even the simplest, basic CSF analyses provide valuable information for a reliable MS diagnosis. The single most informative test is assessment of OB-which is again an integral part of MS diagnostic criteria and should be performed in experienced laboratories, ensuring high analytical standards.

CSF and Prognosis of MS

CSF reflects pathological processes in the brain and spinal cord and thus gives a great opportunity for exploration of novel biomarkers that may track disease course and help in MS prognosis. Many biomarkers with prognostic potential were studied in MS.

CXCL13 is a B cell chemoattractant protein. There is evidence that B cells play an important role in all stages of MS, including the formation of meningeal B-cell follicles which are associated with the disease progression. Clinical studies have shown that CSF CXCL13 correlates with relapses and unfavourable clinical course in MS [24, 25].

Axonal loss is present from the earliest stages of MS and is strongly associated with prognosis of the disease. Neurofilaments are cytoskeletal proteins and consist of three types of chains (light, intermediate and heavy) which differ in size. Neurofilaments are released in axonal injury of any type and can be measured in CSF [26]. Neurofilament light (NFL) is the most studied protein among them. CSF concentrations of NFL are very high in extensive axonal injury, for example in ALS or Creutzfeldt-Jakob disease [27]. Studies in MS showed that CSF NFL concentrations are elevated in patients with recent relapses and are associated with number of

MRI lesions and with gadolinium enhancing lesions. CSF concentrations of NFL predict future relapses, but are not that well associated with disability progression [27, 28]. CFS NFL was also shown to predict recovery after optic neuritis [29]. In our retrospective study including 146 patients with CIS that were followed for nearly 5 years CSF NFL predicted relapses but not EDSS worsening [30]. In addition, some seminal studies demonstrated that CSF NFL normalizes with disease-modifying therapy [31]. For monitoring of therapy response, repetitive lumbar punctures are not feasible but single molecule array immunoassays with high analytical sensitivity already enable NFL measurements in blood. Further developments and broader availability of ultra-sensitive technology could result in simple treatment monitoring in not-so-distant future [27].

Processes of microglial activation and astrogliosis are known to be associated with disability progression in MS. YKL-40 is a glycoprotein which is expressed by astrocytes and microglia. While NFL is more associated with disease activity, YKL-40 may serve as a better marker of disability progression in MS [32]. In our patients with CIS YKL-40 concentrations were in correlation with disability worsening after nearly 5 years of follow-up [30], which is in line with the findings of other studies on YKL-40 prognostic potential [32, 33].

Less inflammatory activity is generally observed in PPMS. It was shown that anti-CD20 drug, ocrelizumab is more effective in a subgroup of patients with active PPMS. Therefore, biomarkers which could predict treatment response are badly needed. In our cohort of nearly 40 patients with PPMS CSF YKL-40 but not NFL was associated with disease duration (negative correlation) and EDSS progression, indicating it might also be used as a prognostic biomarker in PPMS [34].

Glial fibrillary acidic protein (GFAP) is a marker of astrogliosis and scar formation in MS, an important pathological finding in progressive forms of disease. Indeed, CSF GFAP concentrations were found to predict disability progression in early MS [35]. In another study, GFAP in serum but not in CSF was shown to correlate with disease severity in PPMS, which may be explained by direct release of GFAP into the blood compartment [36] or better stability of GFAP in serum during pre-analytical procedures [37]. Furthermore, serum GFAP may become useful in neuromyelitis optica spectrum disorder [38].

Recovery of symptoms, resolution of inflammation and remyelination are also important part of MS, therefore markers of regeneration should also be sought in order to better predict prognosis in an individual patient. Growth associated protein 43 (GAP-43) is a marker of synaptic plasticity and synaptic and axonal regeneration of the central nervous system. Only few studies were performed but they show that CSF GAP-43 is not reduced in early MS and correlates with inflammatory activity of the disease. In patients with progressive MS, on the other hand, CSF GAP-43 was reduced and correlated inversely with disability progression [39, 40].

In summary, there is growing evidence that biomarkers of B cell activity, axonal loss, reactive astrogliosis and microglial activation could assist in MS prognosis or monitoring of therapy response in the near future. NFL and YKL-40 are currently the most promising prognostic biomarkers in MS.

Experience with Newly Implemented Biomarkers in Clinical Routine

In Laboratory for CSF diagnostics, Ljubljana, KFLC quantification is regularly performed since 2018 and is always done when MS is strongly suspected but CSF OB bands are weak or absent. We also encourage our physicians to determine KFLC in patients with early or atypical MS. Yearly our laboratory receives approximately 600 requests for OB testing, in more than half of them, additional KFLC parameters are determined. In our daily clinical practice, we encounter cases with absent OB in CSF and positive KFLC index (2%) and less frequently, patients with a single CSF band and KFLC index below the cut-off (1%), in whom intrathecal inflammation is not ruled-out completely.

CSF NFL in our laboratory is most often determined in patients with atypical and rapidly progressive dementia or in patients with suspected motor neuron disease. However, CSF NFL reflects neuro-axonal loss which is present in many neurological conditions and levels markedly depend on age [41]. Since CSF NFL is 2.5 times higher in control subjects who are >60 years compared to those <30 years (unpublished observation), we established expected values for four different age groups (< 30 years, < 40 years, < 60 years, >60 years). When comparing the magnitude of CSF NFL increase in MS (CIS, RRMS, PPMS) and various neurological diseases in relation to symptomatic control subjects of similar age, we observed on average 1.5 increase in MS patients and 2, 5 and seven-fold higher levels in Alzheimer's disease, frontotemporal dementia and amyotrophic lateral sclerosis, respectively.

With the following examples we would like to emphasize how CSF NFL determination could be helpful in suspected MS.

AJ, 28 years old man had 1 month lasting right leg paraesthesias in January 2020 which again developed in November the same year. EDSS score was 2.0 in February 2021 when lumbar puncture was performed. CSF showed mildly elevated total protein concentration 0.64 g/l and 16 mononuclear cells per microliter. OB were positive in CSF (type 2) and FKLC index was 20.5 (normal <5.9). CSF NFL was 680 ng/l (2.3 higher than age-adjusted expected levels). MRI showed 2 brain and 1 cervical cord lesion and no gadolinium enhancement was seen.

ZS, 29 years old man had 1 month lasting right-sided pins and needles in 2016. He had similar symptoms in September 2020. His EDSS score was 2.0 in January 2021 when lumbar puncture was performed. CSF examination revealed normal total protein concentration 0.43 g/l and 3 mononuclear cells per microliter. OB were positive in CSF (type 2) and FKLC index was 15.2 (normal <5.9). CSF NFL was 8446 ng/l, which is 28 times above the age-adjusted expected levels. Brain MRI showed 9 white matter lesions and no gadolinium enhancement was observed.

Our patients had many similarities: both were young, males, had 2 MS attacks with sensory symptoms and low EDSS score. They had similar duration from symptoms to lumbar puncture and positive CSF OB and elevated FKLC index. The crucial differences were number of MRI lesions and NFL concentrations which were expected for the diagnosis in our first patient, but markedly elevated in the second case. Very high levels in our second patients could not be a consequence of a recent relapse because his symptoms developed 5 months before lumbar puncture. The observed differences found on MRI and in CSF are important for the treating physician and are helpful in the treatment decision making. In a patient with high number of MRI lesions and high CSF NFL, potent medication would probably be selected, whereas in a patient with low number of lesions and relatively normal CSF NFL a platform medication would likely be prescribed. It has to be said that treatment decision could be similar with the MRI results only, therefore more studies are needed to address possible added value of CSF NFL to the MRI findings.

Conclusion

CSF analysis is a valuable tool in the diagnostic work-up in suspected MS. CSF findings have an added value to the MRI data because they reveal inflammatory process in MS and improve differential diagnosis. OB status is again an important part of the McDonald diagnostic criteria. CSF is also a great source of novel biomarkers with differential diagnostic or prognostic potential and some of them are already entering clinical routine.

References

1. Tintore M, Vidal-Jordana A, Sastre-Garriga J. Treatment of multiple sclerosis – success from bench to bedside. Nat Rev Neurol. 2019;15:53–8.
2. Cree BAC, Gourraud PA, Oksenburg JR, Bevan C, Crabtree-Hartman E, Gelfand JM, et al. Long-term evolution of multiple sclerosis disability in the treatment era. Ann Neurol. 2016;80:499–510.
3. Montalban X, Hauser SL, Kappos L, Arnold DL, Bar-Or A, Comi G, et al. Ocrelizumab versus placebo in primary progressive multiple sclerosis. N Engl J Med. 2017;376:209–20.
4. Kappos L, Bar-Or A, Cree BAC, Fox RJ, Giovannoni G, Gold R, et al. Siponimod versus placebo in secondary progressive multiple sclerosis (EXPAND): a double-blind, randomized phase 3 study. Lancet. 2018;391:1263–73.
5. Thompson AJ, Banwell BL, Barkhof F, Carrol WM, Coetzee T, Comi G, et al. Diagnosis of multiple sclerosis: 2017 revisions to the McDonald criteria. Lancet Neurol. 2018;17:162–73.
6. Deisenhammer F, Zetterberg H, Fitzner B, Zettl UK. The cerebrospinal fluid in multiple sclerosis. Front Immunol. 2019;10:726.
7. Petzold A. Applying the 2017 McDonald diagnostic criteria for multiple sclerosis. Lancet Neurol. 2018;17(6):496–7.
8. Huss AM, Halbgebauer S, Öckl P, et al. Importance of cerebrospinal fluid analysis in the era of McDonald 2010 criteria: a German-Austrian retrospective multicenter study in patients with a clinically isolated syndrome. J Neurol. 2016;263(12):2499–504.
9. Solomon AJ, Klein EP, Bourdette D. "Undiagnosing" multiple sclerosis: the challenge of misdiagnosis in MS. Neurology. 2012;78(24):1986–91.
10. Solomon AJ, Bourdette DN, Cross AH, Applebee A, Skidd PM, Howard DB, et al. The contemporary spectrum of multiple sclerosis misdiagnosis: a multicenter study. Neurology. 2016;87:1393–9.

11. Rot U, Mesec A, Pogačnik T. Multiple sclerosis with uncommon cerebrospinal fluid findings. Croat Med J. 2003;44:697–701.
12. Rot U, Mesec A. Clinical, MRI, CSF and electrophysiological findings in different stages of multiple sclerosis. Clin Neurol Neurosurg. 2006;108:271–4.
13. Rot U, Horvat-Ledinek A, Šega-Jazbec S. Clinical, magnetic resonance imaging, cerebrospinal fluid and electrophysiological characteristics of the earliest multiple sclerosis. Clin Neurol Neurosurg. 2008;110:233–8.
14. Deisenhammer F, Bartos A, Egg R, Gilhus NE, Giovannoni G, Rauer S, et al. Guidelines on routine cerebrospinal fluid analysis. Reports from an EFNS task force. Eur J Neurol. 2006;13:913–22.
15. Hegen H, Auer M, Zeileis A, Deisenhammer F. Upper reference limits for cerebrospinal fluid total protein and albumin quotient based on a large cohort of control patients: implications for increased clinical specificity. Clin Chem Lab Med. 2016;54(2):285–92.
16. Andlovic A, Babič M, Accetto S, Rot U. Comparison of two methods for the detection of oligoclonal bands in a large number of clinically isolated syndrome and multiple sclerosis patients. Clin Neurol Neurosurg. 2012;114:659–62.
17. Holmoy T. The discovery of oligoclonal bands: a 50-year anniversary. Eur Neurol. 2009;62:311–5.
18. Freedman MS, Thompson EJ, Deisenhammer F, Giovannoni G, Griemsley G, Keir G, et al. Recommended standard of cerebrospinal fluid analysis in the diagnosis of multiple sclerosis. A consensus statement. Arch Neurol. 2005;62:865–70.
19. Sadaba MC, Gonzalez-Porque P, Massjuan J, Alvarez-Cermeno JC, Bootello A, Villar LM. An ultrasensitive method for the detection of oligoclonal IgG bands. J Immunol Methods. 2004;284:141–5.
20. Duranti F, Pieri M, Centonze D, Buttari F, Bernardini S, Dessi M. Determination of kappa FLC and kappa index in cerebrospinal fluid: a valid alternative to assess intrathecal immunoglobulin synthesis. J Neuroimmunol. 2013;263:116–20.
21. Presslauer S, Milosavljevic D, Huebi F, Abouienein-Djamshidian F, Krugluger W, Deisehammer F, et al. Validation of kappa free light chains as a diagnostic biomarker in multiple sclerosis and clinically isolated syndrome: a multicenter study. Mult Scler. 2016;22:1–9.
22. Leurs CE, Twaalfhoven HAM, Lissenberg-White BI, van Pesch V, Dujmovic I, et al. Kappa free light chains is a valid tool in the diagnostics of MS: a large multicenter study. Mult Scler. 2020;26:912–23.
23. Emeršič A, Anadolli V, Krsnik M, Rot U. Intrathecal immunoglobulin synthesis: the potential value of an adjunct test. Clin Chim Acta. 2019;489:109–16.
24. Khademi M, Kockum I, Andersson ML, Iacobaeus E, Brundin L, Sellebjerg F, et al. Cerebrospinal fluid CXCL13 in multiple sclerosis: a suggestive prognostic marker for the disease course. Mult Scler. 2011;17:335–43.
25. DiSano K, Gilli F, Pachner A. Intrathecally produced CXCL13: a predictive biomarker in multiple sclerosis. Mult Scler J Exp Transl Clin. 2020;6(4):2055217320981396.
26. Teunissen CE, Khalil M. Neurofilaments as biomarkers in mutiple sclerosis. Mult Scler. 2012;18:552–6.
27. Khalil M, Teunissen CE, Otto M, Piehl F, Sormani MP, Gattringer T, et al. Neurofilaments as biomarkers in neurological disorders. Nat Rev Neurol. 2018;14:577–89.
28. Malmestrom C, Haghighi S, Rosengren L, Andersen O, Lycke J. Neurofilament light protein and glial fibrillary acidic protein as biological markers in MS. Neurology. 2003;61:1720–5.
29. Modvig S, Degn M, Sander B, Horwitz H, Wanscher B, Sellebjerg F, et al. Cerebrospinal fluid neurofilament light chain levels predict visual outcome after optic neuritis. Mult Scler. 2016;22:590–8.
30. Emeršič A, Perovnik M, Pišlar N, Pavlič K, Pohar-Perme M, Rot U. Prognostic potential of CSF biomarkers in clinically isolated syndrome. Mult Scler. 2018;24(S2):668–9.
31. Gunnarsson M, Malmestrom C, Axelsson M, Sundstrom P, Dahle C, Vrethem M, et al. Axonal damage in multiple sclerosis is markedly reduced by natalizumab. Ann Neurol. 2011;69:83–9.

32. Novakova L, Axelsson M, Khademi M, Zetterberg H, Blennow K, Malmestrom C, et al. Cerebrospinal fluid biomarkers as a measure of disease activity and treatment efficacy in relapsing-remitting multiple sclerosis. J Neurochem. 2017;141:296–304.
33. Sellebjerg F, Royen L, Sorensen PS, Bang Oturai A, Jensen PEH. Prognostic value of cerebrospinal fluid neurofilament light chain and chitinase-3-like-1 in newly diagnosed patients with multiple sclerosis. Mult Scler. 2019;25:1444–51.
34. Emeršič A, Brecl-Jakob G, Šega-Jazbec S, Horvat-Ledinek A, Čučnik S, Rot U. Cerebrospinal fluid chitinase-3-like protein 1 is associated with disability progression in primary progressive multiple sclerosis. Mult Scler. 2020;26(S3):329–30.
35. Martinez MA, Olsson B, Bau L, Matas E, Cobo Calvo A, Andersson U, et al. Glial and neuronal markers in cerebrospinal fluid predict progression in multiple sclerosis. Mult Scler. 2015;21:550–61.
36. Abdelhak A, Hottenrott T, Morenas-Rodriguez E, Suarez-Calvet M, Zettl UK, Haass C, et al. Glial activation markers in CSF and serum from patients with primary progressive multiple sclerosis: potential of serum GFAP as disease severity marker ? Front Neurol. 2019;10:280.
37. Simrén J, Weninger H, Brum WS, et al. Differences between blood and cerebrospinal fluid glial fibrillary acidic protein levels: the effect of sample stability. Alzheimers Dement. 2022.; [published online ahead of print]; https://doi.org/10.1002/alz.12806.
38. Aktas O, Smith MA, Rees WA, Bennett JL, She D, Katz E, et al. Serum glial fibrillary acidic protein: a neuromyelitis optica spectrum disorder biomarker. Ann Neurol. 2021;89:895–910.
39. Rot U, Sandelius A, Emeršič A, Zetterberg H, Blennow K. Cerebrospinal fluid GAP-43 in early multiple sclerosis. Mult Scler J Exp Transl Clin. 4(3):2055217318792931. https://doi.org/10.1177/2055217318792931. eCollection 2018 Jul-Sep.
40. Sandelius A, Sandgren S, Axelsson M, Malmestrom C, Novakova L, Kostanjevecki V, et al. Cerebrospinal fluid growth-associated protein 43 in multiple sclerosis. Sci Rep. 2019;9(1):17309.
41. Vagberg M, Norgren N, Dring A, Lindqvist T, Birgander R, Zetterberg H, et al. Levels and age dependency of neurofilament light and glial fibrillary acidic protein in healthy individuals and their relation to the brain parenchymal fraction. PLoS One. 2015;10(8):e0.

Chapter 24
A Role of Deep Brain Stimulation in Advanced Parkinson's Disease

Vladimira Vuletić and Valentino Rački

Introduction

Parkinson's disease (PD) is a chronic neurodegenerative disease characterized with motor symptoms and non-motor symptoms. Bradykinesia, rigidity, rest tremor, postural instability are well-known motor symptoms, while pain, autonomic dysfunction, cognitive decline, depression, fatigue, apathy, and sleep disturbances are some of frequent non-motor symptoms [1]. We have few groups of antiparkinsonian drugs, with levodopa being a gold standard, to manage disease symptoms. The medications are very functional in an early stage, called the first honey-moon period, but in the advanced phase, we can witness the shortening of time when there is adequate symptom control. Management of advanced Parkinson's disease is challenging. In that stage, we can see motor fluctuation, severe non-motor symptoms, and insufficient control of motor symptoms with falls, freezing, festination, and sudden ON and OFF periods. Non-motor symptoms, especially neuropsychiatric problems, exert the most influence over quality of life in patients and their caregivers [2]. Proper DBS patient selection and management depends on a multidisciplinary approach that encompasses many specialties, including neurologists, neurosurgeons, neuroradiologists, psychologists, speech pathologists and physiotherapists.

Deep brain stimulation (DBS) has been used in the field of movement disorders for the last 33 years. From previous studies we know that DBS is an effective therapy for Parkinson's disease patients in their advanced state [3]. Subthalamic nucleus is the most widely used target, with individual advantages and disadvantages

V. Vuletić (✉) · V. Rački
Department of Neurology, UHC Rijeka, Rijeka, Croatia

Department of Neurology, Faculty of Medicine, University of Rijeka, Rijeka, Croatia
e-mail: Valentino.racki@uniri.hr

V. Demarin et al. (eds.), *Mind, Brain and Education*,
https://doi.org/10.1007/978-3-031-33013-1_24

253

influencing patient selection, followed by internal globus pallidus (GPi) that also has a beneficial effect [4]. Potential DBS patients are selected using the existing guidelines and the role of an interdisciplinary team is very important. Educated multidisciplinary team reviews each individual's risk–benefit profile for DBS. It is important to mention that DBS is not a cure and it does not stop disease progression and is not the only therapy for advanced PD, therefore, a clear explanation of possible benefits and side effects should be given to patients and their family member during the evaluation process. Their expectations should be realistic and clearly defined. The period after DBS has been lately called as a second honeymoon period by our patients. These advances have led researchers to put DBS among the most important advances in the clinical neurosciences in the past two decades, with ever-growing research. In this review, we will present the role of DBS in advanced Parkinson's disease.

Definition of Advanced Parkinson's Disease and the Indication for DBS

The definition of advanced Parkinson disease is still unclear. There are different definitions based on expert opinions, although it is clear that advanced PD begins when conventional treatment does not provide an adequate level of symptoms control [2]. Some consensus has been met, with a clear focus on both motor and non-motor symptoms, and include disease duration, motor fluctuation with dyskinesia, Hoehn and Yahr staging and specific clinical phenotypes like axial symptoms, cognitive decline or levodopa resistance [5].

Today, all recommendation for adequate management of Parkinson's disease include a multidisciplinary team of health professionals, including the neurologist or geriatrician, Parkinson's disease nurse specialist, physiotherapist, occupational therapist, speech and language therapist, dietitian, clinical psychologist, social worker, urologist, sex therapist, among many others. A multidisciplinary team approach has been shown to improve the quality of life and motor function for people with Parkinson's disease, and is also helpful for their caregivers [6–11]. Additionally, in the DBS team we also have a neurosurgeon, who is crucial to the success of therapy [6]. Finally, we also have studies that have shown beneficial effect of DBS in moderate early PD [12].

The definition of advanced PD is a moving target, but clear guidelines are now emerging on the benefits of DBS in both advanced and early PD patient groups. A recently published guideline using the GRADE methodology by Deuschl et al. had two main research questions and delivered several key recommendations. Most importantly, regarding DBS, there are clear recommendations that it should be preferentially offered to patients suffering from advanced PD with fluctuations bilaterally in the subthalamic nucleus (STN) or GPi, pending proper patient selections. Furthermore, we can consider offering STN-DBS to early PD patients with early fluctuations, as well as those with refractory tremor. The evidence so far does not recommend offering DBS to early PD patients without fluctuation [13].

Effect of DBS in PD

Evidence from previous studies has shown that DBS of either the subthalamic nucleus (STN) or the GPi have beneficial effect on motor fluctuations and dyskinesia associated with advanced PD. Benefits include increased ON time without troubling dyskinesia by a mean of 4.6 hours per day, reducing medications more than 50%, reducing OFF time for 67%, reducing dyskinesias for 70% and increasing quality of life by 50–70% [3]. The exact mechanism is still unknown, while it is thought that stimulation modulates neural circuits, improving transmission of signal and reducing oscillations [14]. DBS makes no major brain lesion, and is both adjustable and reversible. Greatest benefit requires proper adjustments in the whole disease course, which includes often visits to centers of excellence to manage parameter stimulation and adjustment of medical treatments. It is considered as a safe method and adverse events recorded during first 6 months were generally not serious when compared to the groups on best medical therapy [15]. Most commonly, the adverse events are due to operative procedure like intracerebral hemorrhage, infections etc., but cognitive and behavioral complications were infrequent and not significantly different between DBS and medical treatment groups when patients are properly selected [3]. The long term studies have shown persistent effect after 5, 10 and even 15 years [16–18]. Factors that predict benefit of DBS are: preoperative levodopa responsiveness, age, duration of OFF time, dyskinesias and psychiatric symptoms [19]. Comparing different targets, STN-DBS or GPi-DBS are both effective in advanced and early PD, although STN-DBS led to significantly greater improvements compared with GPi-DBS in mean change in the UPDRS motor examination score, disability score, and levodopa equivalent drug reduction in the off-drug phase. However, in the on-drug phase assessment, GPi-DBS was associated with a greater reduction in dyskinesia compared to STN-DBS. Generally it seems that STN as a target is better for more medication reduction, less-frequent battery changes, and has a more favorable economic profile, while GPi is better for more-robust dyskinesia suppression, easier programming, and greater flexibility in adjusting medications. There are also slight differences in the cognitive impact of DBS, with STN-DBS impacting cognitive deterioration more often than GPi-DBS, even though the risks are low for both [20]. In the end, the decision of target (STN or GPi) has to be tailored towards each patients [21, 22].

Research has shown that DBS has beneficial effects on non-motor symptoms at 24-month follow-up [23]. Generally, the impact of DBS is varied, and it can negatively impact cognitive function, especially if patients beforehand already have mild or severe cognitive impairments [20]. On the other hand, beneficial effects of DBS can be seen in autonomic dysfunction, sleep, seonsory function and mood disorder [24]. STN-DBS can improve anxiety in PD patients [25, 26] and impulse control disorders due to a reduction of the dopaminergic drugs after DBS [27]. Consequently, these improvements lead to improved quality of life and treatment satisfaction of patients [28]. There are some investigations in place to find good targets for DBS in PD dementia, such as the stimulation of the cholinergic nucleus basalis of Meynert, although further research is required [29].

Future Perspectives: New Technologies in DBS

The ability to improve advanced PD patient outcomes in intrinsically linked to the advancement of DBS systems. Two key improvements to DBS can be seen in the development of directional and local field potential systems, which have entered routine clinical practice. Research has already shown that directional programming brings many benefits to patients via more flexible stimulation options and can lead to greater therapeutic width and is preferred both by physicians and patients [30]. This is mostly achieved by increasing the threshold for side effects, as the clinical benefits are comparable to conventional DBS [31]. This has been shown in routine practice, with a majority of patients receiving some form of directionality in stimulation for 36 months, mainly for ameliorating side effects [32]. However, longer-term studies are required to see whether the benefits over conventional DBS hold in time.

Another novel approach to DBS programming is using local field potentials for guided stimulations catered to every patient [33]. Local field potentials are biosignals that have been previously used in DBS for confirming correct placements intraoperatively [34]. Furthermore, they are also used for insights into the pathophysiology of the disease from a functional perspective [35]. It's use in clinical practice is still being evaluated, and the greatest promise comes from the ability to reduce initial programming time. Conventional DBS programming is still based on a trial-and-error approach, while personalized measurements of oscillations point to clear targets in each patients. Pilot studies have shown that LFP based programming can streamline the whole process and limit the amount of time spent [36].

Furthermore, looking forward, we can expect further innovation in the field with adaptive closed loop stimulation. Current DBS systems function in an open loop environment, meaning they are programmed beforehand and do not respond to any feedback from the patient. There are several devices in development that could use biomarkers of brain activity and change stimulation settings automatically, leading to the most precise stimulation possible for each patient and each situation [37].

Conclusion

Treatment of advanced PD with invasive methods is becoming more developed each day. DBS has been routinely used for years, and with accumulating experience it is now clear that it should be the first choice for properly selected patients. This is reflected in the most recent guidelines, with STN and Gpi holding strong recommendations for treatment, which lead to a significant improvement to quality of life and functioning.

DBS expert centers are preferentially multidisciplinary and can provide holistic approaches to each patient to facilitate the best possible outcomes. Future perspectives of DBS are bright, with novel systems entering routine clinical practice that offer greater flexibility and effectiveness with less side effects.

References

1. Armstrong MJ, Okun MS. Diagnosis and treatment of Parkinson disease: a review. JAMA. 2020;323(6):548–60.
2. Coelho M, Ferreira JJ. Late-stage Parkinson disease. Nat Rev Neurol. 2012;8(8):435–42.
3. Deuschl G, Schade-Brittinger C, Krack P, Volkmann J, Schäfer H, Bötzel K, et al. A randomized trial of deep-brain stimulation for Parkinson's disease. N Engl J Med. 2006;355(9):896–908.
4. Follett KA, Weaver FM, Stern M, Hur K, Harris CL, Luo P, et al. Pallidal versus subthalamic deep-brain stimulation for Parkinson's disease. N Engl J Med. 2010;362(22):2077–91.
5. Luquin MR, Kulisevsky J, Martinez-Martin P, Mir P, Tolosa ES. Consensus on the definition of advanced Parkinson's disease: a neurologists-based Delphi study (CEPA study). Parkinsons Dis. 2017;2017:4047392.
6. Pedersen SW, Suedmeyer M, Liu LWC, Domagk D, Forbes A, Bergmann L, et al. The role and structure of the multidisciplinary team in the management of advanced Parkinson's disease with a focus on the use of levodopa-carbidopa intestinal gel. J Multidiscip Healthc. 2017;10:13–27.
7. Carne W, Cifu D, Marcinko P, Pickett T, Baron M, Qutubbudin A, et al. Efficacy of a multidisciplinary treatment program on one-year outcomes of individuals with Parkinson's disease. NeuroRehabilitation. 2005;20(3):161–7.
8. Carne W, Cifu DX, Marcinko P, Baron M, Pickett T, Qutubuddin A, et al. Efficacy of multidisciplinary treatment program on long-term outcomes of individuals with Parkinson's disease. J Rehabil Res Dev. 2005;42(6):779–86.
9. van der Marck MA, Bloem BR, Borm GF, Overeem S, Munneke M, Guttman M. Effectiveness of multidisciplinary care for Parkinson's disease: a randomized, controlled trial. Mov Disord. 2013;28(5):605–11.
10. van der Marck MA, Munneke M, Mulleners W, Hoogerwaard EM, Borm GF, Overeem S, et al. Integrated multidisciplinary care in Parkinson's disease: a non-randomised, controlled trial (IMPACT). Lancet Neurol. 2013;12(10):947–56.
11. Martinez-Martin P, Rodriguez-Blazquez C, Forjaz MJ. Quality of life and burden in caregivers for patients with Parkinson's disease: concepts, assessment and related factors. Expert Rev Pharmacoecon Outcomes Res. 2012;12(2):221–30.
12. Deuschl G, Schüpbach M, Knudsen K, Pinsker MO, Cornu P, Rau J, et al. Stimulation of the subthalamic nucleus at an earlier disease stage of Parkinson's disease: concept and standards of the EARLYSTIM-study. Parkinsonism Relat Disord. 2013;19(1):56–61.
13. Deuschl G, Antonini A, Costa J, Śmiłowska K, Berg D, Corvol JC, et al. European Academy of Neurology/Movement Disorder Society - European Section guideline on the treatment of Parkinson's disease: I. Invasive therapies. Eur J Neurol. 2022;29(9):2580–95.
14. Chiken S, Nambu A. Mechanism of deep brain stimulation: inhibition, excitation, or disruption? Neuroscientist. 2016;22(3):313–22.
15. Weaver FM, Follett K, Stern M, Hur K, Harris C, Marks WJ, et al. Bilateral deep brain stimulation vs best medical therapy for patients with advanced Parkinson disease: a randomized controlled trial. JAMA. 2009;301(1):63–73.

16. Rodriguez-Oroz MC, Moro E, Krack P. Long-term outcomes of surgical therapies for Parkinson's disease. Mov Disord. 2012;27(14):1718–28.
17. Castrioto A, Lozano AM, Poon YY, Lang AE, Fallis M, Moro E. Ten-year outcome of subthalamic stimulation in Parkinson disease: a blinded evaluation. Arch Neurol. 2011;68(12):1550–6.
18. Bove F, Mulas D, Cavallieri F, Castrioto A, Chabardès S, Meoni S, et al. Long-term outcomes (15 years) after subthalamic nucleus deep brain stimulation in patients with Parkinson disease. Neurology. 2021;97(3):e254–62.
19. Welter ML, Houeto JL, Tezenas du Montcel S, Mesnage V, Bonnet AM, Pillon B, et al. Clinical predictive factors of subthalamic stimulation in Parkinson's disease. Brain. 2002;125(Pt 3):575–83.
20. Rački V, Hero M, Rožmarić G, Papić E, Raguž M, Chudy D, et al. Cognitive impact of deep brain stimulation in Parkinson's disease patients: a systematic review. Front Hum Neurosci. 2022;16:867055.
21. Odekerken VJ, van Laar T, Staal MJ, Mosch A, Hoffmann CF, Nijssen PC, et al. Subthalamic nucleus versus globus pallidus bilateral deep brain stimulation for advanced Parkinson's disease (NSTAPS study): a randomised controlled trial. Lancet Neurol. 2013;12(1):37–44.
22. Williams NR, Foote KD, Okun MS. STN vs. GPi deep brain stimulation: translating the rematch into clinical practice. Mov Disord Clin Pract. 2014;1(1):24–35.
23. Klingelhoefer L, Samuel M, Chaudhuri KR, Ashkan K. An update of the impact of deep brain stimulation on non motor symptoms in Parkinson's disease. J Parkinsons Dis. 2014;4(2):289–300.
24. Vuletic V, Racki V, Chudy D, Bogdanovic N. Deep brain stimulation in non-motor symptoms of neurodegenerative diseases [Internet]. In: Neurostimulation and neuromodulation in contemporary therapeutic practice. London: IntechOpen; 2019. https://www.intechopen.com/chapters/68407. [cited 2022 Jan 13].
25. Parsons TD, Rogers SA, Braaten AJ, Woods SP, Tröster AI. Cognitive sequelae of subthalamic nucleus deep brain stimulation in Parkinson's disease: a meta-analysis. Lancet Neurol. 2006;5(7):578–88.
26. Castelli L, Perozzo P, Zibetti M, Crivelli B, Morabito U, Lanotte M, et al. Chronic deep brain stimulation of the subthalamic nucleus for Parkinson's disease: effects on cognition, mood, anxiety and personality traits. Eur Neurol. 2006;55(3):136–44.
27. Shotbolt P, Moriarty J, Costello A, Jha A, David A, Ashkan K, et al. Relationships between deep brain stimulation and impulse control disorders in Parkinson's disease, with a literature review. Parkinsonism Relat Disord. 2012;18(1):10–6.
28. Gruber D, Calmbach L, Kühn AA, Krause P, Kopp UA, Schneider GH, et al. Longterm outcome of cognition, affective state, and quality of life following subthalamic deep brain stimulation in Parkinson's disease. J Neural Transm (Vienna). 2019;126(3):309–18.
29. Gratwicke J, Kahan J, Zrinzo L, Hariz M, Limousin P, Foltynie T, et al. The nucleus basalis of Meynert: a new target for deep brain stimulation in dementia? Neurosci Biobehav Rev. 2013;37(10 Pt 2):2676–88.
30. Schnitzler A, Mir P, Brodsky MA, Verhagen L, Groppa S, Alvarez R, et al. Directional deep brain stimulation for Parkinson's disease: results of an international crossover study with randomized, double-blind primary endpoint. Neuromodulation. 2022;25(6):817–28.
31. Dembek TA, Reker P, Visser-Vandewalle V, Wirths J, Treuer H, Klehr M, et al. Directional DBS increases side-effect thresholds—a prospective, double-blind trial. Mov Disord. 2017;32(10):1380–8.
32. Karl JA, Joyce J, Ouyang B, Verhagen ML. Long-term clinical experience with directional deep brain stimulation programming: a retrospective review. Neurol Ther. 2022;11(3):1309–18.
33. Rosa M, Marceglia S, Barbieri S, Priori A. Local field potential and deep brain stimulation (DBS). In: Jaeger D, Jung R, editors. Encyclopedia of computational neuroscience [Internet]. New York, NY: Springer; 2013. p. 1–20. https://doi.org/10.1007/978-1-4614-7320-6_547-1. [cited 2022 Sep 30].

34. Yin Z, Zhu G, Zhao B, Bai Y, Jiang Y, Neumann WJ, et al. Local field potentials in Parkinson's disease: a frequency-based review. Neurobiol Dis. 2021;155:105372.
35. Telkes I, Jimenez-Shahed J, Viswanathan A, Abosch A, Ince NF. Prediction of STN-DBS electrode implantation track in Parkinson's disease by using local field potentials. Front Neurosci. 2016;10. https://www.frontiersin.org/articles/10.3389/fnins.2016.00198. [cited 2022 Sep 30].
36. Fasano A, Gorodetsky C, Paul D, Germann J, Loh A, Yan H, et al. Local field potential-based programming: a proof-of-concept pilot study. Neuromodulation. 2022;25(2):271–5.
37. Frontiers | Deep brain stimulation programming 2.0: future perspectives for target identification and adaptive closed loop stimulation [Internet]. https://www.frontiersin.org/articles/10.3389/fneur.2019.00314/full. [cited 2022 Sep 30].

Chapter 25
Serum Carnitine Levels in Outpatient Huntington's Disease Clinic Population, Brief Report and Review of Literature

Miroslav Cuturic, Ruth K. Abramson, Janice G. Edwards, and Souvik Sen

Introduction

Huntington's disease (HD) is a complex neurodegenerative disorder with autosomal-dominant inheritance. It usually emerges in mid-life, manifested by progressive motor, cognitive and psychiatric deterioration. The genetic defect consists of CAG trinucleotide expansion on the chromosome 4, resulting in polyglutamine chain expansion in the mutant huntingtin protein. Multiple factors have been proposed that lead to the progression of neurodegeneration in HD, including glutamate-induced excitotoxicity [1], oxidative stress [2, 3], and impaired energy metabolism with mitochondrial dysfunction [4, 5]. Beneficial effects of antioxidants in reducing the progression of the illness have been demonstrated in animal models of HD [6].

The amino acid derivative L-carnitine (4-N-trimethylammonium-3-hydroxy-butyric-acid), is an important regulator of lipid metabolism in humans. It is

M. Cuturic (✉) · S. Sen
Department of Neurology, University of South Carolina School of Medicine,
Columbia, SC, USA
e-mail: miroslav.cuturic@uscmed.sc.edu; souvik.sen@uscmed.sc.edu

R. K. Abramson
Hall Institute of Psychiatry, Department of Neuropsychiatry and Behavioral Science,
University of South Carolina School of Medicine, Columbia, SC, USA
e-mail: Ruth.Abramson@uscmed.sc.edu

J. G. Edwards
University of South Carolina School of Medicine, Columbia, SC, USA
e-mail: janice.edwards@uscmed.sc.edu

V. Demarin et al. (eds.), *Mind, Brain and Education*,
https://doi.org/10.1007/978-3-031-33013-1_25

responsible for the transport of long-chain fatty acids into the mitochondrial matrix, facilitating beta-oxidation and energy production via the Krebs's cycle [7, 8]. Levocarnitine has antioxidant properties, acts as a free radical scavenger, and may reduce physical and mental fatigue in elderly [9–11]. High dose levocarnitine treatment in a transgenic mouse model of HD showed illness alleviating effects and neuroprotective properties [12]. Therefore, carnitine is becoming increasingly interesting as a potential therapeutic agent in HD.

In our prior retrospective study of institutionalized HD patients, we found a relatively high (26.1%) prevalence of hypocarnitinemia [13]. In the current review we examined serum carnitine levels and clinical correlates in an outpatient group at our university HD clinic over a period of 3 years. In this retrospective medical record review, our aim was to evaluate the relevance of serum carnitine levels monitoring in outpatient HD population and to compare our findings with the findings from the prior study done on, more advanced, institutionalized patients.

Patients and Methods

This retrospective chart review was approved by the University of South Carolina Institutional Review Board. We reviewed records of new outpatient clients with HD, during a 3-year period, for whom we routinely recommended measurement of serum carnitine levels upon their initial clinic visit. From the patients charts we documented vital statistics, body mass index (BMI), psychiatric and medical diagnoses, medication regimen, laboratory data, Mini Mental State Exam (MMSE) scores, and Unified Huntington's Disease Rating Scale (UHDRS) motor scores [14, 15]. Hypocarnitinemia was defined as serum carnitine levels below the normal laboratory reference values for total and free carnitine (<34 and 25 µmol/L, respectively). We used descriptive statistics to compare patients with and without hypocarnitinemia. The student's t-test was utilized to evaluate certain parameters. In this small seminal study, we did not utilize formal corrections for type I error given that they may be too conservative at this initial stage.

Results

We recommended metabolic screening, including the measurement of serum carnitine levels, for all 34 new HD patients at our clinic for the given period, but only 19 patients (55%) complied with our recommendations. In 5 (26.3%) of the 19 compliant patients, serum carnitine levels were below normal reference range. Two of the five patients with hypocarnitinemia complied with our recommendation for levocarnitine supplementation. The two compliant patients received a low dose supplementation with levocarnitine at 330 mg twice per day, one for three and one for six months, until follow up assessment revealed normalization of the total and free carnitine serum levels. In both patients, serial MMSE scores remained unchanged for the given periods. In one patient serial UHDRS motor score improved from 58 to 48 during the supplementation period of 3 months, while in the other the UHDRS motor score worsened from 6 to 13 during the 6 months of the supplementation. There were no adverse effects reported with levocarnitine supplementation. Of the three patients who did not comply with carnitine supplementation, one dropped out after the initial clinic visit, while two declined the supplementation and subsequently dropped out from follow up.

In 14 subjects (73.7%), the initial carnitine levels were within the normal range. These patients did not have any further follow up of carnitine levels and did not receive any carnitine supplementation during the reviewed period.

In the review of systems, none of the study subjects reported significant weight loss. All of the subjects reported adhering to a balanced diet, without any specific restrictions. Two subjects reported modification of their food consistency, either to soft mechanical or pureed diet, but both of these individuals had normal serum carnitine levels. None of the patients were dependent on tube feedings.

Demographic and selected clinical characteristics of our patients are summarized in Table 25.1, while their nutritional and metabolic parameters are compared in Table 25.2.

Table 25.1 Prevalence of selected demographic and clinical parameters in our Huntington's Disease patients

	Patients with hypocarnitinemia ($N = 5$)	Patients without hypocarnitinemia ($N = 14$)	All ($N = 19$)
Females	3 (60.0 %)	10 (71.4 %)	13 (68.4 %)
African Americans	1 (20.0 %)	1 (7.1 %)	2 (10.5 %)
Caucasians	4 (80.0 %)	13 (92.9 %)	17 (89.5 %)
Valproate therapy*	0 (0.0 %)	4 (28.6 %)	4 (21.1 %)
Hyperlipidemia**	5 (100.0 %)	3 (21.4 %)	8 (42.1 %)
Depression**	1 (20.0 %)	7 (50.0 %)	8 (42.1 %)
DNA confirmed***	3 (60.0 %)	10 (71.4 %)	13 (68.4 %)

N = number of cases
* = number of cases receiving valproic acid
** = most prevalent conditions other than Huntington's Disease
*** = number of cases with Huntington's Disease confirmed by genetic testing

Table 25.2 Comparison of the initial mean values for selected nutritional and metabolic parameters between patients with ($n = 5$) and without ($n = 14$) hypocarnitinemia

	Patients with hypocarnitinemia (SD)	Patients without hypocarnitinemia (SD)	P value
Age	53.8 (11.1)	53.3 (15.6)	0.950
Body Mass Index	24.6 (4.9)	25.7 (6.1)	0.726
Total carnitine, μmol/L	28.2 (5.9)	45.4 (11.8)	**<0.007**
Free carnitine, μmol/L	21.6 (6.2)	32.8 (10.3)	**0.038**
Acyl-carnitine, μmol/L	6.6 (2.5)	11.8 (4.3)	**0.023**
Hemoglobin A1C	5.4 (0.3)	5.4 (0.3)	0.838
Albumin, gm/dl	4.2 (0.5)	4.1 (0.4)	0.653
AST, U/L	26.4 (16.2)	29.7 (16.9)	0.718
ALT, U/L	17.0 (5.1)	25.6 (16.0)	0.266
Creatinine, mg/dl	0.8 (0.1)	0.8 (0.2)	0.788
BUN, mg/dl	12.2 (6.5)	16.2 (4.8)	0.172
Cholesterol, mg/dl	198.0 (49.3)	180.0 (40.3)	0.453
HDL, mg/dl	56.2 (8.2)	49.0 (15.5)	0.354
LDL, mg/dl	128.0 (44.9)	107.0 (32.0)	0.320
Triglycerides, mg/dl	73.2 (23.1)	101.0 (40.5)	0.178
VLDL, mg/dl	12.8 (4.11)	31.4 (27.2)	0.209
UHDRS	29.2 (24.3)	26.7 (20.2)	0.827
MMSE	26.0 (3.7)	25.3 (4.7)	0.783

ALT alanine aminotransferase; *AST* aspartat aminotransferase; *HDL* high density lipoproteins; *BUN* blood urea nitrogen; *LDL* low density lipoproteins; *n* number of cases; *SD* standard deviation; *UHDRS* total motor score of the Unified Huntington's Disease Rating Scale; *VLDL* very low density lipoproteins

Discussion

This review reveals a relatively high prevalence of hypocarnitinemia in an ambulatory, outpatient, HD clinic population. The prevalence of hypocarnitinemia in this review is approximately equal to the prevalence that we found in our prior study of the more advanced institutionalized HD patients, with the values of 26.3% and 26.1 % respectively [13]. These prevalence rates are higher than reported in general and psychiatric populations, but similar to the prevalence reported in critically ill patients at intensive care units, without HD [16–20].

The outpatient population in this review had less advanced HD compared to the institutionalized patients in our prior study, as evidenced by lower UHDRS scores and higher MMSE scores. However, the prevalence and the level of hypocarnitinemia were almost identical between the two reports [13]. This suggests that hypocarnitinemia is not likely a marker of the HD severity, but rather an independent HD epiphenomenon.

Primary carnitine deficiency is a rare autosomal recessive disorder caused by a defect in the plasma membrane carnitine transporter gene SLC22A5 (5q31.2-32), with a prevalence of 1:40,000 in live births, and is not related to HD [18, 21]. The reduction of carnitine serum levels reported here as well as in our prior study were less profound than would be expected with homozygous primary carnitine deficiency [18], suggesting that the carnitine deficiency in our HD patients resulted from secondary depletion. We hypothesize that at the cellular level, accumulation of mutant huntingtin may impair endogenous synthesis of carnitine, as well as the activity of the carnitine membrane transporter, carnitine palmitoyltransferase I or II, all of which would promote carnitine deficiency.

In the general population, carnitine levels are known to be lower in females [22]. In this review, most of our patients were females (68.4%), including the group of patients with hypocarnitinemia (60%, Table 25.1). To the contrary, in our prior study of institutionalized HD patients, most of our subjects were males (65%), even more so in the group of individuals with hypocarnitinemia (83%) [13]. Therefore, we conclude that gender disparities in our reports are the result of the site-specific differences between the studied populations, rather than to gender related prevalence disparities of hypocarnitinemia in HD patients.

Although carnitine deficiency may result from malnutrition [7, 8], the patients with hypocarnitinemia in this review reported no restrictions in diet and their mean BMI did not differ from the rest of the patients (Table 25.2). Chronic renal or hepatic disease may also contribute to secondary carnitine deficiency [23], but none of our patients had any of such diagnoses, all had normal hepatic and renal function laboratory tests, and there was no difference in renal or hepatic parameters when compared to the patients without hypocarnitinemia (Table 25.2).

Although possible role of catabolism and weight loss in the development of hypocarnitinemia was suggested in our prior study [13], in this review the patients with hypocarnitinemia did not report weight loss and their BMI did not differ from the rest of the patients (Table 25.2). It is conceivable that excessive involuntary motor activity and catabolism due to advanced chorea may increase demand for carnitine and promote secondary carnitine deficiency [24]. However, in the current review there was no significant difference in the motor UHDRS scores between the patients with and without hypocarnitinemia.

Chronic valproic acid therapy is a well-established cause of secondary carnitine deficiency in epileptic as well as in psychiatric patients [16, 17, 25]. In our prior study of institutionalized HD patients, valproic acid use was associated with two thirds of the cases of hypocarnitinemia [13]. To the contrary, in our current review of outpatient population, none of the patients with hypocarnitinemia had exposure to valproic acid, suggesting that hypocarnitinemia in the HD population is not necessarily related to valproic acid use and may be related to other etiologies or represent an epiphenomenon of HD itself.

In this current review as well as in our prior study, dyslipidemia was the most prevalent medical diagnosis in the patients with hypocarnitinemia (Table 25.2) [13].

Although in our prior study serum lipid levels were significantly higher in patients with hypocarnitinemia, in the current review the lipid level differences did not reach statistical significance (Table 25.2) [13]. Nonetheless, these findings may suggest a possible relationship between hypocarnitinemia and dyslipidemia in HD patients.

The principal clinical manifestations of carnitine deficiency include metabolic encephalopathy, myopathy and cardiomyopathy, which can be effectively treated by carnitine supplementation [23, 26, 27]. In HD patients, a superposition of metabolic encephalopathy or myopathy may be particularly difficult to differentiate from progression of the illness itself, as both conditions are manifested by cognitive impairment, generalized weakness, imbalance, and behavioral disturbances. In our prior study we observed that carnitine supplementation improved the motor and cognitive symptoms of hypocarnitinemia [13], but in the current review we did not have a sufficient number of patients treated nor sufficient longitudinal data to make such inferences.

Limitations of the Report

The most significant limitation of this retrospective chart review is the small sample size resulting in low statistical power. Consequently, most of the calculated statistical measures are subject to type I and II error, as they may gain or lose significance as the sample size gets larger. In this small review, we did not utilize formal corrections for type I error given that they may be too conservative at this initial stage.

Inter-rater variability of UHDRS scores, MMSE scores, serum handling issues, and natural fluctuations of carnitine serum levels may have additionally affected our results. The retrospective design of this review is another significant limitation. In addition, the findings were restricted only to the patients who were compliant with the recommended blood tests (55%), which can result in either under- or overestimation of the hypocarnitinemia prevalence. Future studies are needed to corroborate the findings.

Conclusion

We found a relatively high prevalence of hypocarnitinemia in an outpatient HD population, which does not differ from the prevalence previously reported among institutionalized HD patients. We conclude that monitoring of serum carnitine levels as a part of metabolic screening in HD patients should be recommended, particularly as HD patients with hypocarnitinemia potentially may benefit from low-dose levocarnitine supplementation. In this revew as well as in the prior study of institutionalized HD patients, dyslipidemia was the most common medical disorder associated with hypocarnitinemia. In comparison to our prior study, this retrospective chart review did not confirm association of hypocarnitinemia with valproic acid use or precipitous weight loss in HD. Future studies are needed to further clarify these findings.

Acknowledgments None.

Conflict of Interest This research did not receive any specific grant from funding agencies in the public, commercial, or not-for-profit sectors. All authors individually, and as a group, declare no conflict of interest.

References

1. Lee W, Reyes RC, Gottipati MK, Lewis K, Lesort M, Parpura V, Gray M. Enhanced Ca(2+)-dependent glutamate release from astrocytes of the BACHD Huntington's disease mouse model. Neurobiol Dis. 2013;58:192–9. [cited 2022 Sept 1]. https://www.sciencedirect.com/science/article/abs/pii/S0969996113001691. https://doi.org/10.1016/j.nbd.2013.06.002.
2. Perluigi M, Poon HF, Maragos W, Pierce WM, Klein JB, Calabrese V, Cini C, De Marco C, Butterfield DA. Proteomic analysis of protein expression and oxidative modification in r6/2 transgenic mice: a model of Huntington disease. Mol Cell Proteomics. 2005;4(12):1849–61. [cited 2022 Aug 30]. https://www.sciencedirect.com/science/article/pii/S1535947620300220. https://doi.org/10.1074/mcp.M500090-MCP200.
3. Klepac N, Relja M, Klepac R, Hećimović S, Babić T, Trkulja V. Oxidative stress parameters in plasma of Huntington's disease patients, asymptomatic Huntington's disease gene carriers and healthy subjects : a cross-sectional study. J Neurol. 2007;254(12):1676–83. [cited 2022 Aug 30]. https://link.springer.com/article/10.1007/s00415-007-0611-y. https://doi.org/10.1007/s00415-007-0611-y.
4. Damiano M, Galvan L, Déglon N, Brouillet E. Mitochondria in Huntington's disease. Biochim Biophys Acta. 2010;1802(1):52–61. [cited 2022 Sept 1]. https://www.sciencedirect.com/science/article/pii/S0925443909001653. https://doi.org/10.1016/j.bbadis.2009.07.012.
5. Sawant N, Morton H, Kshirsagar S, Reddy AP, Reddy PH. Mitochondrial abnormalities and synaptic damage in Huntington's disease: a focus on defective mitophagy and mitochondria-targeted therapeutics. Mol Neurobiol. 2021;58(12):6350–77. [cited 2022 Sept 1]. https://www.springer.com/journal/12035. https://doi.org/10.1007/s12035-021-02556-x.
6. Bicca Obetine Baptista F, Arantes LP, Machado ML, da Silva AF, Marafiga Cordeiro L, da Silveira TL, Soares FAA. Diphenyl diselenide protects a Caenorhabditis elegans model for Huntington's disease by activation of the antioxidant pathway and a decrease in protein aggregation. Metallomics. 2020;12(7):1142–58. [cited 2022 Sept 1]. https://academic.oup.com/metallomics/issue/12/7. https://doi.org/10.1039/d0mt00074d.
7. Rebouche CJ. Kinetics, pharmacokinetics, and regulation of L-carnitine and acetyl-L-carnitine metabolism. Ann N Y Acad Sci. 2004;1033:30–41. [cited 2022 Aug 30]. https://nyaspubs.onlinelibrary.wiley.com/doi/abs/10.1196/annals.1320.003.
8. Evans AM, Fornasini G. Pharmacokinetics of L-carnitine. Clin Pharmacokinet. 2003;42:941–67. [cited 2022 Aug 31]. https://link.springer.com/article/10.2165/00003088-200342110-00002. https://doi.org/10.2165/00003088-200342110-00002.
9. Arockia Rani PJ, Panneerselvam C. Carnitine as a free radical scavenger in aging. Exp Gerontol. 2001;36(10):1713–26. [cited 2022 Aug 31]. https://www.sciencedirect.com/science/article/abs/pii/S0531556501001164?via%3Dihub. https://doi.org/10.1016/S0531-5565(01)00116-4.
10. Bayrak S, Aktaş S, Altun Z, Çakir Y, Tütüncü M, Kum Özşengezer S, Yilmaz O, Olgun N. Antioxidant effect of acetyl-l-carnitine against cisplatin-induced cardiotoxicity. J Int

Med Res. 2020;48(8):300060520951393. [cited 2022 Sept 1]. https://journals.sagepub.com/doi/10.1177/0300060520951393.

11. Malaguarnera M, Cammalleri L, Gargante MP, Vacante M, Colonna V, Motta M. L-Carnitine treatment reduces severity of physical and mental fatigue and increases cognitive functions in centenarians: a randomized and controlled clinical trial. Am J Clin Nutr. 2007;86(6):1738–44. [cited 2022 Aug 31]. https://academic.oup.com/ajcn/article/86/6/1738/4649810. https://doi.org/10.1093/ajcn/86.5.1738.

12. Vamos E, Voros K, Vecsei L, Klivenyi P. Neuroprotective effects of L-carnitine in a transgenic animal model of Huntington's disease. Biomed Pharmacother. 2010;64(4):282–6. [cited 2022 Aug 31]. https://www.sciencedirect.com/science/article/abs/pii/S0753332209001887?via%3Dihub. https://doi.org/10.1016/j.biopha.2009.06.020.

13. Cuturic M, Abramson RK, Moran RR, Hardin JW, Frank EM, Sellers AA. Serum carnitine levels and levocarnitine supplementation in institutionalized Huntington's disease patients. Neurol Sci. 2013;34(1):93–8. [cited 2022 Aug 31]. https://link.springer.com/article/10.1007/s10072-012-0952-x. https://doi.org/10.1007/s10072-012-0952-x.

14. Folstein M, Folstein SE, McHugh PR. "Mini-mental state". A practical method for grading the cognitive state of patients for the clinician. J Psychiatr Res. 1975;12:189–98. [cited 2022 Aug 31]. https://psycnet.apa.org/record/1976-20785-001. https://doi.org/10.1016/0022-3956(75)90026-6.

15. Huntington Study Group. Unified Huntington's disease rating scale: reliability and consistency. Mov Disord. 1996;11(2):136–42. [cited 2022 Aug 31]. https://movementdisorders.onlinelibrary.wiley.com/doi/10.1002/mds.870110204.

16. Cuturic M, Abramson RK, Moran RR, Hardin JW, Hall AV. Clinical correlates of low serum carnitine levels in hospitalized psychiatric patients. World J Biol Psychiatry. 2011;12(1):73–9. [cited 2022 Aug 31]. https://www.tandfonline.com/doi/full/10.3109/15622975.2010.489619.

17. Cuturic M, Abramson RK, Moran RR, Hardin JW. Clinical outcomes and low-dose levocarnitine supplementation in psychiatric inpatients with documented hypocarnitinemia: a retrospective chart review. J Psychiatr Pract. 2010;16(1):5–14. [cited 2022 Sept 1]. https://journals.lww.com/practicalpsychiatry/Abstract/2010/01000/Clinical_Outcomes_and_Low_Dose_Levocarnitine.2.aspx. https://doi.org/10.1097/01.pra.0000367773.03636.d1.

18. Longo N. Primary carnitine deficiency and newborn screening for disorders of the carnitine cycle. Ann Nutr Metab. 2016;68(Suppl 3):5–9. [cited 2022 Sept 1]. https://www.karger.com/Article/Fulltext/448321. https://doi.org/10.1159/000448321.

19. Kelley J, Sullivan E, Norris M, Sullivan S, Parietti J, Kellogg K, Scott AI. Carnitine deficiency among hospitalized pediatric patients: a retrospective study of critically ill patients receiving extracorporeal membrane oxygenation therapy. JPEN J Parenter Enteral Nutr. 2021;45(8):1663–72. [cited 2022 Sept 1]. https://aspenjournals.onlinelibrary.wiley.com/doi/10.1002/jpen.2255.

20. Bonafé L, Berger MM, Que YA, Mechanick JI. Carnitine deficiency in chronic critical illness. Curr Opin Clin Nutr Metab Care. 2014;17(2):200–9. [cited 2022 Sept 1]. https://journals.lww.com/co-clinicalnutrition/Abstract/2014/03000/Carnitine_deficiency_in_chronic_critical_illness.16.aspx. https://doi.org/10.1097/MCO.0000000000000037.

21. Koizumi A, Nozaki J, Ohura T, Kayo T, Wada Y, Nezu J, Ohashi R, Tamai I, Shoji Y, Takada G, Kibira S, Matsuishi T, Tsuji A. Genetic epidemiology of the carnitine transporter OCTN2 gene in a Japanese population and phenotypic characterization in Japanese pedigrees with primary systemic carnitine deficiency. Hum Mol Genet. 1999;8(12):2247–54. [cited 2022 Aug 31]. https://academic.oup.com/hmg/article/8/12/2247/661065. https://doi.org/10.1093/hmg/8.12.2247.

22. Rasmussen J, Nielsen OW, Janzen N, Duno M, Gislason H, Køber L, Steuerwald U, Lund AM. Carnitine levels in 26,462 individuals from the nationwide screening program for primary carnitine deficiency in the Faroe Islands: 50 years of Newborn Screening. J Inherit Metab Dis. 2014;37(2):215–22. [cited 2022 Aug 31]. https://pure.fo/en/publications/

carnitine-levels-in-26462-individuals-from-the-nationwide-screeni.   https://doi.org/10.1007/s10545-013-9606-2.

23. Pons R, De Vivo DC. Primary and secondary carnitine deficiency syndromes. J Child Neurol. 1995;10(suppl. 2):8–24. [cited 2022 Aug 31]. https://journals.sagepub.com/doi/10.1177/0883073895010002S03. https://doi.org/10.1177/0883073895010002S03.

24. Pratley RE, Salbe AD, Ravussin E, Caviness JN. Higher sedentary energy expenditure in patients with Huntington's disease. Ann Neurol. 2000;47(1):64–70. [cited 2022 Aug 31]. https://onlinelibrary.wiley.com/doi/10.1002/1531-8249(200001)47:1%3C64::AID-ANA11%3E3.0.CO;2-S.

25. Saito M, Takizawa T, Miyaoka H. Factors associated with blood carnitine levels in adult epilepsy patients with chronic valproic acid therapy. Epilepsy Res. 2021;175:106697. [cited 2022 Sept 1]. https://www.sciencedirect.com/science/article/abs/pii/S0920121121001509. https://doi.org/10.1016/j.eplepsyres.2021.106697.

26. Breningstall GN. Carnitine deficiency syndromes. Pediatr Neurol. 1990;6:75–81. [cited 2022 Aug 31]. https://www.pedneur.com/article/0887-8994(90)90037-2/pdf doi.org/10.1016/0887-8994(90)90037-2.

27. Nicholson C, Fowler M, Mullen C, Cunningham B. Evaluation of levocarnitine, lactulose, and combination therapy for the treatment of valproic acid-induced hyperammonemia in critically ill patients. Epilepsy Res. 2021;178:106806. [cited 2022 Sept 1]. https://www.sciencedirect.com/science/article/abs/pii/S0920121121002618?via%3Dihub. https://doi.org/10.1016/j.eplepsyres.2021.106806.

Chapter 26
Twelve Year-Follow-up of a Diabetic Patient with Vascular Complications

Vesna Ðermanović Dobrota and Vesna Lukinović-Škudar

Abbreviations

ABI	ankle brachial index
ACI	internal carotid artery
AF	art. fibularis (fibular artery)
AFC	art. femoralis communis (common femoral artery)
AFP	art. femoralis profunda
AFS	art. femoralis superficialis (superficial femoral artery)
AIC	art. iliaca communis (common iliac artery)
AIE	art. iliaca externa (external iliac artery)
AP	art. poplitea (popliteal artery)
AR	radial artery
ATA	art. tibialis anterior
ATP	art. tibialis posterior
CDFI	color Doppler flow imaging
CEA	carotid endarterectomy
CVD	cardiovascular disease
Cx	circumflex branch of coronary artery
DM	diabetes mellitus
DMT1	type 1 diabetes mellitus
DMT2	type 2 diabetes mellitus
DPN	diabetic polyneuropathy

V. Ðermanović Dobrota (✉)
Clinical Hospital Merkur – University Clinic Vuk Vrhovac, Zagreb, Croatia

V. Lukinović-Škudar
School of Medicine, University of Zagreb, Zagreb, Croatia
e-mail: vesna.dermanovic.dobrota@kb-merkur.hr

V. Demarin et al. (eds.), *Mind, Brain and Education*,
https://doi.org/10.1007/978-3-031-33013-1_26

ECG   electrocardiography
EMNG   electromyoneurography
FF bypass femorofemoral (FF) bypass
FP bypass femoro-popliteal (FP) bypass
LAD   left coronary artery
MRI   magnetic resonance imaging
MSCT   multislice computed tomography
PSV   peak systolic velocity
PTA   percutaneous transluminal angioplasty
RCA   right coronary artery
TEE   transoesophageal echocardiography
TrBC   truncus brachiocephalicus (brachiocephalic trunk artery)

Introduction

Diabetes mellitus (DM) is a chronic non-infective disease that has become one of the major health problems worldwide [1, 2]. According to the World Health Organization (WHO) report, 537 million people have been diagnosed with diabetes [3, 4]. About four million people die from diabetic complications every year, and half of them are under the age of 60. Global treatment costs are estimated at 966 billion US dollars annually. It is expected that by year 2045 the number of diabetic patients will be near to 700 million, with an equal number of people in pre-diabetic phase [4]. In 2021, 61 million of people with diabetes have been detected in Europe [3, 4]. In the Republic of Croatia, reports from the Croatian Institute of Public Health indicate that 388 213 adults were confirmed to have been diagnosed with diabetes in 2022, with 58 050 newly diagnosed patients [3]. Considering 50% patients are not aware of their disease, it is estimated that the total number of diabetic patients in Croatia is greater than 500,000 [3, 4]. It is especially important to note that diabetes is diagnosed much more frequently than before in younger patients, aged 45–65, with work ability. Another significant problem is an unavoidable increase of macrovascular and microvascular complications with disease duration and patient age, in spite of mostly satisfactory control of diabetes.

Case Description

Our female patient was born 1965. She was diagnosed with diabetes in 1989, and she started therapy with peroral antidiabetic medicaments. Since 1993 her diabetes she has been regularly monitored in Clinical Hospital Merkur - University Clinic Vuk Vrhovac, Zagreb. Insulin therapy was administered to the patient in 1993, and an insulin pump was implanted in September 2011. In the same period, it was detected that the patient also had hypertension and hyperlipidemia associated with diabetes. The patient is a long-time smoker, but she quit smoking in 2016. In anamnesis, the patient also confirmed that she had a spinal injury in 1978. This resulted in the protrusion of the L5-S1 disc with consequent lumboishialgia on the right. She has fallen on numerous occasions due to weakness in her right leg.

In April 2008, the patient was treated with insertion of two stents into the left subclavian artery, for she suffered from frequent vertigo attacks that led to the patient falling multiple times. In 2010, at Sestre Milosrdnice University Hospital Zagreb, another stent was inserted in the right common iliac artery, leading to claudication, and shortening the walking distance to only 100 m. An angiology specialist found that the patient had ABI of 0.72 on the right side and 1.04 on the left side, index P4 upper arms right 0.72 (previously 0.77), Color Doppler Ultrasonography of leg arteries showed in AFC and AP right monophasic spectra by type of obliterative arteriopathy.

Magnetic Resonance Imaging (MRI) was performed several times. MRI of brain was performed in 1997 and again in 2011, with both scans showing normal results. Leg and arm EMNG in 2011 confirmed chronic distal sensorimotor diabetic polyneuropathy of stage 2a, and chronic compressive lesion of right median nerve in segmental wrist (carpal tunnel syndrome) of moderate grade. ENMG has also shown polytopic radicular lesions of C6 on the right side, C7 bilaterally and L5 and S1 bilaterally.

Color Doppler flow imaging (CDFI) of the carotid arteries performed in 2012 showed a progression of stenosis; right internal carotid artery (ACI) was ca. 45% with peak systolic velocity (PSV) of 178 cm/s and left ACI ca. 50% with increased hemodynamics to PSV 270 cm/s. Status post Percutaneous Transluminal Angioplasty (PTA) of left subclavian artery performed in 2008 was determined. In the same year, the patient was admitted to a hospital for angina pectoris in General Hospital in Bjelovar. In the year 2013 she has been hospitalized for the rupture of the meniscus in the right knee, and she was treated by arthroscopy.

In 2013 the patient started to feel pain and chills in the lower part of the right leg and the walking distance was reduced to 30–40 m. Color Doppler Ultrasonography of leg arteries showed biphasic spectra in AFC, indicating further stenosis of previously dilated common iliac artery (AIC). Superficial femoral artery (AFS) and arteria poplitea (AP) were visible, with biphasic spectra that had low flow values in the distal part of the AP. In the shin, the fibular artery (AF) and the posterior tibial (ATP) showed monophasic spectra while in the anterior tibial artery (ATA) there were biphasic spectra.

In August 2014, an angiography was performed on the patient as well as a dilatation of the right iliac artery with femoro-popliteal (FP) bypass on the right side at the Department of Vascular Surgery, Clinical Hospital Merkur. Control CDFI of the carotid arteries was done in 2014, and results showed the progression of right ACI stenosis of ca. 60%, and velocity of blood flow has increased to PSV 346 cm/s. For that reason, MSCT angiography of aortic arch and carotid vessels was recommended. MSCT angiography of aortic arch and neck blood vessels in 2014 indicated bending of brachiocephalic trunk artery (TrBC) with TrBC stenosis of ca. 50%. Stenosis of both ACI was detected on the proximal site with calcifications and occlusion of both ACI (50–70%), which was more pronounced on the left.

The patient was injured in the car accident in 2014, as a co-driver. She had a blow injury to the head and a whiplash injury of the cervical spine, so she underwent therapy in the ambulance. MRI of cervical spine has shown dorsomedial and dorsolateral osteophytosis with moderate ventral compression at the level of C3–C4, dorsomedial and prevalent left side dorsolateral osteophytosis with compression at the level of C5–C6 and circular bulging of annulus fibrosus. MRI of lumbosacral spine showed degenerative changes of intervertebral disc at L4–L5 with circular bulging of annulus fibrosus, and at L5–S1 degenerative changes of intervertebral disc with incipient lamellar dorsomedial disc protrusion are present. Densitometry has confirmed presence of osteopenia.

Color Doppler Ultrasonography of leg arteries in 2015 showed the progression of the right leg vascular insufficiency - possible occlusion of right common iliac artery (AIC). Ankle - brachial index (ABI) on the right was 0.55 and 0.95 on the left. An MSCT angiography of the aorta was recommended by a vascular surgeon, and an occlusion of right AIC and external iliac artery (AIE) was found, alongside occlusion of right proximal fibular artery (AF).

In the year 2016, control CDFI of carotid arteries has revealed a progression of right ACI stenosis to ca. 65%, with increase of flow velocity (PSV – 403 cm/s). In December, the patient was once more admitted to hospital, at the Department of vascular surgery, with the diagnosis of the critical limb (foot) ischemia on the right side: She was treated by a redo of femoro-femoral (FF) bypass with insertion of prosthesis (Goretex 6 mm), and with thrombectomy.

In 2017 an angiography of the aortic arch and the cervical arteries was performed, which has shown 70% stenosis of the right ACI and 60% stenosis of the left ACI.

Control CDFI of carotid arteries that was done in January 2018 has confirmed right internal carotid artery (ACI) stenosis to 70% with further increase of blood flow velocity (PSV – 597 cm/s), and left ACI stenosis to 60% followed by milder increase of blood flow velocity (PSV – 186 cm/s). The patient was admitted to the hospital, at the Department of vascular surgery, and carotid endarterectomy (CEA) of right ACI was done. The patient is also under the control of an ophthalmologist for photocoagulation as she was diagnosed with diabetic non-proliferative retinopathy with maculopathy. In the same year, the patient was once again hospitalized at the Cardiology Department for the indication of coronarography, but the intervention couldn't be completed, because of prominent vascular tortuosity. Cardiologic examination showed coronary arterial disease. Transesophageal echocardiography (TEE) was performed, with the conclusion that the stent or surgical procedure for revascularization were not indicated at that moment, since the ejection fraction was 75%. The patient was ordered with therapy Preductal MR 2x1 tbl and she was recommended to come for next control examination in 6 months.

Control CDFI of leg arteries performed in 2019 had demonstrated FF bypass occlusion and occlusion of radial artery (AR) on the right forearm, so MSCT angiography of lower extremities and right arm was again recommended. The patient was hospitalized at the Department of Vascular Surgery and in March 2019. MSCT angiography was performed, that had revealed occlusions of FF and FP bypasses on the right side with the occlusion of the left AP. Thus, diagnosed condition presented an indication for the surgical intervention of revascularization of the right leg. Prior to an intervention, another coronarography was done in Clinical Hospital Dubrava, which has revealed changes on the left coronary artery (LAD) with diffuse calcium deposits in its proximal segment rough atherosclerotic changes of circumflex branch of coronary artery (Cx), thin right coronary artery (RCA) at the ostium with rough atherosclerotic changes.

In March of 2019 previously planned intervention for revascularization of the right leg was performed at the Department of Vascular Surgery Clinical Hospital Merkur, where a redo of FF bypass (L-D) was done and control MSCT of abdomen, pelvic and leg arteries angiography showed regular flow.

Arm and leg EMNG performed in 2019. showed chronic distal sensorimotor diabetic polyneuropathy (DPN) of degree 2a in progression, chronic compression of right median lesion in the wrist segment (carpal tunnel syndrome) of moderate degree and chronic polytopic radicular lesions (C6 on the right, C7 bilaterally, L5 and S1 bilaterally).

Control CDFI of carotid arteries performed in 2020. has revealed the condition after right carotid endarterectomy (CEA) – restenosis the right ACI of ca. 45% with PSV 288 cm/s, and stenosis of the left ACI 60% with increased hemodynamics (PSV 257 cm/s).

Brain MRI was performed in July 2020. and it confirmed small microangiopathic lesion in the pons and multiple small lesions of the same ethiology in deep white matter bilaterally frontally. 3D time of flight (TOF) angiography of the brain and neck blood vessels revealed significant stenosis at the origin of the left ACI. Thus, an MSCT angiography of the head and neck was advised.

As MSCT angiography showed significant stenosis of the left ACI, on January 13, 2021 vascular surgeon performed CEA of the left carotid artery, and in August 2021. control CDFI of carotid and vertebral arteries was done.

Electromyoneurography (EMNG) of arms and legs performed in 2021 has shown the further progression of chronic distal sensorimotor diabetic polyneuropathy (DPN) of grade2–3 followed with superposed chronic compressive right median lesion in the wrist segment (carpal tunnel syndrome) of moderate degree and chronic polytopic radicular C6 lesions on the right, and C7 lesion bilaterally and L5 and S1 lesions bilaterally.

Cardiologist's control examination with ECG, and laboratory tests was done in 2021. Instead of Nebivolol tbl. Propranolol was ordered 40 mg 2x1 tbl with recommendation for control examination in 6 months including Holter ECG and new echocardiography examination.

MRI of cervical spine in 2021 has shown degeneration of several intervertebral (iv) discs: C3–4 showed slightly narrowed iv openings; in C5–6 osteophyte complex that mutually compressed nerve roots, and in C6–7, mild bulging of discs was detected, which didn't compromise neural structures.

Vascular surgical examination in 2021 confirmed bilateral restenosis (50–60%) of ACI that was asymptomatic, with walking distance 200 m. Antilipemic therapy was ordered by the vascular surgeon, with addition of acetylsalic acid 100 mg 1 tbl, rivoroxaban 5 mg 2x1/2 tbl, and she was adviced to walk slowly without pause as long as possible with counting of the steps. Control examination was ordered in 6 months, with CDFI of carotid and vertebral arteries.

Control CDFI of carotid and vertebral arteries performed in 2022 showed status post CEA- e ACI on the right done in 2018, with 45% - restenosis and PSV 179 cm/s. Status post CEA- e ACI on the left was detected on 13.01.2021, with restenosis about 45% and PSV 233 cm/s.

An angiology specialist found that the patient had ABI of 0.65 on the right side and 0.75 on the left side. Color Doppler Ultrasonography of leg arteries showed a pervious FF bypass, as well as AFS and FP bypass occlusion. Therapy remains unchanged, as prescribed by the vascular surgeon one year prior.

The patient now complains of occipital and left retroauricular headaches with constant sensation of tingling in the head. She has frequent vertigo with the feeling that her room is turning, and she has postural instability and dizziness with reliance when standing, when lying down or when she lifts her hands. She has also a feeling of falling and drifting to the right, which is more exaggerated in the last 2 months, she feels like she has double vision at rest and when turning her head for more than 3 min; however, these symptoms spontaneously pass. During walking, sense of tumbling becomes more prominent, and it occurs together with swaying to the right. She has tinnitus accompanied with worsening of hearing. She feels progressive bilateral pain in the neck and upper arm with limited mobility on the right side. The patient also complains about pain and feeling of vibrations in the upper arm muscles and paresthesias of the left forearm and 4 and 5 finger, sometimes she describes a feeling of punching in the fingers – like electric strike. She has sense of numbness of fingers of the hands, decreased sense of touch and frequent dropping of objects from her hands, annealing of the fingers of the hands; left fist is sometimes cold; there is tremor of the hand, stronger to the right, sometimes has itching and sweating of left hand skin, while the right hand is dry; (however, the patient is still self-containing and independent in everyday life). She has a burning sensation in the neck and between the shoulder blades, has severe pain in the backs of the right and pain is stronger and more progressive in the right hip. She is frequently woken up by pain, and sometimes the piercing pain spreads along the right leg to her feet. Since last month she feels line-type pain along the back side of the right leg. She has constant paraesthesia all over her right foot. The patient is often falling in her house without loss of consciousness and head injury. She lifts her right leg with her right arm when laying. Her pain in the lumbar area is growing stronger, when sitting, but also when walking, and when bending, the pain is in the type of stabbing. She can walk about 400–500 m on a lighter gait and has to stop due to pain in her right leg, stronger in the calf, but more pronounced on the right ankle and more pronounced paraesthesia of both feet. She has absent sensation of touch of the right groin, in the front of the right lower leg and ankle, and the front halves of both feet, sometimes she feels the heat of the right lower leg and in the toes of the right foot. Rarely she has an annealing feeling in the evening and feeling that the soles of the feet are more often cold and wet, but on touch they are warm. Touch and heat sensations on the right soles are reduced and she doesn't feel the half of the left soles; she more often complains about the feeling the tightening and firing of the front of the left foot, and the numbness and stiffness of all the toes of the left foot and the 4 and 5 fingers of the right. She does not distinguish warm and cold sensation on her feet, she sleeps poorly (3–4 h at night) and she feels that she has to take a walk; cannot find her position. She is taking Ibuprofen 600 mg in granules, alternating it with paracetamol 325 mg/tramadol 37.5 mg 2–3× daily, together with diazepam in tablets (2.2, 5 mg).

Neurological examination has detected the impaired gait with the right leg, and gait on a broad basis, and the patient does not perform gait on the heels and fingers. Romberg test confirms instability with retropulsion and dextropulsion, and this was the condition after surgical treatment of both carotid arteries. Assessment of cranial nerves doesn't show abnormalities. When examining vision, she complains about shimmering spots in the visual field. Coarse muscle strength of the hand is weakened more on the left side, indicating condition after surgical intervention of left wrist, fine movements are clumsier when performed with the left hand; the coordination is normal. Mild flush of hands and both feet is detected; upper extremities reflexes are symmetrical but the reflexes cannot be elicited on the lower extremities; Tinel's wrist sign is positive, more pronounced on the right side; Babinski sign is negative. Vibration sensation is significantly shortened, according to the diabetic polyneuropathy type, and on the feet is absent. Lasegue sign is positive; on the left side it is about 30° and about 40° on the right. The patient can control sphincters.

Present (2022) laboratory findings in serum were: HbA1c 7, 3%, total cholesterol 6, 4 mmol/L, HDL- cholesterol 1, 4 mmol/L, LDL- cholesterol 4, 4 mmol/L, triglyceride concentration 1, 4 mmol/L.

Due to the deterioration of the patient's condition, it was suggested for her to undergo a brain MRI + 3D TOF angiography and ophthalmologic examination because of the visual symptoms, examination of otorhinolaryngologist due to dizziness, control by vascular surgeon as recommended before, control CDFI of carotid and vertebral arteries in 8–10 months, Color Doppler Ultrasonography of leg arteries with strict recommendation of the treatment by physical medicine specialist. Control examination by neurologist with recommendation to patient to bring all older documentation was also ordered.

Current medicament therapy includes: Novorapid 25.95 IU in basal conditions plus 2–4 IU before the meal, Panzoprazol 40 mg 1 tbl, Propranolol 40 mg 3×1 tbl, Diazepam 2–2.5 mg tbl, Norprexanil 10/10 1 tbl, Rosuvastatin 40 mg 1 tbl, Ezetimib 10 mg 1 tbl, ASK pro 100 mg 1 tbl, Rivoroxaban 5 mg 2×1/2 tbl, NTG sprey pp., Mirtazapine 15 mg 1 tbl.

Discussion

Here we present the course of diabetic disease in a patient with type 1 diabetes diagnosed at a young age, in her 30 s, and has developed multiple microvascular and macrovascular complications as well as several neurological comorbidities throughout her life, despite continuous control of diabetes.

In parallel with duration of the disease, diabetes mellitus is unavoidably followed with chronic complications. The number and severity of complications increase with the age of the patient, the duration of the disease and poor glycemic control. The most important macrovascular complications are stroke, myocardial infarction, coronary artery disease and peripheral arteriy disease, while the most important microvascular complications are diabetic retinopathy, nephropathy and neuropathy [2, 5]. All chronic complications of diabetes can be found in patients with both type 1 and type 2 diabetes, although some complications are more common in patients with a certain type. Thus, severe autonomic neuropathy is found much more often in patients with type 1 diabetes, unlike diabetic neuropathy, which is equally common regardless of the type of diabetes [2, 6]. In both types of diabetes, cardiovascular disease (CVD), caused by atherosclerosis with its clinical consequences (coronary heart disease, myocardial infarction and stroke), is the most important cause of illness and death. In patients with diabetes, the risk of death from CVD is two times higher than in people without diabetes [5, 6].

Diabetes mellitus is one of the most important known risk factors for cardiovascular disease, along with hypertension, dyslipidemia and smoking, which can also be influenced with preventive and therapeutic measures. Additionally, DM and CVD share several risk factors, as patients with diabetes have a higher incidence of hypertension, dyslipidemia, and myocardial infarction. Hyperglycemia and insulin resistance, on the other hand, significantly accelerate the development of atherosclerosis and macrovascular complications of diabetes.

Diabetes is an important risk factor for atherosclerosis, regardless of the type of diabetes and development of peripheral arterial disease, another macrovascular complication of diabetes. There is a clear correlation between the degree of development of atherosclerotic plaques and the size of the plaque and the duration of hyperglycemia, which greatly accelerates morphological and functional changes in vessels. Consequently, peripheral arterial disease with gangrene is 8 to 15 times more common in diabetics, especially if they are smokers, with consequent need for amputations [7].

Color Doppler Ultrasonography is a non-invasive, accurate, and cheap technique for detecting early atherosclerotic plaques and changes in arterial walls. The incidence of cardiovascular and cerebrovascular events is significantly higher in patients with atherosclerotic changes of carotid arteries and peripheral artery disease, than in diabetic patients with either one of the aforementioned. Color Doppler Ultrasonography has high sensitivity and repeatability, making it especially important for early clinical diagnosis, prevention, and treatment of atherosclerotic changes of carotid arteries and peripheral artery diseases. Its value becomes even more pronounced when we consider the importance of early detection in delaying the incidence of macroangiopathy and postponing the development of cerebrovascular disease [7].

Lifestyle changes are considered key in the prevention of CVD in patients with DM and pre-DM. Any reduction in weight also reduces the risk of developing pre-DM and DM [1]. Lowering calorie intake causes a drop in HbA1C and improves quality of life. In obese patients with DM, body weight loss greater than 5% improves the regulation of glycemia, lipids, and blood pressure. Moderate physical activity,

150 min/week or more, is also recommended, because it delays the conversion of prediabetes to DM, improves glycemic control and CVD complications. Aerobic and endurance exercises improve insulin activation, glycemic and lipid control, and blood pressure. Patients who consume alcohol less than 100 g per week have a lower risk of MI. Smoking increases the risk for DM, CVD, and earlier mortality [6].

In the treatment of DMT1 and DMT2, a broader and individual approach to treatment is required, based on the reduction of important risk factors for general morbidity and mortality due to the development of micro- and macrovascular complications. A comprehensive and individual approach to pharmacological treatment is desirable, with the application of primary and secondary prevention measures such as blood pressure and serum lipid control, smoking cessation along with walking and physical activity. Beside diabetic therapy, the patient should be regularly examined by a specialist with the aim of early detection of major micro- and macrovascular complications and comorbidities, which are very often found in diabetics [5, 6].

Conclusion

The aim of this case report was to present a patient diagnosed with diabetes in her 30 s, who has developed multiple microvascular and macrovascular complications and comorbidities throughout her life, despite permanent control of diabetes. Complications and comorbidities need to be recognized by clinicians as early as possible, to delay or decrease considerable risk of life-threatening events leading to early mortality.

References

1. Cosentino F, Grant PJ, Aboyans V, Bailey CJ, Ceriello A, Delgado V, Federici M, Filippatos G, Grobbee DE, Hansen TB, Huikuri HV, Johansson I, Juni P, Lettino M, Marx N, Mellbin LG, Ostgren CJ, Rocca B, Roffi M, Sattar N, Seferović PM, Sousa-Uva M, Valensi P, Wheeler DC. 2019 ESC Guidelines on diabetes, pre-diabetes, and cardiovascular diseases developed in collaboration with EASD. The Task Force for diabetes, prediabetes, and cardiovascular diseases of the European Society of Cardiology (ESC) and the European Association for the Study of Diabetes (EASD). Eur Heart J. 2020;41:255–323. https://doi.org/10.1093/eurheartj/ehz486.
2. Tomic D, Shaw JE, Magliano DJ. The burden and risks of emerging complications of diabetes mellitus. Nat Rev Endocrinol. 2022;18(9):525–39. https://doi.org/10.1038/s41574-022-00690-7.
3. Croatian Institute of Public Health CroDiab registry of people with diabetes mellitus. https://www.hzjz.hr/sluzbaepidemiologija-prevencija-nezaraznih-bolesti/odjel-za-koordinaciju-i-provodenje-programa-i-projekata-za-prevenciju-kronicnih-nezaraznih-bolest/dijabetes/. Accessed 11 Jun 2023.
4. International Diabetes Federation. Latest figures show 463 million people now living with diabetes worldwide as numbers continue to rise. Diabetes Res Clin Pract. 2019;157:107932. https://doi.org/10.1016/j.diabres.2019.107932.

5. ADVANCE Collaborative Group, Patel A, MacMahon S, Chalmers J, Neal B, Billot L, Woodward M, Marre M, Cooper M, Glasziou P, Grobbee D, Hamet P, Harrap S, Heller S, Liu L, Mancia G, Mogensen CE, Pan C, Poulter N, Rodgers A, Williams B, Bompoint S, de Galan BE, Joshi R, Travert F. Intensive blood glucose control and vascular outcomes in patients with type 2 diabetes. N Engl J Med. 2008;358(24):2560–72. https://doi.org/10.1056/NEJMoa0802987.
6. Gajos G. Diabetes and cardiovascular disease: from new mechanisms to new therapies. Pol Arch Intern Med. 2018;128(3):178–85.
7. Yang Z, Han B, Zhang H, Ji G, Zhang L, Singh BK. Association of lower extremity vascular disease, coronary artery, and carotid artery artherosclerosis in patients with type 2 diabetes mellitus. Comput Math Methods Med. 2021; 6268856. https://doi.org/10.1155/2021/6268856.

Chapter 27
Autism: Insights from Brain, Mind, and Education

Raphael Béné

Introduction

Autism is a neurodevelopmental condition that groups two significant criteria: difficulties in social communication and interaction, as well as restricted, repetitive patterns of behavior, interests, or activities. In clinical practice, we assess the combination of these two cardinal groups of signs and their impact in different contexts as Autism Spectrum Disorder [1].

However, autistic children, students, and adults also display numerous strengths that can help them achieve fascinating breakthroughs in many areas. Isn't, after all, a good scientist someone devoted to a restricted interest, exploring it by repetitive experiment?

Regarding the history of autism as a clinical entity, it began almost a century ago, when Grunya Sukhareva provided the first accurate clinical description of autistic traits in 1925, after a clinical investigation of six young boys at the faculty of pediatric psychiatry in Moscow [2]. Two years later, in 1927, she described the same clinical signs in five young girls [3]. Only eighteen years later, Leo Kanner published the result of his evaluation of eleven children, eight boys, and three girls [4]. The same year, Hans Asperger described a pattern of behavior and skills in four boys, including "lack of empathy, poor ability to make friends, one-way conversation, strong preoccupation with special interests, and awkward movements." [5].

Another figurehead of autism knowledge was Lorna Wing. In 1980 she launched the concept of the autism spectrum [6] and the denomination Asperger syndrome [7]. Her work resulted in a new delineation of the syndrome, a so-called triad of impairments in social communication, social interaction, and social imagination

R. Béné (✉)
Department of Psychiatry, University Hospital Centre Vaudois (CHUV),
Lausanne, Switzerland

V. Demarin et al. (eds.), *Mind, Brain and Education*,
https://doi.org/10.1007/978-3-031-33013-1_27

281

still referred to as «wing's trial.». Autism as a spectrum, introduced in DSM-III-TR, leads to a considerable increase in its prevalence, among other poorly understood factors.

In 2013, the DSM-5 defined one clinical entity, "Autism Spectrum Disorder," with further specifications regarding intellectual or language impairment, known medical genetic condition or environmental factor, catatonia, and the presence or absence of neurodevelopmental, mental, or behavioral disorder [1]. It is impressive that the clinical criteria described in DSM V are almost identical to Sukhareva's first description nearly a century before.

Insights from the Mind: Cognitive Models

Models explain which processes result in an observed output [8]. In their book "Autism: A new introduction to psychological theory and current debate," Francesca Happé and Sue Fletcher Watson presents a clear overview of different autism cognitive models that they divided into three categories:: "Primary deficit models" (i.e., lack of theory of mind, lack of executive function), "Developmental trajectory model," and "domain-general information processing models." [9].

The first hypotheses in the cognitive science of autism were related to a deficit, or dysfunction, of particular cognitive abilities. On the one hand, Utah Frith, Simon Baron-Cohen, and Alan Leslie formulated the "Theory of Mind" model, in which autistic individuals have difficulties understanding and attributing others' mental states [10]. On the other hand, Damasio and Maurer linked the characteristics of autism to those observed in patients with frontal lesions [11]. They hypothesized that autism leads to planning and controlling behavior challenges due to a deficit in executive functions [12].

New cognitive models switch from deficit-based to social and perceptual origins.

Among these models, two are currently the subject of much interest: the social motivation hypothesis, a developmental trajectory model [for review, see [13]], and the predictive coding theory [for review, see [14]], a domain-general information processing model.

The social motivation theory puts the reward system of the brain in focus. In autistic children, it's as if not being attracted to a social context (less reward) leads to less visual exploration [15]. A study by Amy Klin showed via eye-tracking technology that already six months after birth, autistic children tend to make fewer and shorter gaze fixation on social cues [16]. Also, they display attention deficits for speech but not for non-speech sounds [17]. It implies that autistic's auditory orienting difficulties are not due to sensory impairments and that this deficit may be speech-sound specific. Social motivation refers to the preference to orient to, seek out and enjoy social interactions to create social bonds. The reduction of social interest, and decreasing social motivation, leads to deprivation in social experiences, the underlying support of social learning, and social cognition. A recent study reported that social motivation predicts social skills only in autistic children, whereas cognitive ability predicts social skills in all children [18].

Social motivation theory remains the strongest argument for early naturalistic developmental behavioral intervention in autistic children [19], such as the Early Start Denver Model [20].

The predictive coding theory is a cognitive model of autism that Happé and Fletcher-Watson grouped as an "information processing" model. This model aims to explain how autism changes the way of perceiving, organizing, and responding to input from the surrounding world. The predictive coding model is called Bayesian (based on Bayes' rule for inverse inference). It implies that our brain continuously creates new internal world models by making predictions and comparing them against other perceptual pieces of information [21]. Step by step, we minimize prediction errors or mismatches. Sensory input challenges our brains' predictions of what will happen, and the brain reacts to any mismatch. That's why we experience the world based on our expectations that rely on our prior knowledge [22].

Autistic people's life experience is much more intense and surprising than neurotypical one. Autistic persons have trouble knowing what to expect in daily life. They also show superior visual discrimination [23] and reduced sensitivity to optical illusions [24]. Some aspects of the autism phenotype may manifest different predictions, less influenced by prior knowledge, leading to reduced bias by previous experience.

It has broad consequences in exteroception (how we use sound, light, and discriminatory touch to build models of the external world) [25], proprioception (how we process information from the body to allow a sense of agency) [26], and interoception (The process of sensing signals from the body, like a heartbeat, breathing, hungers) [27] of autistic individuals. However, the predictive coding model of autism still has limited relevance to higher cognitive functions.

With these two new cognitive models, we see a new approach in autism research, not focusing on cognitive deficits but exploring social motivation and predictive abilities. However, none of the cognitive models can explain to a full extent the broad range of difficulties and skills stemming from autism.

Insights from the Brain: Neural Networks

The focus of brain research has shifted from tying a cognitive function to a specific region to connecting it to a network that spans the entire brain. A true revolution in understanding functional connectivity in the autistic brain was the creation of consortia for data sharing, such as ABIDE (Autism Brain Imaging Data Exchange) [28] or ENIGMA (Enhancing NeuroImaging Genetics through Meta-Analysis) [29].

Three major neural networks are of particular interest.

The default mode network (DMN) shows activity when our brain is at rest [for review, see [30]]. The central nodes of this network include the ventromedial prefrontal cortex and the posterior cingulate cortex [31]. This network is crucial in integrating self-monitoring, autobiographical, and associated social cognitive processes [32].

If something grabs our attention, it is via the salience network [33]. It aims to identify the most relevant internal and extrapersonal stimuli to guide attention and behavior. The anterior insula and cingulate cortex belong to the salience network [34].

Suppose the incentives lead to a problem-solving task requiring cognitive engagement. In that case, it will activate the central executive network, essential for active maintenance and manipulation of information in working memory [35].

The salience network drives the switching between the default mode network and the central executive network depending on salient environmental events [36, 37]. It sends signals that regulate behavior, enabling attentional capture, mental flexibility, and the creation of signaling mechanisms that ease access to working memory resources [38].

These three neural networks show different synchronization patterns in autism [39].

Studies consistently demonstrate that the default mode network differs in autism [40]: Its nodes have decreased interaction with other functional systems and enhanced within-network connectivity, particularly between the posterior cingulate cortex and the medial prefrontal cortex.

Salience network hyperconnectivity may also be a distinguishing feature in autistic children. Lucina Q. Uddin found that the salience network demonstrated the highest classification accuracy among all the networks tested. Indeed, the blood oxygen level-dependent signal in the salience network predicted constrained and repetitive behavior scores [41].

Several earlier task-based functional MRI investigations have shown different synchronization patterns in these brain networks while processing social stimuli. In addition, another cortical region, the sulcus temporalis superior, crucial for social cognition and implicated in several steps of social interactions, has decreased activity during task-specific neuroimaging studies in autistic individuals. This node is seen as the gate to the mirror neuron system and connects massively with other regions of the 'social brain' such as the orbitofrontal cortex and the amygdala [42].

Autism is one condition for which the study of the connectome is critical. Even if we still don't have evidence-based clues, due to the heterogeneity of the spectrum and extreme complexity of data processing and analysis, we can target individualized learning techniques, to overtake issues such as sensory overload or executive dysfunction.

The considerable development of imaging techniques, as well as machine-learning analyses [43], will enable pattern recognition, as well as classification other than clinical ones.

Insights from Education: Autistic Students and Autistic Community

Perhaps the most significant shift in approach to autism is the one relying on autistic strengths instead of trying to lessen autistic behavior. For example, repetitive behaviors were seen as targets for therapy. Now, they seem to support autistic individuals in a variety of ways [44]. How do we know it? Because autistics claim it for years. "Don't Mourn For Us," a speech by Sinclair from 1993, focused on autism as a way of being [45].

In the same way, Temple Grandin explains in "The autistic brain," which she wrote with Richard Panek in 2014, why autism and raising autistic children should be less about focussing on weaknesses and more about empowering and nurturing autistic strengths [46].

Autism is not a disease but a condition. Considering a human operating system can help one understand neurodiversity and autism, as Steve Silberman stated in his book Neurotribes [47]. Even if a computer cannot run Windows, it may still be functional. Autistic people may struggle more or less with social communication and executive functions, sometimes requiring substantial professional support. Still, they can also impress with their creativity, focus, honesty, and dedication.

In his new book, "The pattern seekers," Simon Baron-Cohen explains how autistic people have a potent systemizing method of thinking, enabling them to identify If-and-Then patterns in their environment. Most autistic students are known to be logical thinkers, curious, tenacious, persistent problem solvers, and focused. They present various viewpoints and avoid groupthink and non-evidence-based decision-making. They have shaped our society by offering new and innovative approaches to issues that deeply impact human understanding of nature [48]. And yet, little has been done to answer their fundamental questions: Can you help destigmatize autism in society? Can you help us cope with sensory issues? Can you offer alternative ways of communication? If awareness of autism is on the rise, that is not the case regarding the development of accommodations for autistic students, even in wealthy countries. The critiques from the autistic community toward clinicians and researchers are mainly about the fact that autistic priorities and their quality of life never seemed to be the focus of scientific research and budget. Autistic students reported problems interacting in social settings at institutions and underlined a need for better services in higher education meant to enhance their social lives. Jennifer C. Sarrett addresses the need for input from the autism community by gathering the perspectives of autistic individuals on accommodations received [49]. The most frequently cited need was increased autism awareness on campus, sensory-friendly spaces and practices, a disability support group, and more acceptance of self-stimulatory behavior. Another reminder that the true challenge remains tolerance of society toward autistic students.

References

Introduction

1. American Psychiatric Association. Neurodevelopmental disorders. In: Diagnostic and statistical manual of mental disorders. 5th ed. American Psychiatric Association; 2013.
2. Sucharewa GE. Die schizoiden Psychopathien im Kindesalter. Monatsschr Psychiatr Neurol. 1926;60(3–4):235–61.
3. Sucharewa GE. Die Besonderheiten der schizoiden Psychopathien bei den Mädchen. Eur Neurol. 1927;62(3):186–200.
4. Kanner L. Autistic disturbances of affective contact. Nervous Child. 1943;2(3):217–50.
5. Asperger H. Die „Autistischen psychopathen" im kindesalter. Arch Psychiatr Nervenkr. 1944;117(1):76–136.
6. Wing L. The definition and prevalence of autism: a review. Eur Child Adolesc Psychiatry. 1993;2(1):61–74.
7. Wing L. The history of Asperger syndrome. In: Asperger syndrome or high-functioning autism? Boston, MA: Springer; 1998. p. 11–28.

Insight from the Mind

8. Adamson, P. (2012). Stanford encyclopedia of philosophy.
9. Fletcher-Watson S, Happé F. Autism: a new introduction to psychological theory and current debate. Routledge; 2019.
10. Baron-Cohen S, Leslie AM, Frith U. Does the autistic child have a "theory of mind"? Cognition. 1985;21(1):37–46.
11. Damasio AR, Maurer RG. A neurological model for childhood autism. Arch Neurol. 1978;35(12):777–86.
12. Maurer RG, Damasio AR. Childhood autism from the point of view of behavioral neurology. J Autism Dev Disord. 1982;12(2):195–205.
13. Clements CC, Zoltowski AR, Yankowitz LD, Yerys BE, Schultz RT, Herrington JD. Evaluation of the social motivation hypothesis of autism: a systematic review and meta-analysis. JAMA Psychiatry. 2018;75(8):797–808.
14. Sinha P, Kjelgaard MM, Gandhi TK, Tsourides K, Cardinaux AL, Pantazis D, Held RM. Autism as a disorder of prediction. Proc Natl Acad Sci. 2014;111(42):15220–5.
15. Chevallier C, Kohls G, Troiani V, Brodkin ES, Schultz RT. The social motivation theory of autism. Trends Cogn Sci. 2012;16(4):231–9.
16. Jones W, Klin A. Attention to eyes is present but in decline in 2–6-month-old infants later diagnosed with autism. Nature. 2013;504(7480):427–31.
17. Ceponiene R, et al. Speech-sound-selective auditory impairment in children with autism: They can perceive but do not attend. Proc Natl Acad Sci U S A. 2003;100:5567–72.
18. Itskovich E, Zyga O, Libove RA, Phillips JM, Garner JP, Parker KJ. Complex interplay between cognitive ability and social motivation in predicting social skill: a unique role for social motivation in children with autism. Autism Res. 2021;14(1):86–92.
19. Bruinsma YE, Minjarez MB, Schreibman L, Stahmer AC. Naturalistic developmental behavioral interventions for autism spectrum disorder. Baltimore, MD: Brookes Publishing Company; 2020.
20. Dawson G, Rogers S, Munson J, Smith M, Winter J, Greenson J, et al. Randomized, controlled trial of an intervention for toddlers with autism: the Early Start Denver Model. Pediatrics. 2010;125(1):e17–23.

21. Pellicano E, Burr D. When the world becomes 'too real': a Bayesian explanation of autistic perception. Trends Cogn Sci. 2012;16(10):504–10.
22. Van de Cruys S, Evers K, Van der Hallen R, Van Eylen L, Boets B, De-Wit L, Wagemans J. Precise minds in uncertain worlds: predictive coding in autism. Psychol Rev. 2014;121(4):649.
23. Bertone A, Mottron L, Jelenic P, Faubert J. Enhanced and diminished visuospatial information processing in autism depends on stimulus complexity. Brain. 2005;128(10):2430–41.
24. Happé FG. Studying weak central coherence at low levels: children with autism do not succumb to visual illusions. A research note. J Child Psychol Psychiatry. 1996;37(7):873–7.
25. Noel JP, Lytle M, Cascio C, Wallace MT. Disrupted integration of exteroceptive and interoceptive signaling in autism spectrum disorder. Autism Res. 2018;11(1):194–205.
26. Smith R, Badcock P, Friston KJ. Recent advances in the application of predictive coding and active inference models within clinical neuroscience. Psychiatry Clin Neurosci. 2021;75(1):3–13.
27. Marshall AC, Gentsch A, Schütz-Bosbach S. The interaction between interoceptive and action states within a framework of predictive coding. Front Psychol. 2018;9:180.

Neural Networks

28. Di Martino A, Yan CG, Li Q, Denio E, Castellanos FX, Alaerts K, et al. The autism brain imaging data exchange: towards a large-scale evaluation of the intrinsic brain architecture in autism. Mol Psychiatry. 2014;19(6):659–67.
29. Thompson PM, Stein JL, Medland SE, Hibar DP, Vasquez AA, Renteria ME, Le Hellard S. The ENIGMA Consortium: large-scale collaborative analyses of neuroimaging and genetic data. Brain Imaging Behav. 2014;8(2):153–82.
30. Raichle ME. The brain's default mode network. Annu Rev Neurosci. 2015;38:433–47.
31. Smallwood J, Bernhardt BC, Leech R, Bzdok D, Jefferies E, Margulies DS. The default mode network in cognition: a topographical perspective. Nat Rev Neurosci. 2021;22(8):503–13.
32. Spreng RN, Mar RA, Kim AS. The common neural basis of autobiographical memory, prospection, navigation, theory of mind, and the default mode: a quantitative meta-analysis. J Cogn Neurosci. 2009;21(3):489–510.
33. Uddin LQ. Salience network of the human brain. Academic Press; 2016.
34. Menon V, Uddin LQ. Saliency, switching, attention, and control: a network model of insula function. Brain Struct Funct. 2010;214(5):655–67.
35. Collette F, Van der Linden M. Brain imaging of the central executive component of working memory. Neurosci Biobehav Rev. 2002;26(2):105–25.
36. Goulden N, Khusnulina A, Davis NJ, Bracewell RM, Bokde AL, McNulty JP, Mullins PG. The salience network is responsible for switching between the default mode network and the central executive network: replication from DCM. NeuroImage. 2014;99:180–90.
37. Uddin LQ. Salience processing and insular cortical function and dysfunction. Nat Rev Neurosci. 2015;16(1):55–61.
38. Chen AC, Oathes DJ, Chang C, Bradley T, Zhou ZW, Williams LM, et al. Causal interactions between fronto-parietal central executive and default-mode networks in humans. Proc Natl Acad Sci. 2013;110(49):19944–9.
39. Ilioska I, Oldehinkel M, Llera A, Chopra S, Looden T, Chauvin R, Buitelaar JK. Connectome-wide mega-analysis reveals robust patterns of atypical functional connectivity in autism. medRxiv. 2022.
40. Padmanabhan A, Lynch CJ, Schaer M, Menon V. The default mode network in autism. Biol Psychiatry Cogn Neurosci Neuroimaging. 2017;2(6):476–86.
41. Uddin LQ, Supekar K, Lynch CJ, Khouzam A, Phillips J, Feinstein, & Menon, V. Salience network–based classification and prediction of symptom severity in children with autism. JAMA Psychiatry. 2013;70(8):869–79.

42. Bahevalier J, Loveland KA. The orbitofrontal-amygdala circuit and self-regulation of social-emotional behavior in autism. Neurosci Biobehav Rev. 2006;30:97–117.
43. Payabvash S, Palacios EM, Owen JP, Wang MB, Tavassoli T, Gerdes M, Mukherjee P. White matter connectome edge density in children with autism spectrum disorders: potential imaging biomarkers using machine-learning models. Brain Connect. 2019;9(2):209–20.
44. Kapp SK, Steward R, Crane L, Elliott D, Elphick C, Pellicano E, Russell G. 'People should be allowed to do what they like': autistic adults' views and experiences of stimming. Autism. 2019;23(7):1782–92.

Insights from Education: The Autistic Students

45. Pripas-Kapit S. Historicizing Jim Sinclair's "don't mourn for us": A cultural and intellectual history of neurodiversity's first manifesto. In: Autistic community and the neurodiversity movement. Singapore: Palgrave Macmillan; 2020. p. 23–39.
46. Grandin T, Panek R. The autistic brain: Thinking across the spectrum. Houghton Mifflin Harcourt; 2013.
47. Silberman S. Neurotribes: the legacy of autism and how to think smarter about people who think differently. Atlantic Books; 2017.
48. Baron-Cohen S. The pattern seekers: How autism drives human invention. Basic Books; 2020.
49. Sarrett JC. Autism and accommodations in higher education: insights from the autism community. J Autism Dev Disord. 2018;48(3):679–93.

Chapter 28
Co-Creativity with Children and Adolescents

Tatjana Kapuralin

Introduction

Therapeutic work with children and adolescents requires intensive and continuous engagement, consisting of creative interventions and guidance. Co-creativity is a topic that depicts the process of reframing old patterns of behaviour by using creative media. Creativity enables us to connects with our emotions, and co-creativity is the process and a method in which a therapist engages in a play or counselling. While used in various therapeutic methods, co-creativity features frequently in Integrative therapy. Philosophical precept behind it is that human beings, as subjects, are in constant dialogue with themselves and with their environment, accordingly conceiving of the world and life in terms of change and evolution. It opens up new-old ways of perspective, leading us steadily to set the objectives of therapeutic work. Modern life is subject to various challenges, some of which had also been weathered by our predecessors. On the other hand, we are now witnessing the increase in domestic and peer violence, divorces, and consequentially, ruptured families. A serious and complex challenge of the present times is digital media addiction among adolescents. This article will therefore present theoretical overview of Integrative therapy, as well as the concepts of creativity, co-creativity and a brief case study demonstrating practical work with children and adolescents. The article will also briefly reflect on neuroscience, i.e. the main concepts of knowledge acquisition. It endeavours to highlight a key message to all professionals that «heritage», i.e. what is passed on through generations is important, particularly the message . Its other objective is to bring to general awareness the concept of

T. Kapuralin (✉)
University Graduate Social Pedagogue, Integrative Psychotherapist ECP, Family Counseling Center and Private Practice, Pula, Croatia

V. Demarin et al. (eds.), *Mind, Brain and Education*,
https://doi.org/10.1007/978-3-031-33013-1_28

conscious parenting, i.e. to empower parents through the idea that ideal parents or ideal experts on parenthood do not exist, but only those who are willing to continuously learn and improve.

Theory and Perspective of Integrative Therapy

In early 1970s, the American modality of Integrative therapy was established by Richard Erskine. During 1980s, as integrative and Gestalt therapist Kenneth Evans, established European Interdisciplinary Association for Therapeutic Services For Children and Young People (EIATSCYP). Kenneth Evans was also instrumental in the establishment of the Institute D.O.M. in Zagreb, that provides education in Integrative psychotherapy for children and adolescents. The founder of German school of Integrative therapy that emerged in 1960s is Hilarion Gottfried Petzold. Aside from Germany, Integrative therapy is also practiced in Austria, Switzerland, Norway, and to a certain exent in Sweden. In the period between 1950 and 1970, in the former Czechoslovakia, Ferdinand Knobloch combined different theories and methodologies, coming close to the model of Integrative therapy. In this article, however, theoretical approach and the main features of Integrative psychotherapy as developed and taught by Hilarion Petzold will be presented.

Integrative therapy [1] is the model that brings together various theoretical approaches and integrates them into a new, contemporary therapeutic approach. The foundation of integrative therapy is 'Heraclitus' theory', i.e., his philosophy of constant change. With his associates – Johanna Sieper, Hildegunde Heinl, Ilse Orth and others – Petzold set the foundations for a new approach in psychotherapy in the mid-1960s.

Cvetko [1] desribes Petzold's work of bringing together three main strands of psychotherapy: active psychoanalysis (Ferenzi), humanistic psychological approach that combines methods and techniques of psychodrama and Gestalt therapy (Moreno, Perls) and behaviouristic approach, with its methodology directed at the change of behaviour through therapy. Freud's psychoanalysis had set strong foundations for further development of psychotherapy, and integrative therapy also finds its strong roots there, particularly in the active approach to the work with clients (Ferenzi). It was exactly from that basis that integrative therapy derived the concept of co-existence and intersubjective correspondence (constant mutual interchange) that occurs between a client and a therapist. In that way, client is put into a more emancipatory or equal relationship with their therapist. Client is also more responsible for co-creation of the process of therapy and its outcome. Psychotherapy, as the process, thus no longer represents something mystical or incomprehensive for the clients, as they themselves are following the process in all of its phases and are aware of everything that occurs within it.

Another root of integrative psychotherapy is in psychodrama and Gestalt therapy. In its theory, Integrative therapy [1] had built on the concepts of role-play and scene setting (by understanding an atmosphere of therapeutic setting). It also furthered the concept of 'here-and-now' in terms a lifelong growth and development, and experiential activation (as one of the 5 ways of healing within the framework of Integrative therapy). Furthermore, integrative therapy lays down the concepts of emotional orientation through the techniques such as role-play, dramatising, dialogue, emotional catarthic response, body language and other therapeutic interventions. The third starting point of Integrative therapy is increasingly more popular, as it takes into consideration referrals that are covered by medical insurance (though still noti in Croatia). That includes primarily short, effective behavioural therapy that is generally part of final phases of any therapy, and which – in integrative therapy – is referred to as the phase of new orientation. This phase is considered very important, because it is there that old behavioural patterns are abandoned, and new ones adopted.

Integrative therapy [1] relies on several theories, also including cognitive theory. The interpretation is derived from Husserl's phenomenology, originally devised by the French phenomenological philosopher Merleau-Ponty, according to which the man is, first of all, physical and then a social being. As subjects, we are participating in continuous correspondence processes with other subjects, taking place in a context and continuum. It is only after the phenomena and structures we observe are studied, acquainted, understood and explained (hermeneutic spiral) that we can fully grasp the essence, i.e., specific traits of any individual or group. Theory of knowledge is based on Darwin's theory of evolution. The key understanding of that theory is that development is gradual, progressing through a series of steps, whereby each step is emanating from whar was previously known. Newly-emerging forms are always more complex than their predecessors, thus representing evolution. It is defined by Heraclitus' 'Panta rei'.

For Integrative therapy, the science of cosmology is of significant help in understanding the connection between a man and his environment (material and non-material). Man is connected to the cosmos through a non-interrupted correspondence. Through self-reflection and eccentricity, man is able to become aware of his own position of alienation and materialism which, consequentilly, has destructive consequences in the universe (devastation of natural resources, land degradation, climate change). It thus happens that occasionally evolution turns into 'devolution' (Petzold; cited in The Tree of science). Integrative therapy endeavours to raise the awareness of both individuals and the entire society in order to prevent alienation and excessive materialism, which is particularly relevant in the light of unstoppable global changes. Anthropology holds a special place in integrative therapy, primarily anthropological principles of co-existence and co-creation, from which co-creativity emerges. In his article on Creativity [2], Petzold states: »Essentially, we constantly co-exist with others and are constantly co-creative with others. Co-creativity is a phenomenon that occurs as part of interactions and communication processes through coreflexive and coemotional connectedness of different elements (information, material, ideas, etc.), i.e. through connections and complex systems (personal, group, organisational). The product is the creation of new things and the reduction of complexity. Through »creations of higher order«, we can open a new chapter of complexuty that can be acquired and used in configuration of new information. In this way, new »occurrences of co-creativity« emerge, and those newly-created potentials enable greater systems.«

A concept of Integrative therapy with creative media [2] was first mentioned in 1965 (Sieper 1971, Petzold 1972e). Creative media, such as therapeutic painting, music, poetry, bibliotherapy, work with fairytales, nature therapy and meditation are used to generate new experiences. Our body (subject) with its external and internal sensors stores information and creates new experience.

In their article on creativity [3], the authors claim that creativity is when new ideas become reality. Owing to sophisticated brain scanning techniques, modern neuroscience responded to many questions pertaining to that theme, and also explained that creativity, in a certain way, represents »heritage« material which, once adopted, can be further passed on and studied. Human brain supports various forms of creativity, and different brains lead to different types of creativity. This includes two processes, thinking and production/creation.

There is a generally prevalent opinion that only special, talented individuals can be creative, and that only geniuses are gifted. The authors [3] affirm that such claims are only a myth. They also refer to the study conducted by the Exeter University, which concludes that excellence is determined by opportunities, encouragement, training, motivation, but most of all, practice. They also mention George Land and the test he devised in 1968 for the purpose of selection of innovative NASA engineers, which was subsequently used to test creativity of pre-school children. Encouraged by the results, Land used the same test with somewhat older children and adolescents, and came to conclusion that creative behaviour can be learned.

Parents and experts for children and adolescents should, therefore, use creativity in all areas of life (play, school, daily activities) in their interactions with children and young people. Children should spend as much time as possible outdoors (forest kindergardens), because the nature offers endless possibilities for research and creativity. In the current world burdened with uncertainties and effects of modern technology, it is even more of an imperative to use nature therapy in our work with children and teach them about the nature's benefits. In their article

»Green text – the new nature therapies«, Petzold i Homberg [4] present the concept of ecopsychosomatics, which is not only relevant for psychotherapy, but also for environmental medicine and ecology. Ecopsychosomatics is a concept of psychotherapy that connects body, soul, spirit and social behaviour through 'green' exercises (e.g.walking) and 'green' activities (e.g. gardening). It has been proved that it reduces stress, makes people more sensible as to their environment, reduces self-destructive tendencies and supports salutogenesis. When practicing children and adolescents' therapy, we are aware that therapy is not only taking place within confines of therapy room. Ecopsychosomatics assumes deepening of the awareness about 'green' cure and the care for people and nature. It is simultaneously deepening the awareness about protective factors for somatic and spiritual health.

Neuroscience and Creativity

»Modern neuroscience [3] has the privilege to investigate the processes of artistic performance in a healthy brain using modern techniques such as functional magnetic resonance imaging. Not so long-ago scientists could only speculate what brain functions are involved in artistic processes by observing neurological patients « and they arrived to the conclusion that »uncreative people have marked hemispheric dominance and creative people have less marked hemispheric dominance. The right hemisphere is specialized for metaphoric thinking, solution-finding and synthesizing. It is the centre of visualization, imagination and conceptualization, but the left hemisphere is still needed for artistic work to achieve balance by partly suppressing creative states of the right hemisphere and for the executive part of a creative process.... Both brain hemispheres are needed for complete music experience, while frontal cortex has a significant role in rhythm and melody perception. The left hemisphere is important for processing rapid changes in frequency and intensity of tune. Several brain imaging studies have reported activation of many other cortical areas besides auditory cortex during listening to music, which can explain the impact of listening to music on emotions, cognitive and motor processes.«

During treatments and counseling of younger children, we encourage and teach both children and parents to spend time in nature, pay attention to colours, smell the scents, if there is something edible – to try it, to observe if an animal will appear, or to simply read books or poems together. One of accompanying methods is mindfulness tailoured for children. In therapy with preschool children, fairytales, such as the Little Red Riding Hood, and other stories are frequently used. In such instances, however, the Little Red Riding Hood is no longer 'ordinary/simple', but we rather resort to 4D format, which allows us to hear, touch, smell and feel all the things that are narrated. In the end, together we learn a song about the hunter Luke. Though such approach requires more intensive engagement on the part of a therapist, the ultimate effect is significantly stronger, as it carries a deeper sensory feeling, and also neuroplasticity.

Scientist Semir Zeki [3] points out that »all human experience is mediated through the brain and is not solely the product of the outside world«. We now know that »creativity does not just involve a single brain region or even a single side of the brain. Instead, the creative process draws on the whole brain. It is a dynamic interplay of many different brain regions, emotions, and our conscious and unconscious processing system.«

Creativity [3] is the ability of both hemispheres to communicate. A recent study from Duke University reveals that highly creative people have significantly more nerve connections between the right and left hemispheres than less creative people. The authors further point at the neuroscientist of the University of New Mexico who collected the MRI data to assess participants' level of creativity and concluded that the connectedness between the brain hemispheres is of crucial importance.

Differences are visible in the frontal lobe which is the 'control panel' of personality and the ability to communicate. Authors also claim that, despite advances in the neuroscience of creativity, the field still lacks clarity on whether a specific neural architecture distinguishes the highly creative brain. Further, authors also mention the recent study by Beaty and colleagues which revealed that » a person's capacity to generate original ideas can be reliably predicted from the strength of functional connectivity within this network, indicating that creative thinking ability is characterized by a distinct brain connectivity profile which includes three brain regions (i.e. executive, salience and default system). Creative thinking abilitz is characterised by specific brain connectivity profile. Creativity is often confined and even mixed up with artistic talent.«

»We should think of creativity as our intelligence with two facets: crystallized and fluid [3], what can be also said for creativity. Dietrich identified four different types of creativity with corresponding brain activities.« According to him, creativity can either be based on emotions or logical intelligence, and can also be spontaneous or deliberate. In addition to Hermann's typology, other possible modes of creative thinking are convergent and divergent. Convergent thinking assumes that a question has only one right answer and that a problem has a single solution, whilst divergent thinking or 'brainstorming' is where 'downloading' or emptying the brain of a certain topic takes place.«

Our conception of ourselves and the world determines our therapeutic work (teaching, explaining, reflecting, adjusting, evaluating, conceiving and acting again). Petzold [2] claims that differentiation processes grow within the process of creativity, and that such differentiation reduces compexity. They are simultaneously releasing new diversity, forming the basis for creation. The existing patterns are thus reconfigured through self-regulation. »Psychotherapy and art therapy as 'ars curandi et creandi' are intertwined and the strength of integrative art therapy is in opening up preverbal and transverbal spaces without words«. That means that we are gaining a new old during therapeutic process. The work with creative media is inseparable from the social and historical contexts (indivudual and group).

Neuroplasticity and Learning Models

Neuroscientist Vida Demarin, in her book »Healthy brain today – for tomorrow« [5] describes »neuroplasticity or brain plasticity as an increasingly discussed topic among neuroscientists, pertaining to changes occurring in neural pathways and synapses as a consequence of changes in behaviour, environment and neural processes, as well as a consequence of neural tissue injury. Knowledge on neuroplasticity has completely overshadowed previous understandings of brain as a physiologically static organ«.

She further explains that the prevalent opinion amongst neuroscientists of the last century was that of an unchangeable quality of the brain structure. Various studies revealed that changes in the environment can also affect human behaviour and influence cognitive abilities by changing interneuron connections through activation of other neurons (neurogenesis) in hippocampus and cerebellum. Having become aware of the possibility of change of human behaviour, William James introduced the term neuroplasticity in neuroscience in 1890. Another important discovery were mirror neurons. A group of Italian researchers (Giacomo Rizolatti, Vittorio Gallese, Luciano Fadiga i Leonardo Fogassi) published their research on macaque monkeys. They first gave them peanuts and using an electrode placed in one part of monkey's brain, they could monitor which part of the brain was activated when taking peanuts. They came to realization that the same group of neurons becomes active when monkey merely observes someone taking a peanut. Research on mirror neurons helps scientists find the basis of social interactions. These researches are important for us to understand the development of empathy and also gain better insight into autism, schizophrenia and other brain diseases.

Taking into consideration empathy – the awakening of feelings, Damasio [6] also describes that emotional states are defined by countless chemical changes in the body, i.e. internal organs, as well as by the changes in the degree of contraction of striated muscles. These changes are defined as part of neuronal changes in the brain itself. Damasio further describes that emotions as specifically caused transitory changes are essentially feelings of the feeling.

Manfred Spitzer, in his book Digital dementia [7] "Digital dementia" states that one of the most important findings in the field of neurobiology is that the brain is constantly changing by the very fact that we are using it. Perception, thinking, experience, feeling and behaviour all leave behind "traces of memory". In 1908s, these "traces of memory" were only hypotheses. Today, these hypotheses are visible through brain imaging and brain wave technology (fMRI). His long-term psychiatric practice gives him insight into the current state of society and the needs of young generations. His book presents research on the effects of information technology on the brain.

In his book "The Feeling of What Happens" [6], Damasio explains that consciousness is the key to questioning life in good and bad; it is our entry ticket to all knowledge about hunger, thirst, sexuality, tears, laughter, punches and kicks, and to the stream of images we call thought, feelings, words, stories, beliefs, music and poetry, joy and ecstasy. Co-creativity!!!! The bottom line of my thinking as to what makes cocreative personality is to be attracted to diversity, have lively imagination, ability to give and receive, readiness to adapt to someone's style of communication and social interaction, have courage to "work alone in the desert", never be satisfied with what you know, accept that every new problem/decision opens up a new problem/challenge/decision, exceptional pedagogical and analytical skills, readiness for teamwork and learning. The awareness of something is a prerequisite for learning, i.e. co-creation.

Co-creativity, in working with children and young people, represents a joint activity (e.g. if children are playing – you play with them, if they are drawing – you draw with them). Communication flows between activities (children talk between themselves, they say something to a therapist or vice versa). One of the key paradigms of Integrative therapy *I change myself to change my environment and change the world* is very important as a response to numerous demands of the present times. The imperative of lifelong learning is co-creation (lifespan development in context and continuum [1].

Challenges in the Work with Children, Adolescents and (Of Course) Parents

The concepts within Integrative therapy are as follows [1]: correspondence, inter-subjectivity, corporeality, identity, system, context/continuum, figure/background, multiperspective, socio-ecological perception, cognitive, emotional and social learning and finally co-creativity.

A special area of identity awareness are identity pillars. They are important as they represent a reliable instrument for anamnesis and for setting the therapy goal. There are 5 pillars of identity: the first pillar represents body, soul and spirit; the second represents a social network; the third pillar is the area of work and achievement; the fourth pillar represents material values/economic base, while the fifth pillar represents life philosophy and values [8].

Young Children

Co-creative national program "May we also grow together" [9], is implemented as a part of the school system and education for kindergarten groups and in the social welfare of family centers for different groups of parents and children. The population included are pre-school children. The program "Let's grow together" is implemented in kindergartens; "May we also grow together" is intended for parents in challenging parenting situations; "Then let's grow together uno" for foster parents and "Let's continue growing together" for fathers after divorce. These workshops are part of a broader UNICEF program that stems from the concept *The first three (years) are the most important* and are focused on the youngest children (from 3 to 6 years of age) all over Croatia. The main goal of the program is to create a stimulating and solid environment in which parents, together with children, other parents and workshop leaders, exchange parenting experiences, get to know themselves as parents, learn how to relate to a child and about other better ways of treating a child or children.

The first years of life are particularly important for development [9]. The child's brain goes through the most intensive development at that stage. What a child learns and experiences has a huge impact on their opportunities later in life. It is important to point out that the experience about oneself and others from those years is transferred into later life. In the earliest years of a child's life, their emotional safety, i.e. fundamental trust in themselves and others, is established. If a child grows up in an environment without enough love, security and parental care, many synapses in their brain will not be maintained and their brain will be different from a child who grows up in an affectionate environment. Already at the age of three, differences in social and intellectual development are visible. Emotional attachment, early communication and stimulation are just as important for development as are nutrition, health care, hygiene and safety from injury. Apparent wealth, level of informedness, availability of communication via information technology all push us towards the feeling of greater insecurity.

Following the implementation of the program conducted during 2019, the evaluation conducted via the survey of parents confirmed that the green program "We grow together", intended for parents and children with challenges, was implemented in different locations throughout Croatia. A quarter of parents were issued a disciplinary measure by the center for social welfare, and one third were provided professional supervision or were issued a warning. A third of parents attended the program out of the need to have better understanding of child development. During participation in the program, parents reported that they felt more optimistic and fulfilled. The feeling of frustration and anger decreased. They were less stressed, the feeling of guilt was significantly reduced, whilst the feeling of happiness increased. Parents claimed that now they know where and who they can turn to for support. The relationship with other children also improved. At the end of the program, they describe that they have developed a different sensibility towards children. My active role in conducting counseling and therapy is referral to diagnostics (early intervention team), mental health service and clinical treatment (of addiction).

Therapeutic work with children is almost always integrative and based on co-creativity. It, however, requires from therapists to continuously work on themselves, especially because in therapeutic work with children, one can experience a lot of anger (countertransference) towards parents, individuals and the system.

Work with Adolescents

The work with adolescents is particular in its characteristics and requires co-creative approach. The following statements stem from my private practice. A lot of patience is required to establish a meaningful connection, i.e., to build a trust. Adolescents like to have an adult figure by their side, someone who encourages and understands them, learns from them and teaches them how to learn. In this section, I would like to emphasise that a lot of co-creativity has to be found already at the beginning of therapeutic work, in order to understand what is happening with adolescents and how to place that in a context or on a continuum (see Photo 28.1. Teenage loneliness, in Introduction) . Youth of today are faced with new challenges, that are often contradictory. Compared to older generations, they are more enlightened, mature, speak more foreign languages and travel more to foreign countries; yet they still find it difficult to find their way around their own city. They often need assistance to get

Photo 28.1 Teenage
loneliness (private
collection)

from one part of the city to another, or from one city to another (especially in smaller areas). They act as if they know everything about health and conditions such as anxiety, and often live as if they have already been diagnosed and prescribed therapy, which happens not to be the case in reality. They can't be bothered to go out, yet are craving socialisation. They are well familiar with various concepts and terms (thanks to Google), but do not understand them as they they still lack concrete experience. Counseling centre is frequented by adolescents who have been issued special measures by the court, one of which is mandatory counseling. The very definition of mandatory counseling is a big challenge for social pedagogic practice. These consultations require composure and good mental fitness. In private practice (the same instructions apply for parents at home), it has proved as an effective method to involve adolescents in daily activities (especially cooking). Cooking – as an activity and co - creation – is a medium for dialogue. Young people like to be useful and want to show that they know, can and will. It is important for young people that we see them and that we are with them, even if they do not necessarily show it.

Depending on the environment in which it is practiced, the American Counselling Association distinguishes between individual and group counselling. Individual counselling [10] is an opportunity to receive an individual support to a growth experience during different life challenges. It takes place in a confidential environment between a counsillor and a client. It is focused on helping people to define personal goals and teaching them how to achieve those goals. Counselling is carried out with the aim of resolving intrapersonal and interpersonal problems (anger, depression, anxiety, drug abuse, criminality...). In the interpretation of adolescents' life context and continuum pertaining to personal development and processes of separation from the primary family, we use the aforementioned philosophical questions and topics such as, "Who am I"? or "Where am I going?" It is precisely those existential questions that are stimulated through co-creativity.

Group counselling is a great support to adolescents in their growth and development, i.e., co-creativity (should they find themselves within a group). It includes working with the group, i.e., with clients who are faced with same or similar life challenges (except when there is ADHD or extremely dissociative behaviour). In terms of preventing juvenile delinquency, all techniques are aimed at stimulating changes. A special counselling dynamic is needed for adolescents. Informal walks are often useful in order to strengthen a relationship and avoid the counselling to become "boring". Movement always generates information. The difference between counselling and therapy is that during therapy we are not tied to the system and can directly be more co-creative. I myself set the framework of informality, thanks to personal freedom and a range of possibilities that a therapist in private practice (therapy room, kitchen or playroom) can have.

The joint work of adolescents and parents is especially indicated when learning non-violent communication (domestic violence, violence in school, etc.). Together we learn ways of appropriate communication, constructive critical thinking, assertiveness, correction of violent behaviour, as well as the specifics of child and youth development.

Manfred Spitzer [7], among other things, talks about harmful effects of information technology (screens) in the process of education. If, already at kindergarden age, a child spends too much time exposed to screens, as a result, their language acquistion and development of thinking capabilities will be slowed down. In other words, the author states that excessive viewing of TV programs leaves long-term consequences in terms of lower educational capacity. A study conducted in New Zealand in Dunedin region included all children born within one year. When they were five, seven, nine and eleven years old, their parents were asked how much time their children spent watching television? When children were thirteen, fifteen and twenty-one years old, they undertook the survey themselves and were asked how much television they watched between the ages of five and eleven, and between the ages of thirteen and fifteen. When they were twenty-six, a scale from one to four was set (4 meant that they completed university education) and their IQ and social-material status were also measured. The study showed that watching television coincides with lower education. I, therefore, teach parents that there are many ways of parental protection, and one of them is "creative upbringing"; for example, spending a day without information technology (mobile phone, tablet, etc.).

Work with Parents

"Parents also have right for assistance in children upbringing", "Child travels, parent guides", "We will not spoil children with pampering" – these and other similar messages were promoted during the implementation of the program "Growing up together" under the auspices of UNICEF [9]. Different workshops for children and parents are suitable for prevention, as well as for cure, especially when challenges occur (e.g. disciplinary measures issued by the Center for Social Welfare, foster care, specific developmental disorders of children, high-conflict divorces). During therapeutic work with children and young people, conversations are held with parents separately, and together with children. It is advisable to connect with professional service at school (with written consent), since the progress at school is just as important as the progress in therapy. In therapeutic conversations with parents, therapist explains the entire process in detail.

Oaklander points [11] out that regardless of invisible "bonds" and family pressure, things can change transgenerationally. There are many more children's diseases, syndroms or symptoms, but also many, many more different ways of establishing diagnosis and diagnostics. Violet Oaklander explains that special forms of children behaviour are survival strategies, rather than a disease.

In conversations with parents, we describe how – when we were growing up – our parents wanted us to grow up happy (through a choice of adequate profession) and healthy. Regardless of invisible "bonds" and family pressure, things can change transgenerationally. It is, therefore, extremely important to dedicate sufficient time to psychological education on sleep hygene, physical exercise and healthy nutrition of children as the basis for a healthy body, spirit and soul.

Integrative therapist Gitta Hauch [12], in her book "The doctor said it's psychosomatic", in addition to her therapeutic interventions with children and young people, describes in detail her parallel work with parents. She uses an example of baby Ivana, who was hospitalized at the clinic because of cramps and vomiting. After a "chance" meeting at the clinic before Ivana's release from the hospital, a dialogue began. Ivana's mother described her life and incredible battle for her independence and identity. They agreed on the beginning of therapy process. Ivana was released from the hospital only after three weeks. The mother continued her therapy, along with the separation from her primary family and enormous support from her husband, who also started with therapy.

Co-creativity [2] (Petzold 1990) is a phenomenon that occurs within the framework of interactions and communication processes through co-reflexive and co-emotional linking of different elements (information, materials, idea, etc.), or, more precisely, through connections and complex systems (personal, group, organizational). The product is the creation of novel things and reduction of complexity. Through "creations/work of higher order" it is possible to open a new chapter of complexity that can be adopted and used in configuring new information. This is how new "phenomena/appearances of co-creativity" emerge and these newly-generated potentials make systems possible. What follows from the above is that the family – as a system – can be functional again with adequate support.

A Summary of Integrative Therapeutic Work with Ivan

Ivan (the name is changed to protect the child's identity) first came to psychotherapy in April 2017, at the age of 12. At the time, he is a 6th grade pupil at a primary school with adapted program. Before he arrived, accompanied by his father, an informative interview was conducted with both parents. Therapies take place once a week. The parents brought all medical documentation with them and report that the reason for coming is the boy's inappropriate behaviour, i.e. difficulty in controlling anger, outbursts of anger and difficulty in coping with failure at school. The parents describe the boy as socially and emotionally immature, and explain that his treatment of his younger sister is non-affectionate, almost crude. The parents also report that the pregnancy and birth were normal, and that the boy started walking when he was 17 months old. Ivan follows regular curriculum, with adaptations in mathematics and Croatian language. He loves English classes and has good results there. He is not involved in any leisure activities. The boy, his two sisters (one is younger and one older) and parents live together with his grandmother in a small apartment, in a local town. The family moved to that area, and the parents work in the same area. The boy is daily exposed to conflicts between family members. He attended kindergarten for half a year and pre-school for only a few months, as he was finding it difficul to adapt to new environment. Between 2010 and 2014, he was monitored by an orthopedist and wore orthopedic shoes due to congenital deformities of spine, legs and feet. As of 2013, the boy was undertaking regular treatment with a speech therapist. In the latest report, the speech therapist stated that the boy was less inhibited than at an earlier age. He has difficulties in pronouncing sounds and is timid. Images of complex words are unclear to him and he finds it difficult to verbalize what he reads. The speech therapist recommends psychological treatment. After being treated by a clinical psychologist, the boy receives diagnosis, i.e. specific developmental disorder of school skills.

Treatments with the speech therapist continue. In 2014, a control examination by a psychologist and a neurologist was carried out, who described his condition as neuropsychological immaturity, which was the main cause of the boy's learning difficulties. According to the developmental anamnesis, the boy received insufficient preparation for school. Rehabilitative treatment was thus recommended, in order to define adaptations in teaching. However, the treatments do not have a holistic approach, whilst the boy remains constantly exposed to stress due to broken relationships in the family and worsening economic status (mother lost her job). Until that point, the Center for Social Welfare did not require any special measures.

The aim of therapy work is to forge a relationship based on trust, which would enable the boy to more easily cope with emotional difficulties, develop social and communication skills, enhance self-confidence, improve relations with parents and sisters, connect with peers, i.e., empower him in such way that he can change his inappropriate behaviour, and simultaneously encourage the development of his abilities, skills and resilience, all of which are an important protective factor in the phases of development.

First year: At the first session, the boy's posture is bowed down. He is quiet and withdrawn, and answers to all questions with "yes" or "no". At the very beginning of the treatment, pronounced silence and resistance are noted. Defense mechanisms in children serve as material for observation. From the perspective of psychoanalysis [13], defense mechanisms, as a function of ego, become the focus of research, because they are separated from unconscious impulses such as fantasies and images. Defense mechanisms are automatic and unconscious, and as such serve to prevent inputs from the environment to penetrate directly into the consciousness. We get useful information when we observe children playing, drawing, modeling clay or plasticine.

Resistances in psychotherapy play out in different ways; they prevent, reduce or eliminate unpleasant feelings or emotional suffering, and most of all, strong fear – due to the anticipation of danger, shame, guilt, injury to self-love, and sadness, i.e., whenever there is a possibility for preconscious, and especially unconscious content, to come into consciousness [13]. The manifestation of resistance is clearer and more pronounced in adolescents. Showing resistance indicates the process of recognizing some "bad" content within ourselves. In principle, useful information can be obtained by observing children during prolonged periods of rest (e.g. due to illness), though we as professionals are deprived of such possibility. It is evident from the above that we need to spend some time with children and young people, so to be able to learn something from them and they from us. During different phases of psychotherapy, the manifestation of the boy's silence also changed. Considering his restraint, caution and silence, it is difficult to build a "therapeutic alliance". The boy refuses to draw, model the clay or paint a body map. He accepts board games that make him happy and is good in playing them. Gradually developing trust through the play, the boy is becoming more relaxed and talkative (after 5 sessions). Resistances are still present, however. In addition to board games, he was also offered associative cards. With the creation of safe environment, the boy thrives in a way that he now verbalizes his distress, concerns and failures. The first treatment cycle lasted 6 months. He was coming to therapy sessions once a month, which gave him the sense of security and certainty. Whenever his mother came, the dynamic would change and the boy was scared and upset. At the same time, interviews were conducted with the professional service from his school. The boy successfully completed the school year, with certain adjustments and correction of low grades. In other working conditions (private practice), animal-assisted therapy (e.g., with a dog) was advised.After the summer break (**second year**), when he returned to therapy, a big change occurred: he started talking muuch more and was more willing to participate. He was suggested a walk in nature, as a way to relax, and readily accepted it. After the walk, he was offered cards with images of movements on a given topic. He revealed capacity for movement and acting and subsequently joined the drama section at school. In the period of the following 7 months, during our weekly sessions, he accepts to work with figures with which he depicted his family. He also found adequate figures for his school environment. In the continuation of treatment, we created short stories together. The therapist and the boy simultaneously filled out sheets with short stories, and after that – upon his own initiative – he would draw his mom, whenever he got A or D at school. Achievements at school now became his main preoccupation. At the same time, he wants to have friends. A conversation was held with the parents about their inadequate educational methods, i.e., taking photos of him during the episodes of anger and rage outbursts. The parents were explained how to include him in daily activities, such as buying bread, taking out the garbage, bringing wood, etc. After that, mutual trust between them deepened. He feels happy, because he received support. He is also pleased when he is offered to work with associative cards. With redirection of attention, when asked existential question "What if it were?", Ivan enthustically shared that he would love to be a "ninja turtle", because they are not afraid of anything. At the end of the school year, he had low grades, but was no longer crying because of that. Now he wanted to correct them.

After the summer break (**third year**), the boy returns to therapy. He talks about future plans and his wishes for the continuation of education. He grew up and the change is observable. He also describes the loss of his grandmother, and shares how his mother and grandmother used to quarel a lot, and then suddenly stopped. He did not understand why. At therapist's urging to say something to grandmother, he said that he would want to tell her that he loved and missed her. When talking about important events in his life, he specified the following: the birth of his sister, starting with school, grandmother's passing and departure of a schoolfriend at the 4th grade of primary school. Therapist offered to him to draw identity pillars, which was used for their future conversations. Therapy sessions were ongoing for four months. Due to acting out at school (which resulted with material damage), he was hospitalised at the Psychiatric Board for Children and Adolescents at a nearby clinical centre. He was prescribed pharmacotherapy, after which he was having therapy twice a week, within the period of one month. After that, as we were informed, he completed primary school and enrolled a vocational highscool (three-year program), not the one of his choice. He successfully completed the highschool and started working in the same company with his father.

Such complex cases are often seen in the Counselling centres. They require integrative approach and co-creativity in terms of the therapy work with children, with parents and with different systems (school and health). Further multi-sectoral monitoring of development is advised, as well as the continuation of diagnostics and habilitation. It is recommended to continue with integrative therapy, preferably with male therapist, if possible.

Conclusion

Throughout this article, I presented the theory and practice of Integrative therapy in its broadest sense, as a way of describing co-creativity. Co-creativity or joint creation is described as a holistic approach, the dynamics of which assumes that learning and constant changing are everything. It is the approach that not only 'processes' human psyche, but the whole of a person, in all of their dimensions. Deriving from my practical work, I also present various experiences. Co-creativity is the process of personal growth and when viewed from the perspective of neurobiology, a conclusion can be made that the right and left brain hemispheres are equally important. The overview of theory and practice provides insights on bio-psycho-social model, which is very important in interventions within different systems. Co-creativity generates insights into personal resources of knowledge and power, and is a response to challenges of our time and for 'out of the box' thinking. Therapy for children and adolescents requires a particular kind of playfulness, which is often subjected to the criticism of our expertise and professional competence. I do realise that the main task is learning of acceptable behaviour. Co-creativity, among other things, helps us to reduce polarisation between black and white, good and bad. Through Integrative therapy and depiction of creativity from the perspective of neuroscience, I am endeavouring to contribute to therapy work with children and adolescents, because we generally lack experts working with these two groups. Homo homini remedium! What cures is the relationship.

References

1. Maša Žvelc, Miran. Možina, Janko. Bohak: „Psihoterapija", „Psychotherapy" založba Ipsa, Ljubljana 2011, str. 537 – 563.
2. Hilarion G. Petzold, Ilse Orth, Johanna Sieper, Hückeswagen (2008/2010/2019b) „Integrative Therapie mit Kreativen Medien, Komplexen Imaginationen und Mentalisierungen als „intermediale Kunsttherapie" – ein ko-kreativer Ansatz der Krankenbehandlung, Gesundheitsförderung, Persönlichkeitsbildung und Kulturarbeit", Polyloge ISSN 2511-2732, Ausgabe 22/2019. Information: http://www.eag-fpi.com.
3. Creativity—the story continues: an overview of thoughts on creativity, Vida Demarin and Filip Derke, ISBN 978-3-030-38605-4 ISBN 978-3-030-38606-1 (eBook) https://doi.org/10.1007/978-3-030-38606-1.
4. Green texts– 07/2017 Petzold H. G., Hömberg R.: Ecopsychosomatics – An Integrative Core Concept in the „New Nature Therapies", Internet Magazine for Garden, Landscape and Nature Therapy, Forest Therapy, Animal-Assisted Therapy, Green Care, Ecological Health, Ecopsychosomatics http://www.eag-fpi.com.
5. Vida Demarin, »Zdrav mozak danas za sutra«, „Healthy brain today – for tomorrow" Medicinska naklada 2017, str. 43 – 48.
6. Antonio Damasio, „Osjećaj zbivanja" – origin. „The Feeling What Happens", Algoritam Zagreb 2005, str. 19, 274 – 279.
7. Manfred Spitzer, „Digitalna demencija" origin. „Digitale demenz – Wie wir uns und unsere Kinder um den Verstand bringen , Mohorjeva družba Celovec, druga žepna izdaja, 2017, str. 52 – 54, 135.
8. Hilarion G. Petzold, „Transverzale identitat und identitatsarbeit. Die Integrativeidentitatsheone als Grundlagefur eineentwicklungspsyhologischund socalisationstheorethishbegrudendete Personicheirtsheoneund Psichoterapie – Perspektiven „klinischer Socialpsychologie", Polyloge 10/2001, http://www.eag-fpi.com.
9. Ninoslava Pećnik, Branka Starc »Roditeljstvo u najboljem interesu djeteta i podrška roditeljima najmlade djece«, »Parenting in the best interest of the child and support to parents of the youngest children«, UNICEF za Hrvtasku, Zagreb, 2010). https://www.rastimozajedno.hr/rastimo-zajedno-i-mi/.

10. Gabrijela. Ratakajec Gašević i Antonija. Žižak, „Savjetovanje mladih; okvir za provedbu posebne obveze uključivanja u pojedinačni ili skupni psihosocijalni tretman u savjetovalištu za mlade", "Counselling of young people; a framework for the implementation of a special obligation to include in individual or group psychosocial treatment in a counselling centre for young people", Edukacijsko – rehabilitacijska biblioteka, Zagreb, 2017 str. 14 – 22.
11. Violet Ouklander »Put do dječijeg srca« origin. Windows to our children, Školska knjiga Zagreb, 1996, str. 211.
12. Gitta Hauch, »Liječnik je rekao da je psihosomatsko« origin. „Der Doktor hat gesagt, est is psychosomatisch" , Temza, Ljubljana 2008, str. 47 – 48.
13. Anna Freud , „Normalnost i patologija djece" origin."Normality and pathology in childhood", prosvjeta Zagreb2000, 18 - 21.

Chapter 29
The Alps – Adria Section and the Current Trends in Clinical Neurosciences

Leontino Battistin

In the Memorial Book 1961–2010 we summarized the history of the Alps-Adria Section of the Pula Symposia and we listed all the events of this Section that occurred during those years.

During the last ten years we continued on the same way, but, there were also some important news; in fact, in 2009 I was elected as the Chairman of the Research Group on Organization and Delivery of Neurological Services of the World Federation of Neurology (WFN) succeeding to Bosko Barac, and I asked Vida Demarin to act as General Secretary. So, after this fact, I and Vida decided that thereafter at the Pula Congresses the Alps-Adria Meetings should have been done in conjunction with the Research Group on Organization and Delivery of Neurological Services of the WFN.

Then, in 2013 I resigned from the Chairmanship of that Group and I proposed to the Board of the WFN that my place should be taken by Vida Demarin, and my proposal was accepted unanimously; so, from 2013 the Group, now named Applied Research Group, is chaired by Vida Demarin, and she continued to schedule the Alps-Adria Meeting in conjunction with the Research Group on Organization and Delivery of Care.

The programs of the Alps-Adria Section, that, I must underline, was always very well represented by the international presence, and the Meetings in the years 2011–2020 were very rich, and if you look carefully to them you could see that the Organizers always tried to pay attention to the latest news coming from the research field of the various neurosciences branches, like neurochemistry, neuropharmacology, neuroimaging, etc., in the effort to get better ways and discoveries to improve our diagnostic procedures as well as our therapeutical approaches to neurological

L. Battistin (✉)
Department of Neurosciences, University of Padova, Padova, Italy
e-mail: leontino.battistin@unipd.it

V. Demarin et al. (eds.), *Mind, Brain and Education*,
https://doi.org/10.1007/978-3-031-33013-1_29

diseases, but also to the psychiatric ones; this is especially true for the most frequent diseases, like dementia, or Parkinson's disease, or cerebrovascular diseases, or migraine, and in the psychiatric field, schizophrenia and depression.

So, in going through these thoughts I made some reflections, as a Clinician, on the current trends in clinical neuroscience in the present time [1] and I'll participate to you.

The basic question for a Neurologist at the present time is: what are the main lanes of research in clinical neurosciences and which are the main goals of our investigations?

Since the main mission of a clinician is to be of help to the patients, a clinical investigator has to search always to discover etiology and pathology of the diseases in order to be able to reach the best diagnostic and preventive procedures as well as the most efficacious therapies and rehabilitation actions.

So, bearing in mind this guideline for our work as a clinical neuroscientist, we should make a reflection on the "state of the art" of the research in our field in the present time.

I think that a brief historical review of the evolution of knowledges on Parkinson's disease (PD) is the best way to debate the evolution of studies in clinical neurosciences.

In the Fifties neuroleptic drugs (chlorpromazine, reserpine, etc....) were thought efficacious in the therapy of psychoses when the patients were showing the appearance of extrapyramidal symptoms as side effects.

In 1957 Arvid Carlsson [2] showed the deficit of dopamine in the brain of reserpinized rats with extrapyramidal signs; treating the rats with L-Dopa was successful and the extrapyramidal signs disappeared.

In the Sixties and Seventies L-Dopa therapy made the "miracle" for thousands and thousands of Parkinsonian patients throughout the world, passing from the wheelchair to a fairly normal walking.

In the Eighties dopamine agonists added further strength to the dopaminergic therapy of PD and triggered out new biochemical investigations on neurodegenerative diseases.

Thus, in the Eighties it was shown that some parkinsonian patients had cognitive impairments, and that the therapy with anticholinergic drugs that were largely used before L-Dopa was very negative for the cognitive performances. Also, on the base of such observations, the first trial of cholinergic stimulation of dementia started with the use of acetylcholinesterase inhibitors.

In the Nineties a great attention was paid to non-motor symptoms, and especially to the ones that are detectable many years before the appearance of PD. These "preclinical" symptoms were of paramount importance for prevention. Moreover, in the Nineties a great attention was paid to the quality of life of the patients as well as to all the techniques of rehabilitation, both motor and cognitive. As far as prevention is concerned, eating a healthy diet and practicing physical activity have been demonstrated to be key elements for an anti-PD lifestyle [3, 4].

Furthermore, in the Nineties there was an impressive growing of the research in the neuroimaging field with quite important data useful for diagnostic procedures and allowing a deeper knowledge in the physiopathology of PD and related disorders.

In synthesis, the integrated approach of research in PD was by combining clinical observations with molecular neurobiology and neuroimaging both in animal models and in patients.

Nowadays a new approach is growing, that is the so-called "networks connectivity", that is the research on neural connections and their significance in the function of the brain. This approach is certainly a helpful one and it is very much used by the investigators in neuropsychological and behavioral studies. Thus, in this field there was a kind of "explosion" during the last two-three decades in correlating neuropsychological findings with the neuroimaging data (see [5]).

The "network-connectivity" has certainly added new knowledges for a better understanding of physiology of the brain, and particularly in the cognitive and behavioral fields; furthermore, such approach could be useful in the quite complex processes involved in the neurorehabilitation procedures. Now, the point is at which extent such approach, the so-called "connectomics", could add new information and data on the causes of brain diseases and their therapies; a debate on this point could be relevant for all the branches that refer to clinical neurosciences.

As pointed out [6] actually connectomics and molecular neurobiology can be highly complementary on the base of the concept of a hierarchical architecture providing a unified view of the brain morpho-functional organization. As a matter of fact, brain networks, from cellular down to molecular level, are encased one within the other; this means that different connectome levels should be considered namely among brain areas, among brain cellular circuits, within local circuits and within protein circuits; accordingly, connections in the brain and neurochemistry are two sides of the same coin which allow intercellular communications as well as integrative actions. Furthermore, the intercellular communication processes in the Central Nervous System occur via fibers and synapses but also via interstitial fluid channels; thus, not only neurons and astrocytes but also brain interstitial fluid have to be investigated [7].

Within the organization of the nervous system, the cell and its environment can be considered the "centre", and the fibers are the "vehicles" of the messages given by the cells. So, to discover etiology and pathology of a disease we have to study not only connectivity but also cell processes and intercellular communications and what is going wrong in such processes at the different levels of the brain morpho-functional organization. Obviously, if the cell is affected by a pathology, that determines an impairment in the electro-chemical processes, such altered processes can be propagated through the intercellular communication channels to other sites.

Thus, from this broader point of view, it is not clear, at least to me, how investigations based only on neuroimaging data to obtain the so-called "networks-connectivity", could give complete and relevant information on the multi-facet processes that cause cellular pathology and consequently neurological disease, and therefore, I have serious doubts that an approach relying solely on "connectome-imaging" could give final answers to help discover the causes of diseases and their therapy.

This fact makes the difference with the clinical neurobiological integrated to neuroimaging method that has been used in the studying PD and related diseases; such an approach that, as we have seen, was very beneficial for patients and could be defined "translational research".

Actually, I agree with [6, 7] that neurophysiopathology is a big tree with many branches and that the electro-chemical processes at cellular level, neuron and glia cells, the fibers and interstitial fluid modes of intercellular communication as well as the intercellular brain fluid homeostasis, should be considered interrelated branches to be investigated according to an integrated approach. Obviously, in this context, the connectome program could certainly be useful but it should not simply consider as imaging research.

The above-mentioned reflections are on lane with the evaluations of Agnati et al. [7] and of Raichle [8] with the "hyper-network" connectivity and the electromagnetic field allowing integration at the synaptic level; both Authors emphasized the need to base future research in this field on the link with cellular and molecular neurobiology; we are confident that this way of translational research will be the real future.

In the Eighties there were great hopes that introducing spectroscopy research with MR would have been a breaking point to start to do neurochemical research in humans; this fact did not happen; the industries preferred to continue mostly on the neuroimaging approach and therefore they focused on neuropsychological studies, and a great number of neuroscientists followed for various reasons such indication.

I hope that scientific societies and especially the neurological ones, will promote debates and scientific discussions on the above-mentioned points.

References

1. Battistin L. Reflections of a clinician on the current trends in clinical neuroscience – molecular neurobiology and/or connectome? Eur Neurol Rev. 2018;13:16–7.
2. Carlsson A, Lindqvist M, Magnusson T. 3,4 Dyhydroxyphenylalanine and 5-hydroxytryptophan as reserpine antagonists. Nature. 1957;180:1200.
3. Bracco F, Malesani R, Saladini M, Battistin L. Protein redistribution diet and antiparkinsonian response to levodopa. Eur Neurol. 1991;31:68–71.
4. Dam M, Tonin P, Casson S, Bracco F, Piron L, Pizzolato G, Battistin L. Effect of conventional and sensory-enhanced physiotherapy on disability of Parkinson's disease patients. Adv Neurol. 1996;69:551.
5. Sporns O. The human connectome: a complex network. Ann N Y Acad Sci. 2011;1224:109–25.
6. Agnati LF, Fuxe K. New concepts on the structure of the neuronal networks: the miniaturization and hierarchical organization of the Central Nervous System. Biosci Rep. 1984;4:93–8.
7. Agnati LF, Marcoli M, Maura G, Woods A, Guidolin D. The brain as a "hyper-network": the key role of neural networks as main producers of the integrated brain actions especially via the "broadcasted" neuroconnectomics. J Neural Transm. 2018; https://doi.org/10.1007/s00702-018-185.
8. Raichle M. The restless brain: how intrinsic activity organizes brain function. Philos Trans R Soc B Biol Sci. 2015;370:2014017.

Chapter 30
Neurosurgery in the Pula Congresses: A Review at the 60th Jubilee Anniversary of the Congress

Günther Lanner

The International Neuropsychiatric Pula Congresses developed from a small family gathering of neuropsychiatrists with hardly sixty participants in 1961 to a great International Congress with about 500 participants. Responsible for this development were the founder of the Pula Congress Prof. Lopašić (Zagreb) and Prof. Bertha (Graz) until 1964, then Asst. prof. Gerald Gringschl (Graz) until 1984, then the General Secretaries Prof. Helmut Lechner (Graz) and Prof. Bosco Barac(Zagreb) until 2006 and 2008 and since 2008 the General Secretary of the Congress Prof. Vida Demarin (Zagreb).

This International Neuropsychiatric Congress is one with the longest tradition in the field of neurology, psychiatry, and neuropsychiatry. It has also become a tradition of Neuropsychiatric Pula Congresses to promote interdisciplinary collaboration between nowadays independent disciplines neurology and psychiatry with the collaborating and borderline medical disciplines, especially neurosurgery.

Since the Pula Congresses - at the beginning "Symposia" - were founded by Professors Radoslav Lopašić (Zagreb) and Hans Bertha (Graz) in 1961, the neurosurgical lectures and presentations were nearly always included in their programs.

In Academic lectures, in Contributions to the Main themes, in Workshops or in the Round table discussions, neurosurgeons have reported in the 60 last years about their experience in all fields of neurosurgical practice and research.

Already in 1966 at the Kuratorium Meeting of the Pula Symposium Prof.F. Marguth (Munich) was elected into the Board of Trustees as first neurosurgeon. Later on many other neurosurgeons followed him in the Pula Board of Trustees - V. Dolenc (Ljubljana), G. Lanner (Klagenfurt), E. Uhl (Giessen), R. Fahlbusch (Erlangen), B. Richling (Salzburg), M. C. Spendel (Klagenfurt) R. van den Berg (Leuven).

G. Lanner (✉)
Clinic of Neurosurgery Klagenfurt, Klagenfurt, Austria
e-mail: guenther.lanner@a1.net

© The Author(s), under exclusive license to Springer Nature Switzerland AG 2023
V. Demarin et al. (eds.), *Mind, Brain and Education*,
https://doi.org/10.1007/978-3-031-33013-1_30

311

Now there are three experienced neurosurgeons in Pula Kuratorium – the founding honorary member Prof. F. Marguth, Munich, he died 1991, the honorary member Prof. G. Lanner, Klagenfurt and the member of the board of trustees Prof. M. Mokry, Graz.

Professor Frank Marguth, chairman of the University Clinic of Neurosurgery Munich delivered the first neurosurgical lecture in Pula at the *6th "Weekend Symposium"* 1966 (at those times the conference language was still German) about "Surgery in vascular diseases". Moderator of the theme was Professor Reisner, then the Chairman of the University Clinic of Neurology and Psychiatry in Graz.

One year later, at the *7th Neuropsychiatric Symposium (1967)* Prof. Marguth gave a neurosurgical lecture about *"Traumatic intracranial Hematomas"*, followed by Reisner's exposition: "The Late Traumatic Apoplexy". These lectures were followed by a discussion with 13 invited speakers of various affiliations and countries..

In the next annual Meetings during the years 1968 and 1974 many invited neurosurgeons gave their contributions to main themes, such as the Professors Kloss (Innsbruck), Frowein (Cologne), Raimond van Bergh (Leuven), Diemath (Salzburg), Mundinger (Freiburg), Siegfried (Zurich). Kloss and Frowein reported about *"Neurosurgical Aspects in Head Traumas"*, R van den Bergh about *"Intracranial Anastomoses and Vascular Malformations"* – a lecture in memoriam Prof. Hans Bertha, who died 1964, and Diemath, Mundinger and Siegfried about *"New Results in the Stereotactic Treatment of Extrapyramidal Diseases"*.

One of the Main themes of the *15th Neuropsychiatric Pula Symposium (1975)* was *"Cervical and Lumbar Dics Diseases"*. Marguth (Munich) reported about *"Indication and Results of Foraminotomy in Cervical Root Compression"*. He pointed out that in lateral soft disc herniation lateral foraminotomy is indicated. The results after operation were excellent, 95% of the patients were after the operation pain-free, with a normal neurological status.

At the *20th Symposium (1980)* Marguth reported on *"Advances in the Diagnosis and Therapy of Brain Injuries"*. He pointed out that in severe head traumas follow-up examinations of cerebral CT led to a significant progress in the management of posttraumatic intracranial hematomas, cerebral lesions, and posttraumatic brain edema.

One year later, at the *21st Symposium (1981)* Marguth, Stass and Fahlbusch (Munich) reported on *"Neurosurgical Therapy of Intracranial Malignant Tumors"*. They found in glioblastomas without operation a mean survival time of 4 weeks, with operation and antiedema therapy a mean survival time of 4 months, with operation and radiotherapy a survival of 9 months and the same plus chemotherapy the mean survival time was prolonged to 14 months.

One of the main topics of the *29th Symposium 1989* was *"Vertigo and Disturbances of Balance"*. G. Lanner and M.C. Spendel (Klagenfurt) reported at this Congress on " Low *Pressure Hydrocephalus"*. Gait disturbances, mental deterioration and urinary incontinence are the neurological signs of this clinical picture. From their own studies it was pointed out that in accordance to the literature the normal pressure hydrocephalus is determined by the presence of ramp waves, called R waves, described first by Lindsay Symon (Queen Square, London) during continuous measurement of intracranial pressure. An incidence of these R waves of more than 10% of the recording time seems to be a good prognostic sign for operative shunting.

One year later, in *1990 at the 30th Symposium*, one of the main themes was *"Epilepsies in Children"*. Lanner and Lechner (Klagenfurt, Graz) reported about *"Epilepsy Surgery in Children"*, reporting their studies they had performed during their collaboration in Graz. Pharmacoresistant children for more than 2 years were candidates for surgery. In case of temporal lobe epilepsy, often the consequence of birth trauma with following hippocampal gliosis, there was an indication for hippocampectomy during electrocorticography. The reported results were excellent, up to 80% of the children were free of fits or markedly improved.

In *1991* at the *31st Symposium two of the Workshops* were organized on neurosurgical aspects: one dealt with *"Advances in neurosurgery"* in general, a second with *"Progress in diagnosis and therapy of head injuries"*. Many neurosurgeons had participated from Salzburg, Graz and Klagenfurt.

One of the topics in *1993* was *"Rehabilitation in Neuropsychogerontology"*, a common topic of Neurology, Psychiatry and Neurosurgery. Spendel and Lanner (Klagenfurt) reported about *"Neurosurgical Features in Elderly Patients"*. The authors pointed out that because of failure of cerebral compensation; neurosurgical complications following extirpation of tumors or evacuation of intracranial hematomas were more frequent in old age than in other periods of life. The cerebral decompression was often badly tolerated and therefore less invasive operative methods have to be considered in elderly patients, such as radiosurgery and radiotherapy. After evacuation of chronic subdural hematomas the older brain showed poor dilatation due to the reduced mechanisms of adaptation. Complications result from recurring hematoma and pneumatocele in the subdural space.

One topic of the *41st International Neuropsychiatric Symposium in 2001* was Clinical and Social Aspects of Alzheimer´s Disease. Lanner and Spendel reported about *"Neurosurgical Aspects in Alzheimer´s Disease"*. In this disease it is to differentiate between different types of dementia. Sometimes Alzheimer´s disease and vascular dementia may mimic the clinical characteristics of normal pressure hydrocephalus (NPH). In literature and their own studies NPH accounts in 3–6% of cases of dementia, on the other hand the incidence of AD in patients with suspected NPH is 30–50%. NPH was determined by the incidence of Ramp waves during continuous ICP monitoring. In the case of NPH shunt operation is indicated.

At the *42nd Symposium (2002)* Fheodoroff of the Rehabilitation Centre in Carinthia reported together with Lanner about *"Neurorehabilitation after Severe Traumatic Brain Injuries"*. It was pointed out that neurorehabilitation should begin in an intensive care unit, but the patients have to reach first a level of responsiveness to benefit from it.

Neuropathic Pain was one of the main themes of the *44th Symposium (2004)*. Lanner and Spendel reported about *"Spinal Cord Stimulation"* in patients refractory to medication. Since 1977 one of the authors (Lanner) had experience and results with this method already. Stimulation electrodes were implanted in the epidural space percutaneously. After a test phase, a telemetric pulse generator is placed into the subcutaneous pocket in the upper abdominal wall. The reported results were successful, the degree of pain relief was rated in the follow-up as very good or good in 81% of the patients, the level of activity and quality of live increased.

A further topic of the *44th Symposium (2004)* was *New Therapeutic Trends in Neurology*. Lanner and Deinsberger (Klagenfurt) reported about *"Method, Indication and Results of Radiosurgery"* in cerebral metastases, benign tumors, arteriovenous malformations and functional neurosurgery. At the Clinic of Neurosurgery Klagenfurt radiosurgery was used since 1996.

At the *45th Pula Congress (2005)* Lanner and Spendel informed about *"Neuroimaging in Neurosurgery"*. The advances in noninvasive neuroimaging technology over the last decades have been helpful in advancing neurology and neurosurgery in the fields of brain tumors, vascular diseases, brain trauma and spinal lesions. Recent developments in imaging have revolutionized pathomorphological diagnosis in many cases, permitting precise location and clear demonstration of its anatomical relations pre-, intra- and postoperative as well as morphological and functional.

At the *46th International Pula Congress (2006)* one of the main themes was *New Developments in Neurology.* The neurosurgeon Kostron (Innsbruck) reported on *"Modern Management of Malignant Brain Tumors"*. He pointed out that the treatment of malignant gliomas has changed from a single modality to a multidisciplinary approach due to advances in the treatment. Although the prognosis of malignant tumors is in general still poor, the median survival for glioblastoma has been extended from 9 to 15 months, while a 5 year survival is still below 2%. Surgery should be the first and most important step in the cascade of treatment. Adjuvant radio- and chemotherapy is recommended in high grade gliomas following surgery or biopsy. The introduction of Temozolomide is a hopeful substance in the chemotherapy.

Stefan (Erlangen) informed about modern *epilepsy surgery*. He pointed out that the exact presurgical localization of epileptic areas in the brain offer good perspectives for patients with focal seizures undergoing surgery. Non invasive evaluation include intensive video-monitoring and magnetoencephalography, in difficult cases invasive evaluation will be applied with subdural or depth recording, PET, MR-spectroscopy and ictal SPECT. Besides resection of epileptic areas, transactions like callosotomy or multiple pial transection are performed. Other surgical treatments are today vagus nerve stimulation and focal stereotactic radiosurgery.

Dörfler and Richter (Erlangen) reported about *"Coiling as First-line Therapeutic Approach in Intracranial Aneurysms"*. The ISAT study, which was presented in 2002 provided evidence supporting endovascular treatment of first choice. After the ISAT study it should be mandatory that all patients should be seen by a neurointerventionalist together with a neurosurgeon to decide whether the cerebral aneurysm is suitable for coiling or not. Dörfler pointed out that aneurysm therapy had changed in the last years, in specialized centers up to 80% of all aneurysms could be treated via the endovascular approach.

Lanner and Spendel (Klagenfurt) reported about *"Microsurgery of Intracranial Aneurysms"*. In these cases in which the endovascular approach is not possible surgical treatment with exact timing and clipping of the ruptured aneurysms is necessary. After bleeding there is an overall high mortality (25%). Vasospasm and delayed ischemia is the most common cause of death and disability following subarachnoid hemorrhage. The prognosis depends on many factors which include age, Hunt and Hess grad preoperatively and location of the aneurysm.

Spendel und Lanner reported at the *47th Pula Congress (2007)* about *"Operative Decompression in Malignant Brain Infarction"*. In this study the patients were treated by large ipsilateral decompressive hemicraniectomy, durotomy and duraplasty with a minimal size of 12–15 cm. It was found that clinical signs, measurement of intracranial pressure and conventional imaging methods are no significant predictors for early recognition of malignant stroke. The combination of perfusion weighted imaging and diffusion weighted imaging – mismatch provides information on the so-called tissue of risk as a time – independent marker for each patient. The study had shown that early decompressive craniotomy for space occupying cerebral ischemia not only reduced mortality but also significantly improved functional outcome and reduced infarction size. Predictors for poor outcome are increased age (above 50), low preoperative Glasgow Coma Score (under 7) and multi-territorial infarction.

Prof. H.E. Diemath, chairman of the University Clinic of Neurosurgery in Salzburg held at the 47th *Pula Congress* (2007) the Academic lecture about *"Medico – Legal Problems in Neurology and Collaborating Disciplines"*.

At the *48th Pula Congress (2008)* Prof. G. Lanner and M.C. Spendel (Klagenfurt) described the *"Neurosurgical Approach to Patients with Vertigo"*. In general vertigo can be caused by either peripheral or central vestibular disorders. In peripheral lesion vertigo is accompanied by nystagmus and very often by auditory symptoms. Besides of neurotological interventions there are three specific neurosurgical operations – the selective vestibular nerve transection, neuromicrovascular decompression and ventriculo–peritoneal shunt. Selective vestibular transection as an ablative procedure was indicated in patients with unilateral peripheral labyrinthine disease, usually Meniere's disease or vestibular neuritis with an excellent hearing level of the effected ear. In some patients with disabling positional vertigo a vascular compression at the root-entry zone of the cochlear-vestibular nerve is detected in the MR-angiography. In these cases, a microneurovascular decompression at the root entry zone, described first by Jannetta is indicated. In cases of perilymphatic fistulas with vertigo, tinnitus and headache an implantation of ventriculo-peritoneal shunt is necessary.

At the *50th International Neuropsychiatric Pula Congress in 2010* Prof. Lanner (Klagenfurt) reported about the *"Development of Neurosurgery in the last 50 Years"*. Already in the prehistoric times skull drillings were performed. Historical findings had shown that these interventions had been performed with great skill and many of patients had survived. These trepanations were performed to treat traumatic brain injuries, in particular depressed fractures but also for the purpose of liberation from evil spirits, such as epileptic seizures and psychiatric diseases. From the 2nd half of 19th century until 1960 nervous processes could be localized by clinical symptoms already and could be treated by surgery. The birthplace of this pioneer time in neurosurgery was Great Britain. W. Macewen excised 1879 an intracranial meningioma in Glasgow, 1884 R. Godlee and AH. Bennett removed a glioma in Regents Park Hospital London and V. Horsley excised an intradural spinal tumor in National Hospital Queen Square London for the first time.

In the last quarter of the 19th century and at the beginning of the 20th century began in Austria to establish neurosurgery as a medical specialist. Famous names of this time are Theodor Billroth, Anton von Eiselberg and Leopold Schönbauer.

The period from 1960 until now in the development of Neurosurgery is the so called modern Neurosurgery. Prof. Lanner reported about progresses in brain tumor surgery, in neurotraumatology, in functional neurosurgery and about the progress in cerebrovascular and endovascular surgery in this time.

The disciple of L. Schönbauer Prof. Herbert Kraus became in Austria 1964 the first chairman of the Vienna University Clinic of Neurosurgery, Fritz Heppner 1971 in Graz, and Karl Kloss 1972 in Innsbruck.

Spendel, Resch and Lanner (Klagenfurt) informed about the "Microvascular decompression *in trigeminal neuralgia*". They pointed out that microvascular decompression is the therapy of choice especially in patients with good conditions. The key to efficient treatment is to establish whether the pain is indeed trigeminal. 286 patients underwent decompression. Immediately after decompression 99%, up to 3 years 96,9%, up to 6 years 95,1% and up to 10 years after operation 92% were painfree.

Resch, Spendel and Lanner (Klagenfurt) reported also about *"Intrathecal pain therapy with Ziconotide'"*. It was shown that Ziconotide is a new intrathecal option to therapy severe chronic pain. Until now morphine and hydromorphone were the first-line intrathecal drugs with implanted pump systems. With ziconotide an effective alternative to the therapy of chronic pain was available for more than 50% of the patients. Important for the successful therapy was the selection of the patients with evaluation of pain reduction and exclusion of severe side effects before definitive pump implantation.

At the *53rd International Pula Congress* (2013) Lanner and Spendel (Klagenfurt) reported about the *"Surgical Procedures in the Treatment of Trigeminal Neuralgia"*. Their own results of microvascular decompression, radiofrequency rhizotomy and stereotactic radiosurgery were discussed. It could be stated that all these described surgical procedures are complementary methods. Operative treatment requires exactly selection of patients and precise technique of operation.

In *2014* at the *54th Pula Congress* Lanner and Spendel (Klagenfurt) pointed in their studies out, that early endovascular or microsurgical *"Treatment of ruptured cerebral aneurysms"* improve the quality of life of the patients. The prognosis of patients who have suffered a aneurysmal SAH depends on several factors which include age, Hunt and Hess grade and location of the aneurysm.

In 2016 at the *56th International Pula Congress,* one of the main theme at the Stroke Symposium was *"Cerebral and Spinal Malformations"*. G.E.Klein (Graz) and M.Mokry (Graz) reported about Neurointervention and Microneurosurgery in neural malformations,

At the *58th Pula Congress 2018*, at the Stroke Symposium M. Mokry (Graz) reported about "*Neurosurgical Aspects in Hemorrhagic Stroke*".

Many eminent and famous neurosurgeons had attended the Pula Congresses. As a substitution for a vast number of eminent and experienced neurosurgeons from several European countries who had attended the Pula Congress, some well known neurosurgical names we have found in the documents at hand, should be named, at least:

Austria: Prof. Heppner, Prof. Mokry (Graz), Prof. Kraup, Prof. Brenner, Prof. Knosp (Vienna), Prof. Kloss, Prof. Grunert, Prof. Kostron (Innsbruck), Prof. Diemath, Prof. Richling (Salzburg), Prof. Lanner, Dr. Spendel, Dr. Resch, Dr. Deinsberger (Klagenfurt)

Croatia: Prof. Jeličić, Prim. Šurdonja, Prof. Iveković, Dr. Pirker (Zagreb).

Slovenia: Prof. Klun, Prof. Dolenc (Ljubljana), Prof. Vidovič (Maribor)

Germany: Prof. Marguth (Munich), Prof. Schürmann, Prof. Pernetzky (Mainz), Prof. Fahlbusch, Prof. Richter, Prof. Dörfler (Erlangen), Prof. Frowein (Cologne), Prof. Mundinger, Prof. Siegfried (Freiburg), Prof. Lorenz (Frankfurt), Prof. von Wild (Münster), Prof. Lanksch (Berlin), Prof. Schmiedek (Mannheim), Prof. Schwab (Heidelberg)

Hungary: Prof. Pasztor, Prof. Vajda (Budapest), Prof. Merei (Pecz), Prof. Bobest (Szombathely).

Belgium: Prof. van den Bergh (Leuven)

This partial review on neurosurgical participation had to present my conviction that the International Neuropsychiatric Pula Congresses, which have this year their 60th Jubilee anniversary have not been only of high scientific quality and of great importance in the fields of Neurology and Psychiatry, but as well as of the collaborating disciplines, especially Neurosurgery. They preserve in their programs the basic ideas of medicine for collaboration of scientists and professionals from different specialties and from various nations, countries, and regions.

Chapter 31
The "Psychopathology Summer School" Within the INPC

Karl Bechter and Martin Brüne

The Course of Psychopathology Summer Schools at Pula Congresses – from the 1st PSS in 2012 to 2022

The International Neuropsychiatric Congress (INPC) has a long-standing tradition of covering both neurological and psychiatric topics in one meeting. The breadth of the scientific presentations at INPC has been innovative ever since. Bridging the gap between the two "brain" disciplines in medicine has recently become not just "modern", but essential to the benefit of both. The 60th anniversary of the INPC took place, as always, at Verudela, a beautiful peninsula in the vicinity of Pula.

On the occasion of the 52th INPC in 2012, the European Psychopathology Summer School (PSS) was included in the programme for the first time. It was established under the patronage of the European Psychiatric Association (EPA), then coordinated by Prof. Henning Saß, Aachen, Germany, Prof. Karl Bechter, Ulm/Günzburg, Germany, and Prof. Francesco Benedetti, Milan, Italy. The topics of the first Pula Psychopathology Summer School (PPSS) included quite diverse contributions, thus, very much in the spirit of the INPC. Karl Bechter spoke about "Differentiating biological and psychological factors by clinical approach", Henning Saß presented "The impact of phenomenology for psychiatric diagnosis", Tilo Kircher's (Marburg, Germany) two talks were entitled "Brain plasticity due to

K. Bechter (✉)
University of Ulm, Department Psychiatry and Psychotherapy II, Günzburg, Germany
e-mail: Karl.Bechter@bkh-guenzburg.de

M. Brüne
LWL University Hospital Bochum, Clinic for Psychiatry, Psychotherapy and Preventive Medicine, Division of Social Neuropsychiatry and Evolutionary Medicine, Ruhr University Bochum, Bochum, Germany
e-mail: martin.bruene@ruhr-uni-bochum.de

V. Demarin et al. (eds.), *Mind, Brain and Education*,
https://doi.org/10.1007/978-3-031-33013-1_31

319

psychotherapy in patients with schizophrenia" and "Introduction of a new scale for formal thought disorder", and Johannes Schröder (Heidelberg, Germany) presented "Assessment and interpretation of neurological soft signs in schizophrenia". The two final talks by Werner Strik (Berne, Switzerland) were about "Symptom dimensions in psychosis: neurobiology and clinical relevance", and by Francesco Benedetti about "Sleep and sleep disturbances". Part of the agenda of the PPSS has been its interactive nature, with opportunities of young clinicians and researchers to interact with the speakers and discuss the different topics in much greater detail than is usually the case in conferences and congresses.

It has to be gratefully acknowledged that the chair of the organising committee of the INPC, Professor Vida Demarin, Zagreb, Croatia, has always been very supportive in including the PPSS in subsequent INPCs. Without her support and encouragement, the PPSS would not have been possible to continue.

In the following year (2013), the PPSS, organised by Karl Bechter and Francesco Benedetti, reached out to cover innovative topics that usually lie outside the mainstream of psychiatry, namely "Evolutionary Psychiatry" presented by Martin Brüne, Bochum, Germany, and "Autonomic nervous system dysfunction in mental illness" by Karl-Jürgen Bär, Jena, Germany.

In 2014, the 3rd European Psychopathology Summer School, organised by Karl Bechter, Francesco Benedetti and Martin Brüne combined an Inaugural Symposium about "Evolutionary Psychopathology" and "Sports therapy in Psychiatry".

More recent PPSS, organised by Karl Bechter and Martin Brüne, accommodated a diverse range of topics, including research into "Social Brain" dysfunction in psychosis and severe personality disorders, immunological issues pertaining to psychopathological conditions, psychotherapy of severe mental illness, and (in 2018) a "Young Researchers Track" with junior clinicians and scientists presenting their projects and findings. Along with establishing the PPSS as an integrative part of the INPC, speakers have become more international. While initially, the PPSS was a German-Italian endeavour, recent PPSS hosted speakers from Austria, Croatia, Germany, Italy, and The Netherlands.

The focus in the forthcoming yearly (except 2020 due to COVID 19 pandemic) PPSSs until 2022 varied including biological underpinnings of psychopathology, e.g. the cerebral correlates of cognitive reserve or lasting environmental consequences of psychological insults onto behavior and altered biology, e.g. the surprising relation between trauma and pain, or the role of the autonomous nervous system and neurological soft signs in schizophrenia, or the role of infections on altered behavior or the role of immune alterations in severe mental diseases, including the recent diagnoses of autoimmune encephalitis and autoimmune psychosis, and not least the various respective treatment approaches, for example immune treatments in biological approaches, also training, and psychotherapeutic approaches in various diseases with detailed discussion of case presentations prepared by participants, also of ethical aspects for example in treatment of refugees.

Thanks to Professor Vida Demarin, the PPSS, for the first time, has been granted an extra time slot to include six speakers on the occasion of the 60th anniversary of the INPC.

As research and science critically depend on curiosity, novelty, and rigour, as well as opportunities to "think out of the box", the future of the PPSS lies in its embeddedness in and inspiration by the general spirit of the INPC. That is, a unique social and natural environment characterised by openness, acceptance, lively discussion about interdisciplinary and/or controversial topics --- wherever appropriate, be it on the huge deck of the "Park Plaza Histria", on the beach, or in one of the beautiful bars and restaurants overlooking the Adriatic Sea.

We wish the INPC, and the PPSS, all the best to thrive and continue to deliver the important messages that clinics and science can go together, that Neurology and Psychiatry need interdisciplinary exchange with one another, and that both research and patient care are highly rewarding challenges that need young people to take on.

Some Thoughts on Desired Future Developments of Psychopathology

On the background of our own research including the experiences with colleagues through the PPSS in Pula Congresses we present some thoughts about desired future developments of psychopathology in modern psychiatry.

The editors of the influential Journal "Psychopathology" stated recently "The convergence of complementary concepts from phenomenology, psychodynamics, experimental psychopathology, neuroscience, and developmental psychopathology corresponds to a reorientation of *Psychopathology* as an integrative platform for scientific exchange." and accordingly structured the journal into four thematic fields: *Phenomenological Psychopathology, Experimental Psychopathology, Neuropsychology and Neuroscience*, and *Developmental Psychopathology and Youth Mental Health Psychopathology* [1]. Such statement demonstrates the broadness and complexity of modern psychopathology research, still searching an improved role beyond the most relevant though rather "solipsistic" diagnostic role even in the most recent diagnostic system s DSM-5 and ICD-11, both having not achieved a direct combination with/inclusion of paraclinical findings including laboratory data, neuroimaging results and other paraclinical methods. Such direct inclusion remains a so-called "clinicians" role. Herein the role of personal presentation and evaluation in exchange with other experienced clinicians and researchers apparently has an important role, which can be realized in psychopathology schools like the Pula summer school.

Beyond such practical approaches, we saw that such personal interactive exchange provides opportunities to integrate new basic approaches like evolutionary psychiatry sights and the new field of immunopsychiatry and respective new emerging treatments into daily practice by discussing single cases in very detail and outcome over the longer term.

Another aspect was the critical discussions to combine traditional psychopathology not least the Kurt Schneider "School", celebrated over years in many Pula Congresses by Schneider's renown pupil Gerd Huber with prominent worldwide influence on schizophrenia research. Such discussions are enhancing the personal insights and critical sights importantly incuding for example relevant later misunderstandings, for example such about Schneider's first-rank symptoms as often discussed by Gerd Huber, recently independently recognized to be incorrectly resp. overinclusive evaluated and partly integrated into the DSM system, having lost validity due to imprecise adoption [2].

Thus personal encounter and place for discussion is an invaluable opportunity at such school like the Pula Psychopathology Summer School.

For the interested reader we include some selected additional references covering only part of the presentations/presenters [3–14].

References

1. Fuchs T, Herpertz SC, Kaess M, Lenzenweger MF. 125 years of psychopathology. Psychopathology. 2022;55(3–4):127–8. https://doi.org/10.1159/000525084.
2. Nordgaard J, Henriksen MG, Berge J, Siersbæk NL. Associations between Self-Disorders and First-Rank Symptoms: an empirical study. Psychopathology. 2020;53(2):103–10. https://doi.org/10.1159/000508189.
3. Wagner G, Herbsleb M, de la Cruz F, Schumann A, Köhler S, Puta C, Gabriel HW, Reichenbach JR, Bär KJ. Changes in fMRI activation in anterior hippocampus and motor cortex during memory retrieval after an intense exercise intervention. Biol Psychol. 2017;124:65–78.
4. Vai B, Serretti A, Poletti S, Mascia M, Lorenzi C, Colombo C, Benedetti F. Cortico-limbic functional connectivity mediates the effect of early life stress on suicidality in bipolar depressed 5-HTTLPR*s carriers. J Affect Disord. 2020;263:420–7.
5. Ramseyer F, Ebert A, Roser P, Edel MA, Tschacher W, Brüne M. Exploring nonverbal synchrony in borderline personality disorder: a double-blind placebo-controlled study using oxytocin. Br J Clin Psychol. 2019;59:186.
6. Poletti S, Melloni E, Mazza E, Vai B, Benedetti F. Gender-specific differences in white matter microstructure in healthy adults exposed to mild stress. Stress. 2020;23(1):116–24. https://doi.org/10.1080/10253890.2019.1657823.
7. Pazen M, Uhlmann L, van Kemenade BM, Steinsträter O, Straube B, Kircher T. Predictive perception of self-generated movements: commonalities and differences in the neural processing of tool and hand actions. NeuroImage. 2020;2020:206.
8. Maxeiner H-G, Marion Schneider E, Kurfiss S-T, Brettschneider J, Tumani H, Bechter K. Cerebrospinal fluid and serum cytokine profiling to detect immune control of infectious and inflammatory neurological and psychiatric diseases. Cytokine. 2014;69(1):62–7. https://doi.org/10.1016/j.cyto.2014.05.008.
9. Markser VZ. Sport psychiatry and psychotherapy. Mental strains and disorders in professional sports: challenge and answer to societal changes. Eur Arch Psychiatry Clin Neurosci. 2011;261(Suppl 2):182–5.

10. Kong L, Herold CJ, Cheung EFC, Chan RCK, Schröder J. Neurological soft signs and brain network abnormalities in schizophrenia. Schizophr Bull. 2020;46(3):562.
11. Brüne M. Textbook of evolutionary psychiatry and psychosomatic medicine: the origins of psychopathology. 2nd ed. Oxford: Oxford University Press; 2016.
12. Bechi M, Spangaro M, Agostoni G, Bosinelli F, Buonocore M, Bianchi L, Cocchi F, Guglielmino C, Bosia M, Cavallaro R. Intellectual and cognitive profiles in patients affected by schizophrenia. J Neuropsychol. 2019;3(13):589–602.
13. Bär KJ, Markser VZ. Sport specificity of mental disorders: the issue of sport psychiatry. Eur Arch Psychiatry Clin Neurosci. 2013;263(Suppl 2):205–10.
14. Bechter K. Interactions of biological and psychological factors in psychopathology – a remaining challenge. Neurol Psychiatry Brain Res. 2011;17(2):37–8. https://doi.org/10.1016/j.npbr.2011.06.001.

Chapter 32
'Mind & Brain' International Neuropsychiatric Pula Congress - Reflections on 60 Years of Tradition

Vida Demarin, Hrvoje Budinčević, Darko Marčinko, Osman Sinanović, Anton Glasnović, Vladimir Rendevski, and Ninoslav Mimica

The Role of Psychiatric and Psychodynamic Factors in the Course of Psychiatric Symposiums at INPC Pula (Darko Marčinko)

The objective of psychiatric presentations and investigations during a series of INPC psychiatric symposiums has been the role of psychiatric and psychodynamic factors in the course of different psychiatric disorders. Due to huge mental health risks, the number of people who need psychiatric help is going to continually increase due to fear of uncertainty, death, loss of job, drastic changes in lifestyle, stigmatization, isolation, separation from family and beloved persons, etc. New knowledge has led to more holistic treatment in psychiatry and psychological medicine. It is very important to improve public awareness of prevention and intervention strategies by providing daily updates about surveillance and active cases on social media. A psychodynamic understanding of human pain and psychological problems should be

V. Demarin (✉)
International Institute for Brain Health, Zagreb, Croatia

H. Budinčević
Department of Neurology and Neurosurgery, Faculty of Medicine, J.J. Strossmayer University of Osijek, Osijek, Croatia

Department of Neurology, Sveti Duh University Hospital, Zagreb, Croatia

D. Marčinko · A. Glasnović · N. Mimica
School of Medicine, University of Zagreb, Zagreb, Croatia

O. Sinanović
University of Tuzla, Tuzla, Bosnia and Herzegovina

V. Rendevski
University Clinic for Neurosurgery, Medical Faculty, Ss. Cyril and Methodius University in Skopje, Skopje, North Macedonia

© The Author(s), under exclusive license to Springer Nature Switzerland AG 2023
V. Demarin et al. (eds.), *Mind, Brain and Education*,
https://doi.org/10.1007/978-3-031-33013-1_32

useful in the process of communication between health professionals and the general population. Psychodynamic approaches to working with problematic behaviour focus on several different factors that may influence or cause symptoms, such as early childhood experiences (particularly level of attachment to parents), current relationships and the things people do (often without being aware of it) to protect themselves from upsetting thoughts and feelings that are the result of experiencing a stressful event (defence mechanisms). Unlike cognitive-behavioural theory, psychodynamics places a large emphasis on defence mechanisms and the unconscious mind, where upsetting feelings, urges, and thoughts that are too painful for us to directly look at are housed. Even though these painful feelings and thoughts are outside of our awareness, they still influence our behaviour in many ways. Psychological resilience is the ability to mentally or emotionally cope with a crisis or to return to pre-crisis status quickly. Public and global mental health promotion may be predicated on the theory of salutogenesis and three key interrelated terms: empathy, coherence and resilience (series papers of Miro Jakovljević and Darko Marčinko).

Our basic psychodynamic investigation is related to the possible relationship between pathological narcissism and suicidal behaviour linked to the complex emotion of shame. Our study examined the relationships between pathological narcissism, shame and suicidal symptoms in psychiatric outpatients. Consistent with earlier investigations pathological narcissism exhibited a stronger relationship with acute mental distress manifested as suicidal symptoms. Body and character shame proved to be a significant mediator of this association, while the mediating effect of behaviour shame was not confirmed. This investigation also addresses how and why shame can generate an internal subjective experience that evokes suicidal behaviour in individuals with pronounced pathological narcissism. Narcissistic rage and shame-based aggression can lead to suicide if turned against the self. Kernberg described suicide associated with malignant narcissism as a vehicle of omnipotent wishes for sadistic control. Shame is the veiled companion or underside of narcissism. The inability to express shame is an important factor of vulnerable ego and is frequently related to pathological narcissism. Reluctance or inability to share feelings and thoughts of shame with others can force loneliness and isolation and predispose the suicidal process. Our results should be explained by the fact that pathological narcissism accompanied by body and character shame led to affective dysfunction, manifested as depression which can lead to suicidal behaviour. Suicide is a way to reassert pride and reclaim control in narcissistic patients. Envy (frequently presented in narcissistic patients) may lead to shame as it implies inferiority of the subject and superiority of the object of envy. Shame emphasizes weakness, vulnerability, and the likelihood of rejection— so much so that its acknowledgement often generates more shame which frequently leads to suicidality. Our results confirmed the theory of Lansky, that tolerance of shame goes beyond the tolerance of the pain of the effect per se and should be considered also as unbearable mortification and a signal of incipient social annihilation and auto-destruction. Regarding the clinical implications of our results, we may emphatically or intuitively detect and clarify shame and label it as such, which may be helpful when working with difficult patients such as narcissistic or suicidal. Assessing the risk of suicide in narcissistic patients can be challenging for the clinician.

In a series of the book published by Medicinska naklada and presented during INPC psychiatric symposiums (Suicidology, 2011; Narcissistic Personality Disorder: Diagnostic Contribution, 2013; Eating Disorders: From Understanding to Treatment, 2013; From Violence to Dialogue, 2014; Mourning, 2014; Personality Disorders: Real People, Real Problems, 2015; Psychoanalytic Models of Communications, 2016; Hysteria, 2017; Civilization and Its Discontents in 21st Century- psychodynamic approach, 2018; Psychodynamic of Love and Hate; 2019), our team emphasized the link between psychopathology of personality with different and difficult personality and eating disorders patients.

In Pula, We Make Bridges Between Neurology and Psychiatry, Between Established Scientists and Young Ones (Osman Sinanović)

Pula neuropsychiatric congresses have always been a place of good atmosphere and friendship, a place of possibilities for contacts, and an exchange of experience and ideas. At the same time, Pula meetings have made very specific bridges between neurology and psychiatry. These bridges, today when we have a lot of narrow fields in both, neurology and psychiatry, are very important.

In the last 10 years Pula international neuropsychiatric congresses were guided by academician Vida Demarin in the role of the secretary general. She introduced a new symbolic title for the Pula meeting – Mind and Brain. With her enthusiasm and energy Pula, as the oldest medical European meeting is still a very important and „young" international meeting in the field of neurology and psychiatry with participants not only from the Croatia and region but also from Italy, Austria, Hungary, Russia, Greece etc.

In the last 10 years, I tried to extend one of the main goals of the member of the kuratorium (scientific advisory board/board of trustees) – the stimulation of young colleagues from my former Departments of Neurology and Psychiatry (University of Tuzla) to come to Pula with good posters (two of them, Biljana Kojić and Sanela Zukić were winners in neurology), and to be on disposal to our Vida for the Pula good scientific and friendly spirit.

For this 60 years of International Neuropsychiatric Pula Congresses Memorial Book, according to the suggestion of academician Vida Demarin a prepared one review from the Mind and Brain field with tittle Functional neurological disorders. They may also be called psychogenic, non-organic, somatoform, dissociative or conversion symptoms. The most common FNSs are non-epileptic attacks and functional weakness. These are common in neurology and general medical practice, especially in emergencies, where they can be mistaken for epilepsy, stroke or another neurological disease. These disorders are typical neuropsychiatric disorders and a typical bridge between neurology and psychiatry, but unfortunately very often in practice no man land.

I am sure that the Pula meeting will be important and successful in the following years.

New Contents of the Congress and Impressions from the Position of Participants and Organizers (Anton Glasnović)

The Neuropsychiatric Congress in Pula has always had a special place among neurologists. The very mention of that congress aroused the feelings of an intimate meeting of world-renowned experts who met in the beautiful ambience of Pula and discussed the most serious neurological and psychiatric topics. Precisely for this reason, this congress has stood the test of time and is one of the longest-running in the region. Few similar gatherings can be proud of such a long uninterrupted experience, but this one can be proud also of simultaneous concentration of knowledge, experience and a relaxed atmosphere.

An important feature of this gathering is the presence of young doctors and scientists, who with their energy give new value to the whole event. They got a lot of knowledge and experience from their mentors, and in return, they gave them vitality, so it is always a real pleasure to see how from year to year young people become more experienced, and how older people get more energy and will to transfer their knowledge.

All this would not have been possible without academic Vida Demarin, who with her energy, knowledge, charm and experience, and most of all with her undisputed authority, managed to preserve this congress for so many years. Also, without her associates, of which the most prominent is Assoc. Prof. Hrvoje Budinčević, this congress would be much more difficult to run due to a large number of participants, but also the large number of lectures that sometimes have to take place in parallel. Nevertheless, despite all the resistance that can arise from the organization of such a project, academician Demarin and her team manage to surprise every year with new and fresh content and presentation of the latest scientific achievements to the audience and participants of the congress.

One encounter managed to bring additional novelty and freshness to this congress. As I remember it was in 2015 when I approached academician Demarin on my initiative and offered her cooperation. We didn't know each other best then, but we immediately saw that we were thinking similarly, and she gladly accepted my proposal to add this new value to Congress. Also, on my initiative, in addition to several working versions, the Congress was given a new name, which is now "Mind and Brain", since academician Demarin recognized the most modern aspirations in the neuroscience community.

As early as the following year, 2016, psychoanalysis appeared in congress for the first time. Until then, the focus was predominantly on neurology and psychiatry, and psychoanalysis was only occasionally part of the sessions regarding psychotherapy. However, given the recent use of the term neuro psychoanalysis and research conducted in this field, the Congress in Pula is one of the few "mainstream" neuropsychiatric congresses that felt that this aspect of brain and mind research is necessary to fully cover the field of mind and brain. By thinking about who would be the best speaker on this topic and who would introduce Congress to a new era, we simply decided on the best, that is, the one who coined the term - Prof. Mark Solms. His engagement at the Congress heralded a new era, and mainstream neurologists, psychiatrists, and psychoanalysts listened enthusiastically to all of his lectures that year. His charm, but also his knowledge and experience in conveying information to the listener were at a top level, and the interest of the audience was astounding. The great lecture hall was full! It was followed by other lecturers on the topic of the connection between the psyche and the brain as an organ, and finally, a discussion developed which showed how in a pleasant and uncompetitive atmosphere one can come to the knowledge that is the cutting edge of current science.

In 2015., as an introduction to this new moment of the congress, and as an introduction to the fruitful cooperation with the Croatian Medical Journal, an essay entitled "Psychoanalysis has its place in modern medicine, and neuro psychoanalysis is here to support it" by Anton Glasnović, Goran Babić and Vida Demarinwas published. In it, we have given a short overview of how the time has come in which psychoanalysis is once again becoming a scientific discipline. At the neuro psychoanalytic session that year, it was shown that only when cooperating, and working together, we can achieve our goal of embracing the mind and brain as a whole and not just two different and separate fields of research...

On the other hand, psychoanalysts themselves, who do not have their congress in Croatia, but only occasional gatherings, have found a haven here and a place where they could exchange experiences with each other, and also with professionals that are complementary to them. Psychoanalysis is a profession in which psychoanalysts are mostly in their session rooms, with their patients and after a long day find it difficult to find time to socialize with colleagues. And when they relax, they tend to get away from psychoanalysis. This 2016., almost all Croatian psychoanalysts appeared at the Congress, as well as candidates and educators, which is another proof of its quality and organization. The scene in which psychoanalysts sit at the same table with other medical professionals was simply stunning.

The following 2017, Mark Solms came again as one of the key speakers and presented his concept of evidence that neuro psychoanalysis is indeed a science, and not just something other than science, as Karl Popper claimed. Again, it was interesting to hear how this new scientific discipline came about, from the mid-1990s until today when neuro psychoanalysis has become a relevant scientific field. Likewise, it was interesting to see how many medical professionals other than psychoanalysts participated in that particular session.

Furthermore, the Croatian Medical Journal proved to be a good partner to the Congress that added another value to the congress, and that is the session within the Congress entitled "How to read and write the scientific paper". From the beginning, this session offered participants very valuable knowledge and tools for writing scientific articles. And what is more, this knowledge and tools were given by those who in their everyday practice review and prepare scientific manuscripts for publication. Over the years, the almost entire editorial staff of the Croatian Medical Journal has participated in this session, from language editor Antonija Paić, and manuscript editor Hrvoje Barić, to editors-in-chief Srećko Gajović and Anton Glasnović. Topics covered ranged from IMRD article structure, statistical methods, study design, open questions related to scientific publishing and the like. The topics of predatory journals, databases in which it is published and differences in the conditions for obtaining scholarships and grants in different countries based on published scientific articles were also covered. The hall where the session usually takes place was slightly smaller, but even more intimate, so the participants could ask all possible questions that interested them without shame and any bad feelings. Regular discussions on certain topics continued even after the end of the scheduled time for the session.

From the axis of academics Demarin and myself, another part of the congress has developed that had not been represented to that extent until then, and that is the neuroscientific session related to neuroinflammation. This part was led and organized by Prof. Srećko Gajović and I am sure that a few more words will be said about it in other sections. But let's just mention that in that part of the Congress in 2019, when it was first held, the world's greatest experts on experimental stroke participated, primarily from Germany.

But after the professional parts of the congress were over and after the participants and lecturers had a little rest, an equally important part would follow, and that is a social gathering. This part is perhaps the most important part of the whole Congress because in an informal atmosphere one could continue to discuss all the important things that could not be discussed in the official space. Cooperation on future projects was agreed upon here, which were presented at the official part of the program the following year. Here, experiences from clinical settings were exchanged and new insights into the hitherto unclear matter were gained. Also, after such social gatherings, he would return home empowered and refreshed.

All of the above is the result of a long tradition and sufficient robustness of the Congress to add some new or some old issues to its already existing contents (although this is a rare case). Therefore, in the hope that the Congress will continue to grow, or at least remain to this extent, I would like to congratulate all the collaborators and organizers of this great and valuable international gathering on its 60th anniversary.

Pula and My Memories (Vladimir Rendevski)

The international Neuropsychiatric congress „Mind & Brain", held in Pula, Croatia, is one of the symposia with the oldest tradition in the Balkan area. It has a multidisciplinary approach, focusing on bridging between neurology and psychiatry. The congress is supported by The World Federation of Neurology and the European Academy of Neurology. The congress is also accredited according to the Regulations of the European Accreditation Council for Continuing Medical Education (EACCME®) and the Croatian Medical Chamber. Every year, it comprises around 400 highly reputable lecturers from all over the world, allowing seeing the newest achievements of their expertise.

Due to these reasons, the International Neuropsychiatric congress „Mind & Brain" have become also a tradition for me and my working group (comprising of M-r Boris Aleksovski, molecular biologist and, Dr Ana Mihajlovska Rendevska, radiologist), and we try to attend it every year, showing our results from our last studies. Last year, at the 59th Mind & Brain congress, 2019, we also received the award for second place for the best scientific poster on the topic "Advanced 3D Modeling For Prediction And Quantification Of The Perihematomal Brain Edema Formation After Intracerebral Hemorrhage: Implications Of Biochemical, Radiological And Clinical Variables". This served as a motivation for us to continue with our work and present ourselves with even better studies for this year's congress.

We are looking forward to attending this year's 60th anniversary of Mind & Brain.

Pula in My Mind (Ninoslav Mimica)

When I was asked by Professor Vida Demarin to send a contribution for this Anniversary Book, I was thinking (for a few moments) about what kind of contribution would be the most appropriate to write. As I am two years younger than INPC I was not able to speak about how it all started, and as I was not a part of Kuratorium, either not a neuropsychiatrist I couldn't tell something from that view. Although there was a kind possibility to write a psychiatric article or publish some piece of research, I choose the possibility to share with you what Pula Congress means to me, and why is it so.

Let's go back to 1994, to my first visit to Pula Symposium. At that time, I was in my last year of psychiatry specialization, having already 5 years of clinical experience working with psychiatric patients in Psychiatric Hospital Vrapče, Zagreb, Croatia. My boss and mentor was Prof. Vera Folnegović-Šmalc, neuropsychiatrist involved in many research projects, including genetic study in collaboration with American scientists. In connection to this, I have already been in the USA and was trained at Columbia University in New York. So, in the Workshop at 34th INSP entitled: "Genetic aspects of psychiatric diseases", I give a presentation on "The importance of homogeneous groups in psychiatric genetic research" [1]. Besides this, I was also co-author of a poster named "The application of clomipramine in the therapy of resistant depression" [2, 3].

Although I have already been active in several international Congresses [4–10] e.g. this was not my first active presentation at an international professional meeting, it was a great surprise for me that such kind of international high-quality activities was performed on our ground for so many years. Immediately I promise myself to come next year, and every other year, because I realise that this was one of the oldest professional meetings in Europe, in which I can in my home country Croatia meet experts from European countries and discuss various topics. I admire the fact that, despite the war against Croatia, the idea of meeting neurologists and psychiatrists together never stops. The 32nd Neuropsychiatric Symposium took place in Graz, Austria in 1992, but the next year (1993) came back to Pula, Croatia [11]. Next year, as the war was still present in Croatia, we present the poster about the occurrence and intensity of depression in displaced persons and refugees with PTSD [3, 12]. One year later, in 1996, after 5-years of wartime, we try to change the topic and present the poster on patients with eating disorders and treatment with antidepressants [3, 13]. One year later I came also to Pula and present a poster about the pharmacoepidemiologic study of antimanic agents [3, 14]. At that time, the main venue for Pula Congress was still Hotel Brioni. All participants enjoy this accommodation very much, but we all know that due to more participants and parallel sessions we will in the future need to move to another, bigger hotel.

After four years in a row of attending the INPC, I skip two years and come back in the year of the second millennium with a presentation about some bioethical aspects of genetic research in psychiatric patients [15]. On 42nd INPC Prof. Vera Folnegović-Šmalc and I gave a presentation about the tolerability and safety of novel antipsychotics [16, 17]. And then in 2003 group from Vrapče present a poster on extrapyramidal symptoms in patients treated with typical and atypical antipsychotics [3, 18].

I have had a big gap of 8 years in attending the Pula Congres, and I came back in 2011 with forensic aspects of informed consent in Alzheimer's and pharmacoeconomic aspects of memantine treatment. Perhaps the reason for not visiting Pula sooner lies in the fact that I change my field of work, from schizophrenia to Alzheimer's, and need some time to recognise that INPC is the perfect match for bridging people from neurology and psychiatry in dementia research and clinical practice [19, 20]. In 2012 I gave a lecture on the evaluation of testamentary capacity in deceased older persons [21].

I was honoured to organise 2018, on 58th INPC a Mini-symposium on the topic of dementia [22, 23], and in 2019 a satellite on the connection between depression, dementia and delirium [24, 25]. This year, on the 60th Anniversary of INPC I will organise and speak about contemporary hospital treatment and care for people with dementia in Croatia [26].

In conclusion, I discovered Pula Congress 26 years ago, while I was still specialization in psychiatry. In the first part of my career, I was studying schizophrenia, and psychoses and that was my field of interest which I was observing and presenting in Pula. But, when I skipped to the field of dementia, there was no need to change the conference, because Pula was also the perfect platform for this field. All together now is my twelfth visit to the Pula meeting, till now I have been participating in 17 presentations, either poster or oral. Pula was always for me the place to remember, a great opportunity to present my work, meet colleagues from all around and discuss various topics. I admire, love and recommend INPC, and will come again for sure. See you all in Pula.

References

1. Mimica N. The importance of homogeneous groups in psychiatric genetic research. Neurol Croat. 1994;43(Suppl 1):27–8.
2. Rušinović M, Kozarić-Kovačić D, Mimica N, Makarić G, Folnegović-Šmalc V. The application of clomipramine in the therapy of resistant depression. Neurol Croat. 1994;43(Suppl 1):56.
3. Mimica N, Jukić V (ur). Knjiga postera stručnjaka psihijatrijske bolnice Vrapče 1978–2006. Zagreb: Psihijatrijska bolnica Vrapče; 2006.
4. Vojnić-Zelić D, Krapac L, Mimica N. Gonarthrosis in a sample of the older population. First European Conference on the Epidemiology of Rheumatic Diseases. Dubrovnik, Yugoslavia, 16–19 September 1990. Period Biol. 1990;92(Suppl 2):38.
5. Folnegović-Šmalc V, Folnegović Z, Kozarić-Kovačić D, Mimica N, Makarić G. 30-year follow-up of a representative sample of hospitalized schizophrenic patients in Croatia. VIth Biennial European Workshop on Schizophrenia. Badgastein, Austria, January 26–31, 1992. Abstracts, pp. 2.
6. Folnegović Šmalc V, Kittner B, Makarić G, Mimica N. The use of measurement scales to diagnose and assess the psychic condition of pentoxyphiline-treated MID cases. 15. Donausymposion für Psychiatrie. Regensburg, Germany, 7–10 October 1992. Abstracts, pp. 130.
7. Folnegovic-Smalc V, Kittner B, Makaric G, Mimica N, Rößner M. Pentoxifylline treated MID patients; Results of a European double-blind placebo-controlled multicentre study. 26th Danube Symposium for Neurological Sciences. Innsbruck, Austria, September 30–October 2, 1993. Abstracts, pp. 19.
8. Folnegović Z, Hrabak-Žerjavić V, Folnegović-Šmalc V, Mimica N. Hospital incidence of schizophrenia in Croatia over a 25-year period. VIIth Biennial European Workshop on Schizophrenia. Les Diablerets, Switzerland, 23–28 January, 1994. Schizophr Res. 1994;11(2 - special issue):99–100.
9. Mimica N. Catatonic schizophrenia in a representative sample of schizophrenics in Croatia. VIIth Biennial European Workshop on Schizophrenia. Les Diablerets, Switzerland, 23–28 January, 1994. Schizophr Res. 1994;11(2 - special issue):100.
10. Folnegović-Šmalc V, Kozarić-Kovačić D, Mimica N, Makarić G, Folnegović Z. Family care of schizophrenics. 7th European Symposium of the Association of European Psychiatrists (AEP) Section Committee "Psychiatric Epidemiology and Social Psychiatry". Quality of Life and Disabilities in Mental Disorders. Vienna, Austria, April 7–9, 1994. Abstracts, pp. 157.
11. Barac B, Demarin V. 50 years – International Neuropsychiatric Pula Congresses Memorial Book 1961 – 2010. Zagreb: Udruga za neuropsihijatriju; 2010.
12. Folnegović-Šmalc V, Henigsberg N, Jukić V, Makarić G, Mimica N. Occurrence and intensity of depression in displaced persons and refugees with PTSD. 35th International Neuropsychiatric Symposium. Pula, Croatia, 14th–17th June 1995. Neurol Croat 1995;44(Suppl 1):47.
13. Folnegović-Šmalc V, Folnegović Z, Makarić G, Mimica N, Škrinjarić LJ, Rušinović M, Kocijan-Hercigonja D. Antidepressives in the treatment of patients with eating disorders. 36th International Neuropsychiatric Gerald Grinschgl Symposium in Pula. Pula, Croatia, 5th–8th June 1996. Neurol Croat. 1996;45(Suppl 2):57–58.
14. Makarić G, Folnegović-Šmalc V, Mimica N, Rušinović M. Antimanic agents: a pharmacoepidemiologic study. 37th International Neuropsychiatric Gerald Grinschgl Symposium in Pula. Pula, Croatia, May 28–June 1, 1997. Neurol Croat. 1997;46(Suppl 1):46.
15. Folnegović-Šmalc V, Uzun S, Mimica N. Genetic research in psychiatric patients - some bioethical aspects. 40th International Neuropsychiatric Pula Symposium. Pula, Croatia, June 21–24, 2000. Neurol Croat. 2000;49(Suppl. 2):23–24.
16. Mimica N, Folnegović-Šmalc V. Tolerability of novel antipsychotics – similarities and differences. 42nd International Neuropsychiatric Pula Symposium (INPS). Pula, Croatia, May 29–June 1, 2002. Neurol Croat. 2002;51(Suppl 1):104.

17. Folnegović-Šmalc V, Mimica N. Safety – are there any differences among novel antipsychotics. 42nd International Neuropsychiatric Pula Symposium (INPS). Pula, Croatia, May 29–June 1, 2002. Neurol Croat. 2002;51(Suppl 1):104.
18. Folnegović-Šmalc V, Jukić V, Uzun S, Kozumplik O, Mimica N, Makarić G, Mimica N, Rušinović M. Extrapyramidal symptoms in patients treated with typical and atypical antipsychotics. 43rd International Neuropsychiatric Pula Symposium (INPS). Pula, Croatia, 18 – 21 June, 2003. Neurol Croat. 2003;52(Suppl 2):90.
19. Mimica N. Inform consent for clinical trial obtained from person with Alzheimer's disease – forensic aspects. 51st International Neuropsychiatric Pula Congress (INPC), Pula, Croatia, June 15–18, 2011. Acta Clin Croat. 2011;50(Suppl 2):78.
20. Benković V, Mimica N, Stevanović R. Pharmacoeconomic modelling of Alzheimer's disease – asssessment of memantine in treating moderate to severe Alzheimer patients. 51st International Neuropsychiatric Pula Congress (INPC), Pula, Croatia, June 15–18, 2011. Acta Clin Croat. 2011;50(Suppl 2):104.
21. Mimica N. Evaluation of testamentary capacity in deceased older persons. 52nd International Neuropsychiatric Pula Congress (INPC), Pula, Croatia, June 20–23, 2012. Acta Clin Croat. 2012;51(Suppl 1):33.
22. Mimica N. Why Alzheimer's disease became World's health priority? 58th International Neuropsychiatric Congress, Pula, Croatia, May 25–27, 2018. Abstract Book 2018, pp. 32.
23. Kušan Jukić M, Mimica N. The role of psychiatrist during the progression of Alzheimer's disease. 58th International Neuropsychiatric Congress, Pula, Croatia, May 25-27, 2018. Abstract Book 2018, pp. 35.
24. Mimica N. Depression, dementia, and delirium - continuum or comorbidity. 59th International Neuropsychiatric Congress, Pula, Croatia, May 30–June 2, 2019. Abstract Book 2019, pp. 75.
25. Kušan Jukić M, Mimica N. Pathophysiology, clinical picture, and treatment of delirious state in persons with dementia. 59th International Neuropsychiatric Congress, Pula, Croatia, May 30–June 2, 2019. Abstract Book 2019, pp. 78.
26. Mimica N. Contemporary hospital treatment and care for people with dementia in Croatia. Mind&Brain – 60th International Neuropsychiatric Congress. Main Theme: Bridging between Neurology & Psychiatry. Pula, Croatia, May 28–31, 2020. Abstract Book 2020 (in press).

Chapter 33
'Mind & Brain' International Neuropsychiatric Pula Congress – The Event Which Direct Future Career

Hrvoje Budinčević, Filip Derke, Luka Filipović-Grčić, Ana Sruk, Franka Rigo, Ana Filošević, Milan Petrović, Rozi Andretć Waldowski, Maida Seferović, Petra Črnac Žuna, Filip Mustač, and Vida Demarin

Youth at International Neuropsychiatric Pula Congress (Luka Filipović-Grčić)

There can be no talk about the youth and the future if the past is forgotten and discarded as something useless. Even if modern problems require modern solutions, the past can at least offer consolation and reinforce the determination for success by

H. Budinčević
Department of Neurology and Neurosurgery, Faculty of Medicine, J. J. Strossmayer University of Osijek, Osijek, Croatia

Department of Neurology, Sveti Duh University Hospital, Zagreb, Croatia

F. Derke (✉)
Department of Neurology, Dubrava University Hospital, Zagreb, Croatia

Department of Psychiatry and Neurology, Faculty of Dental Medicine and Health, JJ Strossmayer University of Osijek, Osijek, Croatia
e-mail: filip@mozak.hr

L. Filipović-Grčić
University Hospital Center Zagreb, Department of Radiology, Zagreb, Croatia

A. Sruk · M. Seferović · P. Črnac Žuna
Department of Neurology, Sveti Duh University Hospital, Zagreb, Croatia

M. Seferović
Department of Neurology, Zabok General Hospital and Croatian Veterans Hospital, Zagreb, Croatia

F. Rigo · A. Filošević · M. Petrović · R. Andretć Waldowski
University of Rijeka, Rijeka, Croatia

V. Demarin
International Institute for Brain Health, Zagreb, Croatia

F. Mustač
Department of Psychiatry, University Hospital Centre Zagreb, Zagreb, Croatia

V. Demarin et al. (eds.), *Mind, Brain and Education*,
https://doi.org/10.1007/978-3-031-33013-1_33

reminding us of the obstacles it had overcome. The International Neuropsychiatric Congress in Pula with its 60 year long tradition represents the oldest Croatian international congress. Its tradition dates back to the years following the Second World War. The foundations were laid already in 1951, during the encounter of Austrian neurologist Gerald Grinschgl and Croatian neurologist Arnulf Rosenzweig. Their joint efforts resulted in what was to become the first symposium, and then called Weekend Symposium of the Neuropsychiatric Society of the University of Graz. This first symposium held in Pula in 1961 fostered international collaboration among neuropsychiatrists, however, it brought together the representatives of only three countries, namely, Austria, West Germany and Croatia (Yugoslavia). By today standards, it may seem exactly as insignificant as a weekend symposium of a single Austrian society sounds; however, one has to remember that it was a time only 15 years after the greatest and most destructive conflict in human history, a conflict in which the homelands of the participants were on opposing sides. Moreover, all three countries saw their own rebirths following the war as Austria declared a second republic in 1945 and regained full independence in 1955, West Germany reestablished its sovereignty in 1949 following the allied occupation with East Germany remaining within the Eastern Bloc, Yugoslavia, in turn, was a communist state which founded the Nonaligned Movement in 1961 as a way of strengthening its relations with western and non communist countries. The idea of a united Europe was still in its infancy; only four years had passed since the establishment of the European Economic Community. It took great patience and tact to organize this kind of event in an era of financial and travel restrictions and omnipresent political distrust. As the years went by, the symposium grew, attracting more and more experts from different countries and reaching over 1000 participants before the war in Yugoslavia. It continued through and despite the war. In 2004 it was renamed and became the International Neuropsychiatric Pula Congress (INPC) [1]. Throughout the 60 years the congress remained committed to its aims and ideals, paying close attention to the signs of time, constantly adapting and reinventing itself, both in form and content. Following this tradition, in 2016 the first student session was organized, providing a frame for cooperation and familiarization between medical students coming from different universities. The students, under- and postgraduate alike, as well as residents, have always been welcome at the INPC, usually taking part in poster sessions, but also in various other forms. Especially worth mentioning is the Young Psychiatrists section, which gathers psychiatry residents to discuss hot topics and exchange experiences. This chapter will focus on students as the youngest participants at the INPC.

The medical profession has always demanded that its protagonists be not only professionals, but also researchers and scientists. In the light of this tradition, medical students all around the world are encouraged to broaden and update their knowledge through medical journals and other publications, a task that has been greatly simplified with the spread of the internet in the last 20 years. They are also expected to design, produce and present scientific works with help from their mentors. They may present their work on special student congresses, symposia, national or international congresses. Unfortunately, to my knowledge, the difference between the above-mentioned events, their significance, rules of conduct and work preparation are not taught at any point during medical school education in Croatia. The INPC, as one of only few international congresses in Croatia, presents an excellent chance to experience and learn about the rules, functioning and opportunities offered by such event. The Kuratorium has proven very forthcoming in securing special reduced fees for the students, encouraging them to make another step closer to becoming full members of the scientific community. However, congress is not just about lectures and presentations, but provides an opportunity for networking with colleagues of different ages and specialties and learning from them in a less formal context. Another appealing aspect of the INPC is the fact that it thematises neurology, psychiatry and neuroscience, topics interesting not only to medical students, but to a large array of other future professionals ranging from artists to educational rehabilitators. They too have expressed interest and took part in INPC on different occasions in order to expand their knowledge of neuroscience and its potential application in their respective fields. It is thanks to the INPC that young people of so diverse orientations have the opportunity to discuss neuroscience from its theory to practical implementation. They are hereby provided with a setting for learning from one another, for presenting their work and testing what they know thus far. Provided they don't forget what they learnt there, this kind of sessions may bring about fruitful collaboration on projects and ultimately yield deeper understanding and more effective cooperation between the future professionals and their respective fields. Let us therefore hope this tradition will live on and continue to attract new, enthusiastic young people ready to contribute to this atmosphere of humanism and mutual respect.

The INPC played yet another important role in strengthening the ties between student societies for neuroscience from four Croatian schools of medicine, namely Zagreb, Rijeka, Split and Osijek. In every medical school in Croatia, student societies for neuroscience count among the oldest, largest and most prolific societies. They organize numerous lectures and workshops, attracting large numbers of students from other branches of the university. The central event every year is the Brain awareness week, during which the students and their respective institutions organize public lectures and exhibitions, inviting and welcoming the general public in their facilities with the aim of improving the public understanding of the brain, its health and disorders. Societies also have their individual projects, such as Student Congress of Neuroscience NeuRi, hosted by Student society for neuroscience from Rijeka, congress Practical Knowledge for Students, held by Student society for neuroscience from Split, and Gyrus Journal, published by Student society for neuroscience from Zagreb. The members of the societies are constantly encouraged to develop their writing and presentation skills, initially through student congresses and journals, and afterwards on international congresses and scientific journals. Working within the frames of their schools of medicine, the societies may sometimes find themselves a bit isolated and lacking information about one another. Therefore, the INPC represents a great opportunity to meet a great deal of students' needs. It enables the representatives of societies to meet and share ideas for future projects, to solidify mutual support, arrange student exchange and to present their own achievements and scientific works at the congress. Furthermore, at the INPC they have a chance to meet internationally renowned scientist and experts and invite them to give talks and hold lectures at their universities. Surely, all of them cherish the times spent in Hotel Histria Plaza as a dear and lasting memory.

A congress with a tradition so long and rich as the INPC often inspires the younger generations to draw inspiration from that tradition and build upon it. One such case is the association between art and neuroscience. Indeed, the interest neuroscientists have in different art forms, their impact on brain health and their neurophysiological roots reflected on this congress as well, thus „Psychiatry and Arts "was the topic of the symposium already in 1962. Following this tradition, the students have organised a student session themed Neuroscience and Art in 2016. The session gathered both psychology and medicine students in an interesting debate on the progress of the understanding of the intricate connections of arts and neuroscience over the course of the last century. The session covered a multitude of topics, going into depth of neurophysiological explanations of human experience of art works and presenting the current implementation of art in therapy of various neurological and psychiatric disorders. The following year saw the presentation of another amazing project involving art and neuroscience, namely The aestheticization and rehumanization of public space: art as therapy. The project was devised by Melinda Šefčić, a professional artist, and developed within the CreArt project 2017, a multidisciplinary collaboration involving the Croatian Association of Fine Artists, the Department of Ethnology and Cultural Anthropology of the Faculty of Humanities and Social Studies of the University of Zagreb, the Institute of Ethnology and Folklore Research (within their project City-making: space, culture and identity supported by the Croatian Science Foundation) and the University Hospital Centre Zagreb. This exciting venture served as a perfect showcase of what can be achieved when one understands the neuroscientific background of art and the effects it produces in humans. For medical students listening to the lecture it was an inspiring display of how otherwise relatively dull neuroscientific theory can be used and applied in everyday life. It broadened the views of this generation, making them aware that neuroscience is not just something confined to laboratories and clinics, but rather something that is all around us, something one can detect almost everywhere if one looks close enough.

As COVID-19 pandemic grips the world at the moment this chapter is being put to paper, one can does not escape the overwhelming number of webinars and other forms of video communications tackling this crisis, which seem to have sprung out overnight. A communication form already used by the scientific community has suddenly become our only way of communication. It provides a chance to listen and participate in the discussions of some of the best experts in the field from the comfort of one's home. It is incentive to ponder the consequences that may have on scientific communication once all restrictions declared due to the pandemic will be lifted. Some already prophesy the end of congresses and symposia as we know them, as this alternative is incomparably cheaper. Nevertheless, it underestimates the human need for personal contact and face-to-face socialising, a matter not easily substitutable via video communication. Furthermore, congresses represent a stimulus for the economy through travel and accomodation. It is unlikely they will die out any time soon. However, what one might expect is a general increase in scientific communication, as well as an extent of its accessibility. The INPC, hence needs not fear, as it possesses all the qualities necessary for a successful continuation of its 60-year-old tradition. By constantly reinventing itself and offering a comprehensive and diverse experience, with particular attention paid to the youth, it will continue to attract new participants and guarantee the return of those who had already attended it.

From Poster Presentation to Professional and Personal Development or Fulfilment (Ana Sruk)

In June 2007, I got the opportunity to attend my first-ever congress in neurology. I was especially excited because, just like nowadays, 47th International Neuropsychiatric Pula Congress was the congress with the longest tradition in our part of the world. Earlier during my internship, I had started to work on epidemiological study in the field of epilepsy with professor Ivan Bielen, who later become my mentor during residency in neurology. Attending the Congress gave me another memorable reason for a thrill because professor Bielen had decided that I should have my first poster presentation of our results. I remember very well how nervous I was back then and the jitters I had preparing every little detail for the poster presentation. From that point of view who could imagine that ten years later I would win a poster price award or that I would give a lecture with the same leaders and experts who I was admiring so much? I was especially fond of professor Vida Demarin who made an extraordinary impression immediately with her enormous knowledge and social interaction, the same as with a charismatic and affectionate personality. It could clearly be seen how many young colleagues` professional development she had influenced and for how much of us the role model she would come to be. Over the years, Mind&Brain INPC in Pula always has been my favorite congress to attend. It brings together people from other parts of the world who are likeminded, allowing to build professional network depending on the person and their field of interest. It is a great ambience to expand knowledge, to enhance your own work, to discuss some problems, tips and tricks, and to ask presenters questions about their work. Nevertheless, it is a great occasion to present interesting findings from your own research or ideas to other colleagues or to exchange information and knowledge with a variety of people from similar, related or completely different areas of work. Sometimes, given that Congress covers various areas of neurology and psychiatry, through the flow of knowledge, you may learn a lot of new multidisciplinary information. To outline, an interactive atmosphere at Mind&Brain INPC in Pula is always intellectually and mentally beneficial. The presence at the Congress could be crucial for professional and personal development or fulfilment. No less imporant, it is a great way to expand a circle of people with a productive mindset, as well as connect with old and new friends.

Drosophila's Self-Administration and Behavioral Sensitization to Methamphetamine (Franka Rigo, Ana Filošević, Milan Petrović and Rozi Andretć Waldowski)

In our poster at the 2019 Mind and Brain congress we presented FlyCafe assay - a new method which we developed for measuring self-admnistration of psychostimulants in Drosophila melanogaster [2]. Self-administration is a behavioral phenotype characteristic for addiction and develops as a result of the neural plasticity caused by substance abuse. Addictive drugs act on the rewarding mechanism in the brain, which stimulates the motivation for repeated drug use. Drosophila is a good model organism to study addictive behavior because of the high percentage (70%) of genetic similarity with disease-causing genes in humans.

The Capillary Feeder (CAFE) assay, although originally designed to measure food consumption, was used in our laboratory to measure self-administration and preferential consumption of psychostimulants by offering the flies choice between food with or without added methamphetamine. We improved CAFE assay in a way that it allows for precise quantification of different parameters at the level of individual fly using Drosophila Activity Monitoring System (DAMS). In FlyCafe we measure the amount of food that flies consume form the capillaries, their locomotor activity and the percentage of time they spend close to the food source. In our poster we showed that all these parameters are correlated with consumption of methamphetamine food, and most strikingly, that flies prefere consumption of food with added methamphetamine.

Recent experiments provided us with the new insights. We discovered that there are differences in preferential self-administration on the individual level and we identified two subpopulations of wild-type flies - flies with high preference for methamphethamine and flies with low preference. We confirmed differences between these two groups by finding that high preferring flies show higher locomotor activity and spend more time close to the capillary containing food with methamphetamine.

Another behavioral phenotype related to addiction is behavioral sensitization, characterized by significant increase in the locomotor activity after exposure to the two doses of volatilized methamphetamine of the same concentration. In our lab we measure behavioral sensitization using FlyBong, a method we also designed ourselves.

Although behavioral sensitization has been used extensively to elucidate the genetic basis and molecular mechanisms of neuronal plasticity, it is arguable how, if at all, it contributes to development of addiction. We tested if a development of locomotor sensitization to methamphetamine affects voluntary self- administration, in order to investigate how two drug-associated phenotypes influence one another. Our results showed that the development of locomotor sensitization to methamphetamine significantly reduces the preferential self-administration of methamphetamine. This suggests that locomotor sensitization has long-lasting effects of on brain functioning and can affect other types of addiction-related behaviors.

To better understand genetic basis of methamphetamine-induced behavior we used circadian mutant in period (per01) gene, as it is known that per gene has a direct role in the regulation of dopaminergic reward circuitry. Using FlyCafe and FlyBong assay we showed that per01 mutants do not develop preference for methamphetamine nor do they develop locomotor sensitization to volatilized methamphetamine, further emphasizing the neuronal connection between behavioral sensitization and preferential voluntary self-administration of methamphetamine [2].

Pula Congress Place Where I Have Learnt, Made Connections, and Enjoy in Interesting Lectures (Maida Seferović)

It was a great honor and pleasure to attend the 57th International Neuropsychiatric Congress in Pula in 2017, where I had the opportunity to learn a lot about new knowledge in the field of neurology and psychiatry. I also had the opportunity to meet colleagues from other cities and countries with whom I was able to exchange experiences in daily neurology clinical practice, and in treating patients with stroke and other cerebrovascular diseases.

I was especially honored to have had the opportunity to present a poster entitled „Dyslipidemia in subclinical hypothyroidism requires assessment of small dense low density lipoprotein cholesterol (sdLDL-C)", for which I won the 2nd prize [3].

The aim of my research was to evaluate the lipid profile in patients with subclinical hypothyroidism (SHypo) in comparison to controls and to determine the association of SHypo and dyslipidemia in attempt to find importance of small dense low-density lipoprotein cholesterol (sdLDL-C) in atherosclerosis. This study involved 100 women, aged 30 to 70 years that were divided in subgroups according to their age. According to the values of levels of thyroid hormones they were divided into euthyroid (control) group (n = 64) and (newly discovered) subclinical hypothyroidism (SHypo) group (n = 36). Results showed that changed lipid profile, as well as elevated triglycerides and sdLDL-C was observed in the group with subclinical hypothyroidism compared to the control group. So we concluded that it is important to determine serum lipid levels, especially serum sdLDL-C levels at an early stage of subclinical hypothyroidism, since they represent atherogenic LDL particles and are better indicators for dyslipidaemia in subclinical hypothyroidism and the development of atherosclerosis with potential complications, such as cardiovascular and cerebrovascular diseases [3].

My research and poster were the basis of discussion with many colleagues from other clinics and countries and these discussions gave me an insight into their work.

These kinds of events give us an opportunity for networking and possible new collaborations, and I find them very valuable. I highly appreciate the invitation to the 57th International Neuropsychiatric Congress and I would like to thank the organizers, especially Academician professor Vida Demarin, F.C.A., for the excellent organization of the congress, with very interesting and useful experienced lecturers.

I would recommend this Congress to my colleagues, as I believe this is a proper place to hear about things the other neurologists do and about the patient treatment in other hospitals.

My Experience with the 'Mind & Brain' International Neuropsychiatric Pula Congress (Petra Črnac Žuna)

I first participated at the „Mind & Brain" International Neuropsyciatric Pula Congress in June 2016, as a first year neurology resident. I was looking forward to participate in the international meeting with the longest tradition in this part of the world I've heard so much about. The main topics of the 56th Mind & Brain INPC were neuroinflammation, neuropsychoanalysis and neurorepair. The Congress gathered the most prominent experts in their respective fields – the list of speakers was quite impressive. They brought their expertise and experience, shared valuable insights, engaging the audience in constructive, fruitful and open exchanges, both through the formal and the informal, social programme of the Congress. The lectures and workshops were often followed by lively discussions, leading to an effective, productive and successful meeting. The level and diversity of topics discussion were outstanding.

The programme of the congress included several important neurologic and psychiatric conditions, and presented the current knowledge and know-how about the diagnosis and treatment, as well as potential uncertainties, dilemmas and challenges in clinical practice.

The meeting was very well organized, both logistically and in terms of its contents, making the participation in lectures and workshops easy to follow, rewarding and pleasurable.

I had the opportunity to actively participate at the poster session of the meeting and to present the study results of our research group at the Department of Neurology of the University Hospital 'Sveti Duh'. We analyzed the effect of prior antithrombotic therapy in acute stroke in patients with non-valvular atrial fibrillation [4]. I was very proud to win the first prize for the best poster. I am thankful to my mentor Assist.prof. H. Budinčević for all the guidance and for the opportunity to participate in the research and to present the results.

Last but not least, the social program of the Congress is a truly valuable experience, gathering and connecting people from different backgrounds and different countries, in the setting of the beautiful Croatian seaside, with delicious food and great music.

It was a privilege to attend a congress with a central role in the promotion of the most recent insight of both neurology and psychiatry, their similarities and differences and their common path towards a more comprehensive perception of the brain's hidden depths. I am very grateful for the opportunity to meet respected international experts, to communicate with them in person and learn about their professional path, challenges and goals, including one of the organizers of the meeting, academician Vida Demarin, who truly is a lively example of an extraordinary mind, as an intellectual, a leader and a teacher.

The Mind & Brain International Neuropsychiatric Pula Congress is just the right mix of a diverse set of speakers, subjects, and participants from different medical fields to successfully endorse international and interdisciplinary cooperation to achieve an integrative insight into the brain and mind and their interactions.

Pula Congress is a Perfect Congress for Young Residents and Researchers (Filip Mustač)

I am extremely honored that you have given me the opportunity to express my delight with the award I received last year. Last year's "Mental health problems in medical students" poster discussed the issues such as a sense of apaty, irritability, nervousness and sleeping difficulties, as well as the habits of taking anti-anxiety medication in the population of medical students; that is, it compared two generations of medical students over a 10-year period. The results of this study, as presented in this poster, showed younger generations to be less anxious and irritable and taking less anti-anxiety medication. The conclusion could be made that the younger generations have probably adopted more mature and healthier habits. These results, highlighting a positive trend, should certainly be encouraged and further explored for the purposes of having even more mature and healthier population in the future [5]. All of this is most certainly conducive to increased optimism.

On the other hand, the 60th anniversary of the International Neuropsychiatric Congress in Pula is an evidence that speaks to the quality and longevity of this Congress. Such longevity is surely the result of hard and laborious work in all of the segments of the organization. This longevity is most certainly also owed to the quality of the presentations, learnings and discussions made at the Congress itself. This award has great importance for me, giving me the strength and additional motivation to pursue scientific work.

I would like to thank all of my mentors and colleagues for their efforts, cooperation, generosity and creativity which has resulted in passionate and devoted efforts in collaborative scientific projects.

Likewise, I believe that everyone involved in the organization of this Congress, this 60th Anniversary should give the wind beneath their wings. There certainly are reasons for optimism! Congratulations!

References

1. Barac B, Demarin V. Six decades of the pula neuropsychiatric meetings--from neuropsychiatry to borderlands of neurology and psychiatry brain and mind. Acta Clin Croat. 2015;54(4):500–8.
2. Rigo F, Filošević A, Andretić Waldowski R. Quantification of psychostimulant self-administration in drosophila melanogaster. Abstr book (Intern Neuropsychiatr Congr, Online). 2019;1:125.
3. Seferovic Saric M, Jurasic M, Sovic S, Kranjcec S, Kovacic S, Bogoje Raspopović A, et al. Dyslipidemia in subclinical hypothyroidism requires assessment of small dense low density lipoprotein cholesterol (SDLDL-C). Abstr book (Intern Neuropsychiatr Congr, Online). 2017;1:99–100.
4. Črnac P, Friedrich L, Budinčević H, Bielen I, Demarin V. The effect of prior antithrombotic therapy in acute stroke in patients with atrial fibrillation. Abstr book (Intern Neuropsychiatr Congr, Online). 2016;1:82.
5. Mustač F, Majer M, Mužić R, Jureša V, Musil V. Mental health problems in medical students. Abstr book (Intern Neuropsychiatr Congr, Online). 2019;1:121–2.

If you have any concerns about our products,
you can contact us on
ProductSafety@springernature.com

In case Publisher is established outside the EU,
the EU authorized representative is:
Springer Nature Customer Service Center GmbH
Europaplatz 3, 69115 Heidelberg, Germany

Printed by Libri Plureos GmbH
in Hamburg, Germany